Modified Cyclodextrins

Modified Cyclodextrins

Scaffolds and Templates for Supramolecular Chemistry

Christopher J Easton

Australian National University

Stephen F Lincoln

University of Adelaide

Imperial College Press

ICP

Published by

Imperial College Press
203 Electrical Engineering Building
Imperial College
London SW7 2BT

Distributed by

World Scientific Publishing Co. Pte. Ltd.

P O Box 128, Farrer Road, Singapore 912805

USA office: Suite 1B, 1060 Main Street, River Edge, NJ 07661

UK office: 57 Shelton Street, Covent Garden, London WC2H 9HE

Library of Congress Cataloging-in-Publication Data
Easton, Christopher J.
    Modified cyclodextrins : scaffolds and templates for
supramolecular chemistry / Christopher J. Easton, Stephen F.
Lincoln.
        p.   cm.
    Includes bibliographical references and index.
    ISBN 1-86094-144-3
    1. Cyclodextrins.   2. Macromolecules -- Synthesis.   I. Lincoln.
Stephen F.   II. Title.
QD321.E27   1999
547'.7815--dc21   99-19665
                                                                        CIP

British Library Cataloguing-in-Publication Data
A catalogue record for this book is available from the British Library.

This book is printed on acid-free paper.

Printed in Singapore by Uto-Print

Preface

Through modification, the native cyclodextrins are effective templates for the generation of a wide range of molecular hosts. This makes it possible to tailor a cyclodextrin host to a particular guest, to meet specific requirements in the host-guest complex, and opens the way to diverse new areas of supramolecular chemistry. Metallocyclodextrins, rotaxanes and catenanes, and surface monolayers of modified cyclodextrins are now accessible. The native cyclodextrins serve as scaffolds on which functional groups and other substituents can be assembled, with controlled geometry. This results in substantially improved molecular recognition and procedures for chemical separation, including enantiomer discrimination, through guest binding. Access to the gamut of functional groups greatly expands the utility of cyclodextrins in chemical synthesis and affords catalysts which mimic the entire range of enzyme activity. Modifications to the cyclodextrins also lead to a wide range of photochemistry of cyclodextrin complexes, through which the enhancement of guest reactivity occurs. In addition, light harvesting molecular devices and photochemical frequency switches may be constructed. In solution, modified cyclodextrins have been used to construct molecular reactors, and molecular, temperature and pH sensors. At surfaces, they form semipermeable membranes and sensor electrodes. Such innovative fields of chemistry, only made possible through modifications to the native cyclodextrins, are the subject of this book.

To Kylie

Acknowledgements

We gratefully acknowledge the support of our wives, Robyn and Chris, and families, throughout our careers and particularly during the preparation of this manuscript. They are a constant source of motivation and inspiration. We are also indebted to enthusiastic colleagues and research associates who continually stimulate our interest in this area of science and who are too numerous to name individually. However, special thanks are due to James Kelly for his generous assistance with this project.

Christopher J. Easton   *Stephen F. Lincoln*

Contents

CHAPTER 1
MODIFIED CYCLODEXTRINS

CYCLODEXTRIN STRUCTURE AND COMPLEXATION BEHAVIOUR

1.1. Introduction

The first reports of the naturally occurring cyclodextrins (CDs) appeared in the late nineteenth and early twentieth centuries [1,2]. Since then the study of CDs has gained increasing momentum to the point where Szejtli commented that more than 15,000 CD-related papers appeared in the literature by the end of 1997 [3]. Apart from their intrinsic interest, the fascination with these cyclic oligosaccharides has largely arisen from the homochiral doughnut-like annular structures of the smaller CDs and their ability to complex other molecules within their annuli. This complexation exhibits size and chiral discrimination between guests to varying degrees, and in some cases the guest may undergo reactions in the annulus at an increased rate or yield different product ratios from those arising in the free state, while in other cases the guest reactivity may be decreased. Inevitably this has led to an extensive range of chemical modifications of the natural CDs to capitalise on and enhance these characteristics while retaining the ability to complex guests within the CD annulus. The basic structures of the natural CDs have proven to be remarkably resilient during such modifications and have permitted the design and synthesis of modified CDs to achieve specific outcomes. Hence, the natural CDs may be viewed as molecular platforms on which molecular scaffolds of increasing sophistication and designed for specific roles may be erected. It is a recounting of the many facets of the modification of the natural CDs and their outcomes about which this book is primarily concerned.

Before such a recounting can proceed it is necessary to gain a basic understanding of the chemistry of the natural CDs, an understanding which this first chapter seeks to establish. Accordingly, it discusses the structures of the naturally occurring CDs and the stoichiometries, equilibria, kinetics and mechanisms of complex formation and the interactions controlling them. These characteristics of the natural CDs carry through to their modified analogues which are discussed in detail in the chapters which follow. Such has been the plethora of natural CD modifications that it is now possible to systematise strategies for achieving desired modifications and this forms the content of the Chapter 2. The modified CDs and their structural complexing characteristics provide an evolving sequence through Chapters 3 to 8 where, in general terms, the modifications progress from those based on a single CD to those with two covalently linked CDs to CD rotaxanes, catenanes and polymers, and finally to those forming monolayers. In parallel with this broad structural progression, the complexation, catalytic, photochemical and other characteristics of modified CDs are discussed.

While this book is predominantly concerned with the chemistry of modified CDs as a research area, it should be remembered that increasingly natural and modified CDs are found in a range of industrial applications. Thus, several thousands of tonnes of natural CDs are produced annually, and their major usage in 1996 was found in the foods and cosmetics (~73%), pharmaceutical (~15%), chemical (~10%) and pesticide (<1%) industries, and in analytical chemistry (1%) [4]. Readers wishing to learn more about the practical applications of natural and modified CDs are referred to the range of monographs and reviews available [3-14].

1.2. Natural Cyclodextrins and Their Structures

Naturally occurring CDs are homochiral cyclic oligosaccharides, the most common of which are composed of 6, 7 or 8 α-1,4-linked D-glucopyranose units. They are produced, together with some larger CDs and some linear oligosaccharides, through the degradation of starch by the enzyme CD glucosyltransferase. The CDs composed of 6, 7 and 8 glucopyranose units are usually referred to as αCD, βCD and γCD (**1.1a** in Fig. 1.1), respectively, are the most plentifully produced and extensively studied [6,8,15-31]. For the larger CDs with the number of glucopyranose units extending from nine to fourteen [26,30-33], the Greek prefix follows the alphabet from δCD to ιCD as the number of D-glucopyranose units increases. Obviously, this nomenclature cannot extend to the CDs composed of 100 or more such units obtained through the action of a disproportionating enzyme on amylose [34]. However, the vast majority of publications on natural and modified CDs describe research based on αCD, βCD and γCD, and retain the CD nomenclature which is also retained throughout this text. (The CDs are also known as cycloamyloses, so that αCD, βCD, and γCD are named cyclohexa-, cyclohepta- and cyclooctaamylose, respectively, or CA6, CA7 and CA8 where the number indicates the number of glucopyranose units in the macrocycle [26].)

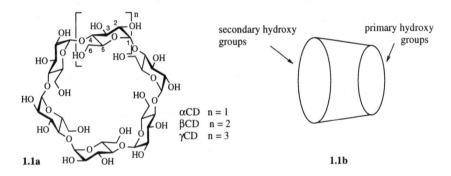

Fig. 1.1. Schematic illustrations of αCD, βCD and γCD (**1.1a**). A truncated cone (**1.1b**) is frequently used to represent a native or modified CD. When a substituent is shown at the narrow end of the cone, it indicates that it replaces one of the C(6) hydroxy groups, while a substituent shown at the wide end of the cone indicates that it replaces either a C(2) or a C(3) hydroxy group.

Many crystallographic X-ray studies of CDs and their complexes have been reported from which a comprehensive picture of their structures has been generated [16,17,19,20,24,26,27]. These crystalline state structures also provide a valuable guide to CD interactions in solution. The αCD, βCD and γCDs have doughnut-like annular structures with wide and narrow hydrophilic ends delineated by O(2)H and O(3)H secondary and O(6)H primary hydroxy groups, respectively, and hydrophobic annular interiors lined with H(3), H(5) and H(6) hydrogens and O(4) ether oxygens. Crystallographic X-ray studies show that each glucose unit adopts a fairly rigid $^4C_1$ chair conformation. The only conformational freedom that exists is in the rotation of the C(6)-O(6) groups and constrained rotation about the glucosidic C(1)-O(4)-C(4) link. Usually the C(6)-O(6) bonds are directed away from the centre of the CD annulus, such that the torsion angle O(5)-C(5)-C(6)-O(6) is (-)-*gauche*, although hydrogen bonding between a guest molecule in the annulus and the O(6)H group may turn the C(6)-O(6) bonds towards the centre of the annulus, such that the torsion angle O(5)-C(5)-C(6)-O(6) becomes (+)-*gauche*. Neutron diffraction studies show that the CD structure is stabilised in the solid state by intramolecular hydrogen bonding between the secondary hydroxy groups of adjacent glucose units [20,26]. The O(4) oxygens delineating the macrocyclic hexagons, heptagons and octagons of αCD, βCD and γCD, respectively, are almost coplanar with < 0.25 Å deviation from the common plane. However, substantial deviations from planarity are seen in the O(4) nonagon of the nine glucopyranose unit δCD which forms a boat-like ellipse [30]. Molecular modelling shows that this distorted structure is a result of the steric strain induced by the larger ring size of δCD, and that further increases in size with increase in the number of glucopyranose units results in a departure from the doughnut shape of αCD, βCD and γCD. Thus, the solid state structures of the ten and fourteen unit εCD [31,33] and ιCD [33] are twisted as a result of a ~180° rotation of two diametrically opposed glucopyranose units about the glucosidic link so that the ring of intramolecular hydrogen bonds present in αCD-δCD is broken. At the ~180° rotation site two adjacent glucopyranose units are oriented *trans* to each other while the other units retain a *cis* orientation. The structure of the 26 unit υCD describes a figure of eight where each half consists of two left handed single helical turns with six glucopyranose moieties per repeating unit [35].

Apart from being the most extensively studied of the natural CDs, αCD, βCD and γCD also represent the platforms for modifications ranging from the substitutions exemplified by that of a hydroxy group proton by a methyl group and the replacement of a hydroxy group by an amine group, to major single and multiple hydroxy group substitutions where the substituent mass may approach that of the natural CD. Nevertheless, the substituted CD's annulus usually retains a central importance in the properties of the modified CD, and accordingly some of the properties of αCD, βCD and γCD are presented in Table 1.1. The solubilities of the CDs vary in an irregular manner [26,36], and the relatively low solubility of βCD provided an early reason for its modification to improve its solubility as its annulus is of a size particularly suitable for the complexation of drug molecules, and its more soluble modified forms yield a range of drug complexes which are more soluble than the drugs in their free states [14]. The annular diameters and volumes increase substantially from αCD to γCD and this provides a crude discrimination between molecules entering the annulus to form complexes. The partial molar volumes are much larger than the annular volumes and arise from the doughnut-like overall shape of the CDs [26,37]. The homochirality of the CDs is reflected by their α_D values [6,17,26]. The final parameters in Table 1.1 are the pK_as assigned to O(2)H and O(3)H in each CD [38], and which have an important role in the CD mediation of the hydrolysis of a range of guests inside the CD annulus as discussed in Chapter 4.

Table 1.1. Some Characteristics of αCD, βCD and γCD.

CD	αCD	βCD	γCD
Number of glucopyranose units	6	7	8
Molecular weight (anhydrous)	972.85	1134.99	1297.14
Solubility per dm^3 H$_2$O at 298.2 K^a	14.5	1.85	23.2
Annular diameter measured from the C(5) hydrogens, Å^b	4.7	6.0	7.5
Annular diameter measured from the C(3) hydrogens, Å^b	5.2	6.4	8.3
Annular depth from the primary to the secondary hydroxy groups, Å^b	7.9 - 8.0	7.9 - 8.0	7.9 - 8.0
Annular volume, Å^3 ^{c}	174	262	472
Partial molar volumes, cm^3 mol^{-1} ^{d}	611.4	703.8	801.2
α_D, deg ^{e}	+150.5	+162.5	+177.4
pK_a O(2)H and O(3)H at 298.2 K^f	12.33	12.20	12.08

[a]References 26 and 36. [b]Measured from Corey-Pauling-Koltun models. [c]References 6 and 26. [d]Reference 26. [e]References 6, 17 and 26. [f]Reference 26.

1.3. Cyclodextrin Complexation Processes

Much of the interest in natural and modified CDs arises from their ability to include or encapsulate a substantial part of a guest molecule or ion inside their annuli to form complexes usually described as either inclusion complexes, host-guest complexes or simply as complexes. These complexes are unusual in that only secondary bonding occurs between the CD and the guest (G), yet their stability can be quite high depending on the nature of the CD and G. The stoichiometry of the complexes is usually encompassed by the ratios: 1:1, CD.G; 1:2, CD.G$_2$; 2:1,

$CD_2.G$; and 2:2, $CD_2.G_2$, characterised by the sequential stability constants K_{11}, K_{12}, K_{21}, K_{22} and K_{22}', respectively.

$$K_{11} = \frac{[CD.G]}{[CD][G]}, \; K_{12} = \frac{[CD.G_2]}{[CD.G][G]}, \; K_{21} = \frac{[CD_2.G]}{[CD.G][CD]},$$

$$K_{22} = \frac{[CD_2.G_2]}{[CD.G_2][CD]}, \; \text{and} \; K_{22}' = \frac{[CD_2.G_2]}{[CD_2.G][G]}.$$

Quite often complexes of different stoichiometries may coexist in solution so that more than one stability constant characterises the system as shown in Eqs. (1.1) and (1.2). Although the equilibria characterised by K_{22} and K_{22}' produce the same complex, $CD_2.G_2$, this is achieved through different routes and there is no necessity for K_{22} and K_{22}' to be of similar magnitude.

$$CD + G \; \xrightleftharpoons{K_{11}} \; CD.G \; \underset{-\,G}{\overset{K_{12},\,+\,G}{\rightleftharpoons}} \; CD.G_2 \; \underset{-\,CD}{\overset{K_{22},\,+\,CD}{\rightleftharpoons}} \; CD_2.G_2 \qquad (1.1)$$

$$CD + G \; \xrightleftharpoons{K_{11}} \; CD.G \; \underset{-\,CD}{\overset{K_{21},\,+\,CD}{\rightleftharpoons}} \; CD_2.G \; \underset{-\,G}{\overset{K_{22}',\,+\,G}{\rightleftharpoons}} \; CD_2.G_2 \qquad (1.2)$$

As expressed above, K_{11}, K_{12}, K_{21}, K_{22} and K_{22}' are concentration stability constants rather than true stability constants where activities of the equilibrium participants are employed instead of concentrations. However, concentration stability constants are usually determined in CD complexation studies in solutions where a constant ionic strength (I) is maintained by a supporting electrolyte which does not participate in the complexation process.

Stability constants do not directly indicate the structure of the complexes formed. Thus, if it is assumed that a ditopic guest (AB) with differentiated complexing ends (A and B) enters the CD annulus through the secondary face alone, there still exists the question as to which end of the guest enters first (Fig. 1.2). When isomeric complexes (CD.AB and CD.BA) are formed, where either end of the guest enters the annulus first, they are characterised by the microscopic stability constants, K_1 and

K_1', respectively. If a second CD adds, the microscopic stability constants, K_2 and K_2' apply to the formation of $CD_2.AB$. Generally, the complexes CD.AB and CD.BA are not separately detected and K_{11} and K_{12} are related to K_1, K_1', K_2 and K_2' through: $K_{11} = K_1 + K_1'$, and $K_{12} = zK_2K_2'/K_{11}$, where $z = K_2/K_1 = K_2'/K_1'$ [23,25]. Should the guest, AB, enter the CD annulus through both the primary and the secondary faces additional complexes may form, and other microscopic stability constants arise, as is also the case should $CD.(AB)_2$ form.

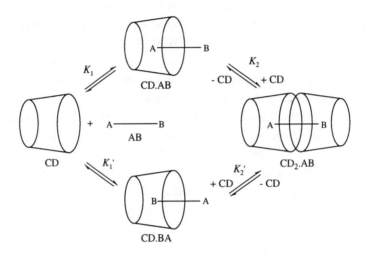

Fig. 1.2. The complexation of a ditopic guest, AB, by a CD through the secondary face only.

The majority of CD complexation studies have been carried out in aqueous solution and as a consequence considerable attention has been focused on the change in hydration of the CD and the guest in the complexation process [23,25,28]. Depending on the identity of the CD, differing numbers of water molecules (xH_2O) occupy its annulus. Similarly, differing numbers of water molecules (yH_2O) hydrate the guest depending on its identity. Upon formation of a 1:1 CD.G complex, the guest displaces water molecules from the CD annulus while also losing part of its own hydration shell so that several water molecules (zH_2O) enter the bulk water as shown in Eq. (1.3).

$$CD.xH_2O + G.yH_2O \overset{K_{11}}{\rightleftharpoons} CD.G.(x + y - z)H_2O + zH_2O \qquad (1.3)$$

While the change in the number of water molecules associated with the CD and guest on formation of the complex is unknown, it, together with the entry of the guest into the CD annulus, are considered to be major factors affecting the stability of the complex. In addition, hydrogen bonding between the guest and the CD appears to stabilise complexes, and it may be that conformational changes in either the CD, the guest or both also contribute to complex stability. Within this broad outline of stabilising factors, more specific effects have been invoked which include hydrophobic effects, release of "high energy" water from the CD annulus, relief of conformational strain in the uncomplexed CD, ion-dipole, dipole-dipole and induced-dipole-dipole interactions, and London dispersion forces. While there is little doubt that the quite strong secondary bonding found in CD complexes results from the summation of multiple weak interactions, their individual identification is less certain. Accordingly, only the salient aspects of these interactions are considered below.

The majority of the thousands of CD complexes reported involve a guest incorporating a hydrophobic moiety which resides in the vicinity of the hydrophobic region of the CD annulus. This feature is thought to arise through the "hydrophobic effect" whereby non-hydrogen bonding entities are excluded from the bulk water hydrogen bonded structure to form aggregates, in this case CD complexes, which thereby experience a stabilising effect. Water occupying the CD annulus cannot form the same hydrogen bonding interactions as bulk water, and is sometimes referred to as "high energy" water. The entry of a guest into the CD annulus expels some or all of this "high energy" water which enters the bulk water structure, a transition which is sometimes thought to contribute to the stability of CD complexes. However, neither the number of water molecules occupying the CD annulus, nor the number expelled on complexation, has been determined in solution and accordingly much of the interpretation of solution studies is based on observations of the crystalline state. One X-ray crystallographic study of $\alpha CD.6H_2O$ shows two water molecules hydrogen bonded to each other lying within the αCD annulus, almost on its axis, with the water molecule closest to the primary face hydrogen bonding to two O(6) hydroxy groups [39]. A second study [40] of $\alpha CD.6H_2O$ finds one water molecule within the annulus, while a third [41] finds 2.57 water molecules statistically distributed over four sites within the annulus. There appears to be a consensus that 2 to 3 water molecules occupy the αCD annulus in aqueous solution and are expelled on complexation of a guest.

Another observation pertinent to the complexation process is that in the first of the above X-ray crystallographic studies of αCD.6H$_2$O one of the glucopyranose rings is tilted with respect to the other five whereas this is not so in simple αCD complexes in the solid state [39]. Calculations show this puckered macrocyclic ring conformation to be strained by comparison with that in which all six glucopyranose units are equivalent and accordingly it has been postulated that relief of this strain on complexation may stabilise the complex [42]. However, in the third of the above solid state studies of αCD.6H$_2$O, significant puckering of the macrocyclic ring is found to be absent and strain energy is diminished [41]. This observation of different macrocyclic ring conformations for αCD in the solid state is consistent with solution NMR [43-45] and Raman [46] studies which show significant flexing of the macrocyclic ring and changes in the C(5)-C(6) rotamers to occur in natural and modified CDs, as do molecular mechanics and molecular dynamics calculations [47-55]. The βCD annulus accommodates 6.5 water molecules statistically distributed over eight sites, and the γCD annulus accommodates 7.1 water molecules distributed over fourteen sites, as shown by X-ray crystallographic studies of βCD.12H$_2$O and γCD.13.3H$_2$O [56,57]. The macrocyclic rings of βCD and γCD show little puckering akin to that sometimes observed for αCD, and as a consequence the relief of strain energy for the larger CDs is generally not considered significant in stabilising their complexes.

1.4. Enthalpy-Entropy Compensation in Cyclodextrin Complexation

The enthalpy and entropy of complexation, ΔH^o and ΔS^o, respectively, of more than 2000 natural CD complexes and more than 100 modified CD complexes have been determined. In the overwhelming majority of cases, ΔH^o is negative and ΔS^o is usually negative, and as a consequence CD complexation is said to be "enthalpy driven" [25,28,58-60]. In the "classical" hydrophobic interaction between nonpolar molecules the structure of the neighbouring water is a critical factor, both ΔH^o and ΔS^o are positive, and the interaction between the nonpolar molecules is said to be "entropy driven". The latter observation could be taken to indicate that hydrophobic interactions are unimportant in stabilising CD complexes. However, CDs are quite polar and it is only the annulus interior which is hydrophobic. "Nonclassical" hydrophobic interactions between semipolar molecules may have either a negative ΔH^O or ΔS^O [25,61,62], and consequently it appears to be of dubious value to attempt to deduce the CD complex stabilising interactions in too great a detail from the magnitude and sign of ΔH^O or ΔS^O.

An extensive range of CD complexations shows an enthalpy-entropy compensation where, for a given CD, a plot of ΔH^O against ΔS^O is linear for all guests and has a positive slope [18,23,25,28,59,60]. This observation has generated considerable mechanistic discussion about the formation of CD complexes. Before entering this discussion it is salutary to briefly review the thermodynamics of 1:1 CD complex formation through Eq. (1.4):

$$\ln K_{11} = -\Delta G^O/RT = -\Delta H^O/RT + \Delta S^O/R \qquad (1.4)$$

In many cases, ΔH^O and ΔS^O are determined from a plot of $R\ln K_{11}$ against $1/T$ which yields ΔH^O from the slope and ΔS^O from the intercept. Two issues arise from this. First, the errors in ΔH^O and ΔS^O are closely correlated so that an error in the determination of one is compensated for by the error in the other so that a range of errors in the studies of the same system could produce a linear relationship between ΔH^O and ΔS^O. Second, there is no thermodynamic requirement for a correlation to exist between the magnitudes of ΔH^O and ΔS^O. Thus, an experimental correlation between ΔH^O and ΔS^O represents an extrathermodynamic relationship. The range of spectroscopic and other techniques used in the determination of ΔH^O and ΔS^O from the temperature dependence of K_{11} are all subject to the first of the above cautionary comments concerning errors [23]. It also appears that calorimetric methods which determine K_{11} and ΔH^O directly might be subject to a similar caution because of the possibility of a correlation of error between these parameters. However, the existence of a real correlation between ΔH^O and ΔS^O can be tested for by appropriate statistical analysis [25,63,64] and subsequent discussion concentrates on ΔH^O and ΔS^O correlations established in this way.

The linear plot of ΔH^O against ΔS^O for 1:1 CD complex formation has a positive slope, β, which is a temperature, referred to as the "compensating" or "isoequilibrium" temperature (Eq. (1.5)). It may be argued that because ΔH^O is related to the energy of the secondary bonding interactions in the CD complexation process, an increasingly negative ΔH^O in a series of complexations reflects an increase in the strength of these interactions, a decrease in the number of degrees of freedom and a correspondingly more negative ΔS^O and *vice versa*. Thus, the observation of an enthalpy-entropy compensation is consistent with the CD complexations to which it applies sharing similar interactions. This is an important observation, but there is considerable debate as to what further information about the CD complexation process can be inferred from such an enthalpy-entropy compensation. Substantially, the debate revolves around the contention that the interrelation of Eqs. (1.5) and (1.6) means that variations in ΔH^O and ΔS^O convey no additional information to that derived from variations in ΔG^O.

$$\delta \Delta H^O = \beta \delta \Delta S^O \tag{1.5}$$

$$\delta \Delta G^O = \delta \Delta H^O - T \delta \Delta S^O \tag{1.6}$$

Nevertheless, mechanistic conclusions have been drawn from the linear plots of $T \Delta S^O$ against ΔH^O for 524 αCD, 488 βCD and 58 γCD 1:1 complexes formed with a wide range of different guest molecules [28]. In broad terms these linear relationships (Eq. (1.7)) indicate that the dominant stabilising factors within each set of CD complexes are similar, and that a change in $T \Delta S^O$ is compensated for by a change in ΔH^O as the guest changes, as indicated by Eq (1.8). Eq. (1.7) shows that the overall entropy change is made up of a term, $T \Delta S^O$, proportional to the enthalpy change and a term independent of it, $T \Delta S_0{}^O$, and further that when $\Delta H^O = 0$ the complex is still stable provided that $T \Delta S_0{}^O$ is positive. Eq. (1.9), derived through Eqs. (1.6) and (1.8), shows that α represents the entropic contribution decreasing the enthalpic stabilisation of the CD complex such that only a $(1 - \alpha)$ portion of $\delta \Delta H^O$ contributes to increasing complex stability. It has been deduced that α arises from conformational change and that $T \Delta S_0{}^O$ arises from the extent of dehydration occurring on complex formation, and that these are the dominant effects determining CD complex stability. For αCD, βCD and γCD, α increases in the sequence: 0.79, 0.80 and 0.97, respectively, and $T \Delta S_0{}^O$ increases in the sequence 8, 11 and 15 kJ mol^{-1}, respectively. These increases in α and $T \Delta S_0{}^O$ are thought to be consistent with increases in conformational change and dehydration occurring on complexation. This is in accord with the expected increase in flexibility and an increase in the number of water molecules in and around the annulus as CD size increases.

$$T\Delta S^O = \alpha\Delta H^O + T\Delta S_O{}^O \tag{1.7}$$

$$T\delta\Delta S^O = \alpha\delta\Delta H^O \tag{1.8}$$

$$\delta\Delta G^O = (1 - \alpha)\delta\Delta H^O \tag{1.9}$$

1.5. Examples of Cyclodextrin Complexation

The homochirality and variation in size of the αCD, βCD and γCD annuli provide opportunities for both chiral [21] and size [18,25,28] discrimination in complexation, as indicated by changes in complex stability as the identity of either the CD or the guest is varied. A wide range of neutral and ionic guests are complexed [65] as is frequently reported for aliphatic [66-75], aromatic [75-84], amino acid [85-94], drug [14,95-104], and dye, indicator and related [105-113] guests, and less frequently reported for sugar [114], inorganic anion [115-118], cation [119], noble gas [120] and fullerene [121-123] guests. A selection of examples which illustrate some of the factors affecting natural CD complex stability are now discussed in more detail, and many examples of modified CD complexes are discussed in the chapters which follow.

The stereochemistry and stability of a CD complex can be profoundly affected by the structure and charge of the guest as is illustrated by the complexation of 4-nitrophenols and -phenolates by αCD (Table 1.2) [124]. Predominantly, these guests enter nitro-substituent first through the secondary face of αCD as shown by ^{1}H NMR studies. This is in accord with 2,6-dimethyl-4-nitrophenol and -phenolate and the 4-nitrophenol and -phenolate analogues forming stable complexes with αCD, while 3,5-dimethyl-4-nitrophenol and -phenolate are inhibited from forming αCD complexes because of steric hindrance to the nitro-substituted end entering the αCD annulus. The less hindered 2-methyl-4-nitrophenolate forms a weak αCD complex, but its phenol analogue does not. When complexes form in this series, neither the phenol nor the phenolate end of the guest enters the αCD annulus first, and the phenolates complex more strongly than the phenols. Possible reasons for both of these observations are that either head-to-tail alignment of the dipole of the guest with that of αCD [125], or hydration of the phenol and phenolate groups inhibit complexation, or both.

Table 1.2. Stability Constants for 1:1 αCD Complexes of 4-Nitrophenols and 4-Nitrophenolates.[a]

Guest	$K_{11}/dm^3\ mol^{-1}$	Guest	$K_{11}/dm^3\ mol^{-1}$
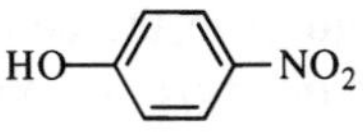 HO—⟨⟩—NO$_2$	1.9×10^1	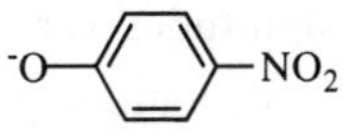 $^-$O—⟨⟩—NO$_2$	2.5×10^3
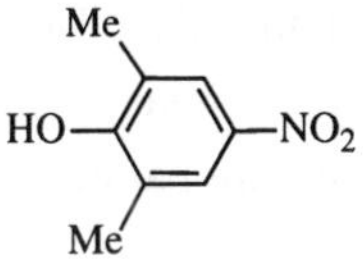 HO—⟨Me⟩—NO$_2$ (Me)	5.7×10^2	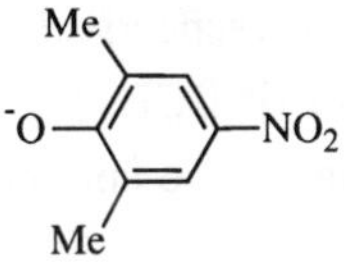 $^-$O—⟨Me⟩—NO$_2$ (Me)	1.06×10^3
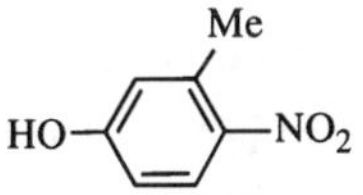 HO—⟨Me,Me⟩—NO$_2$	b	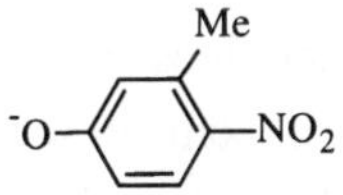 $^-$O—⟨Me,Me⟩—NO$_2$	2.4×10^1
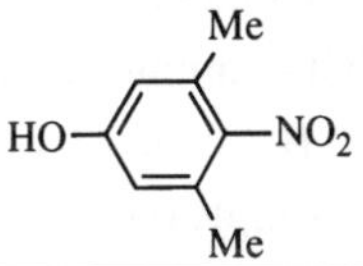 HO—⟨Me,Me⟩—NO$_2$	b	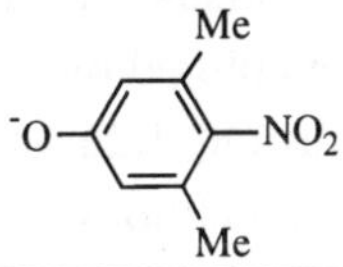 $^-$O—⟨Me,Me⟩—NO$_2$	b

[a]Determined spectrophotometrically at pH 6.5 and pH 11.0, respectively, in aqueous phosphate buffer ($I = 0.5$ mol dm^{-3}) at 298.2 K. [b]No complex detected.

It is of interest to briefly review some of the literature on the dipole moments of CDs and their guests. Calculations based on crystallographic structures arrive at dipole moments for αCD in the range 12-20 D depending on the method of calculation [125-127]. The calculated values of the dipole moment of αCD in its 4-nitrophenol complex are 13.5 D and 12.4 D when the CNDO/2 MO and MM2 routines are used, respectively. These dipole moments align approximately with the αCD annular axis with its positive pole at the primary face, where the nitro group of the guest in the αCD.4-nitrophenol complex is also located. This is the structure of αCD.4-nitrophenol found in solution [128] and the crystalline state [24,27,129]. The dipole moment of 4-nitrophenol (5.0 D) has its negative pole adjacent to the nitro group, from which it is seen that the dipole moments of αCD and the guest are aligned head-to-tail in the complex. Benzoic acid (1.5 D) and 4-hydroxybenzoic acid (1.5 D) have their negative poles adjacent to their carboxylic acid groups and align head-to-tail with the dipole moment of αCD in their complexes. These calculations and observations are in accord with dipole-dipole interactions being significant in orienting guests CD complexes and stabilising them. However, the magnitude of these interactions is subject to uncertainty given the dependence of the calculated CD dipole moments on the assumptions in the calculations. Thus, CNDO/2 calculations show the magnitude of the dipole moment to increase in the sequence: $\alpha CD < \beta CD < \gamma CD$, and that these magnitudes vary with the nature of the guest in the complex [127]. However, AM1 calculations yield dipole moments of 7.06, 2.03 and 2.96 D for αCD, βCD and γCD, respectively [130]. Related AM1 calculations also show that the dipole moment of βCD changes from 2.9 D when the primary hydroxy groups are perpendicular to the annular axis, to 14.9 D when they are parallel to the axis [131].

High CD complex stabilities are observed under favourable conditions as is found for the βCD complexes of some of the steroids shown in Table 1.3 [132]. They also show how the stereochemistry of guest molecules can have a marked effect on complex stability as exemplified by the formation of the 1:1 βCD complexes of the steroids with *cis* AB fusion, **1.2a** and **1.2b**, and with *trans* AB fusion, **1.2c** and **1.2d**. Under the same conditions, lithocholic acid (**1.2a**) is complexed 20 times more strongly by βCD than is its isomer 5α-cholanic acid 3β-ol, **1.2c**. Molecular modelling indicates that this stronger complexation of lithocholic acid is a result of the *cis* AB fused structure causing the A ring to better fit the wider secondary end of the βCD annulus than is the case with the *trans* AB fused 5α-cholanic acid 3β-ol such that the latter is more exposed to solvent. In both cases, the B, C and D rings thread through the annulus. The same trend in stability is seen for **1.2b** and **1.2d** consistent with the effect of *cis* AB fusion being important for the complexation of **1.2b** also. The variation of K_{11} with solution composition for the complexation of **1.2a** is consistent with its aggregation which competes with its complexation. Thus, the true magnitude of K_{11} is probably greater than 1.17×10^6 dm^3 mol^{-1} which appears to be the highest K_{11} reported for a βCD complex to date. None of **1.2a**-**1.2d** show appreciable complexation by the smaller αCD.

Adamantane-1-carboxylic acid (**1.3a**) and bicyclo[2.2.1]heptane-1-carboxylic acid (**1.4a**) and their conjugate bases (**1.3b** and **1.4b**) provide interesting contrasts in their complexation by αCD and βCD [133] (Table 1.4). While αCD forms both 1:1 and 2:1 complexes with both **1.3a** and **1.4a**, βCD only forms 1:1 complexes which are much more stable than the αCD complexes as a consequence of the better fit of the guest to the βCD annulus. The anionic **1.3b** and **1.4b** form 1:1 complexes of

Table 1.3. Stability Constants for 1:1 Complexes of βCD and Some Steroids.[a]

Steroid	$K_{11}/$ dm^3 mol^{-1}
1.2a	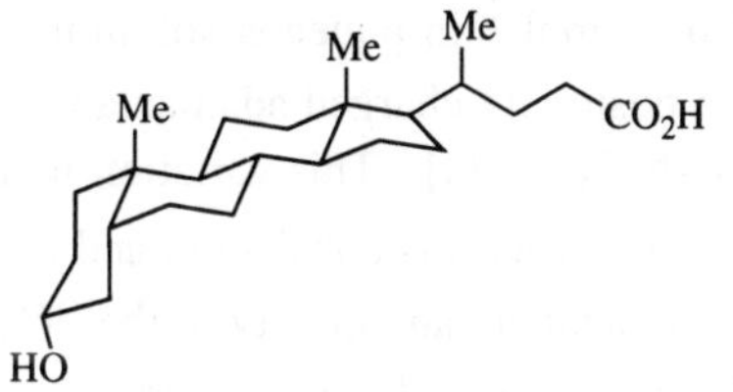 1.17×10^6 [b] 2.01×10^5 [c] 2.77×10^5 [d] 4.83×10^4 [e]
1.2b	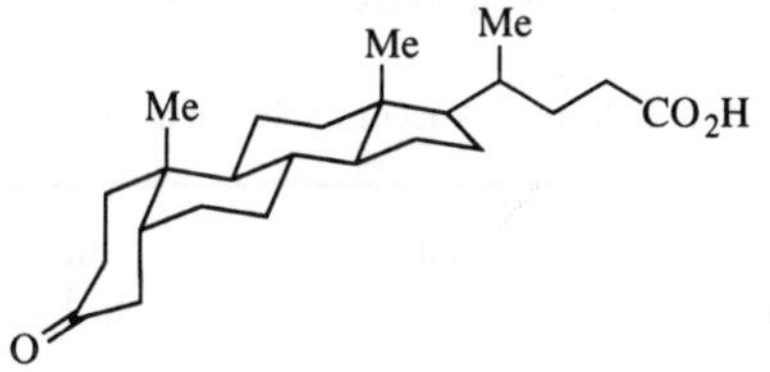 2.40×10^5 [c]
1.2c	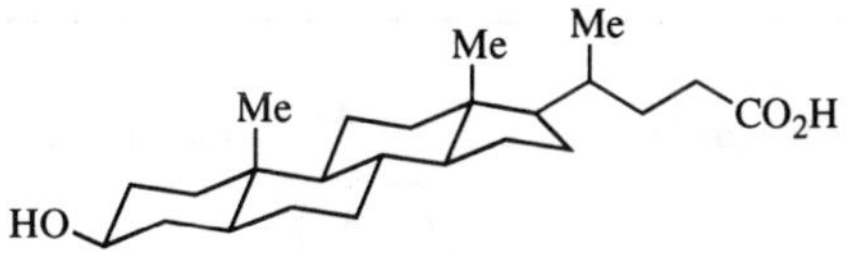 1.07×10^4 [c]
1.2d	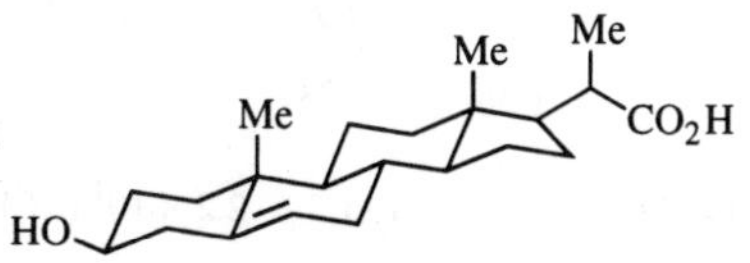 1.67×10^3 [c]

[a]Determined calorimetrically in 0.2 mol dm^{-3} pH 9.0 NaHCO$_3$ - Na$_2$CO$_3$ buffer in water/dimethyl sulfoxide solution at 298.2 K.
[b]2.0×10^{-3} mol dm^{-3} βCD, 5.0×10^{-5} mol dm^{-3} steroid, 1% dimethyl sulfoxide.
[c]2.0×10^{-3} mol dm^{-3} βCD, 1.0×10^{-4} mol dm^{-3} steroid, 10% dimethyl sulfoxide.
[d]2.0×10^{-3} mol dm^{-3} βCD, 5.0×10^{-5} mol dm^{-3} steroid, 10% dimethyl sulfoxide.
[e]5.0×10^{-3} mol dm^{-3} βCD, 5.0×10^{-4} mol dm^{-3} steroid, 10% dimethyl sulfoxide.

similar stability to those of their conjugate acid analogues with αCD and form no 2:1 complexes. There is a significant decrease in stability in the analogous βCD complexes. The formation of αCD.**1.3a** and αCD$_2$.**1.3a** reveals cooperativity in the complexation of the second αCD through K_{21} being substantially larger than

K_{11}. Cooperative complexation also occurs in $\alpha CD_2.\mathbf{1.4a}$. The complexation of several other similar carboxylic acid guests also shows cooperativity, and provides a major rationale for the syntheses of covalently linked CDs which are discussed at length in Chapter 6. At a lower ionic strength, conductivity measurements yield K_{11} = 1.05×10^5, 3.3×10^4 and 1.9×10^3 dm^3 mol^{-1} in aqueous solution for the 1:1 βCD complexes of **1.3a** and **1.3b** and the positively charged adamantane-1-aminium where the -CO$_2$H of **1.3a** is replaced by -NH$_3^+$ [134]. This variation in stability is quite small when the charge variation in these guests is considered and is consistent with the closeness of fit of the hydrophobic adamantane moiety to the βCD annulus being the dominant factor determining the stabilities of these complexes.

Table 1.4. Stability Constants[a] of αCD and βCD Complexes in Aqueous Solution at 298.2 K.

Guest	$K_{11}{}^b$ dm^3 mol^{-1}	$K_{21}{}^b$ dm^3 mol^{-1}	Guest	$K_{11}{}^c$ dm^3 mol^{-1}
1.3a (CO$_2$H)	1.3×10^2 (αCD) 3.0×10^5 (βCD)	4.2×10^2 (αCD)	**1.3b** (CO$_2^-$)	1.4×10^2 (αCD) 1.8×10^4 (βCD)
1.4a (CO$_2$H)	3.1×10^1 (αCD) 7.0×10^3 (βCD)	1.57×10^2 (αCD)	**1.4b** (CO$_2^-$)	2.2×10^1 (αCD) 8.3×10^2 (βCD)

[a]Determined microcalorimetrically. [b]In 0.1 mol dm^{-3} sodium acetate at pH 4.05. [c]In 0.05 mol dm^{-3} sodium borate at pH 8.5.

The change in environment from full hydration to partial dehydration and entry into the hydrophobic CD annulus can substantially modify the chemical behaviour of the guest. This is exemplified by the change in pK_a of the carboxylic acid group of **1.3a** from 6.90 to 6.6 in αCD.**1.3a** and to 6.2 in βCD.**1.3a**, and for **1.4a** from 4.68 to 5.89 in αCD.**1.4a** and to 5.88 in βCD.**1.4a** [133].

The structures of CD complexes in solution are the subject of much discussion. It has been widely assumed from the earliest complexation studies that a major portion of the hydrophobic part of a guest occupies a site within the hydrophobic region of the CD annulus on the basis that guests possessing substantial hydrophobic character tend to complex more strongly than those which do not. Increasingly sophisticated computational [55] and NMR studies [45] have brought more certainty to the structure determination of CD complexes in solution. Thus, the ROESY ^{1}H NMR technique identifies through-space dipolar interactions between the protons of a guest inside the CD annulus and the protons of the annulus interior from which the position of the guest within the annulus can be accurately determined. Generally, computed structures show substantial agreement with those deduced from such NMR studies.

Usually there is strong evidence that entry of a substantial portion of the guest into the CD annulus occurs on complexation. However, examples of complexes where this appears not to be a dominant interaction have been reported as shown by the βCD complexation of the bile pigment (4Z,15Z)-bilirubin-IXa where the enantioselective binding is dominated by hydrogen bonding between the guest carboxylate groups and the βCD secondary hydroxy groups [135]. Similar hydrogen bonding between the (P)- and (M)-1,12-dimethylbenzoic[c]phenanthrene-5,8-dicarboxylate enantiomers and βCD appears to stabilise diastereomeric complexes. Those of the (P)- and (M)-enantiomers are characterised by $K_{11} = 2.2 \times 10^3$ dm^3 mol^{-1} and 1.87×10^4 dm^3 mol^{-1}, respectively, which are quite large despite the expectation that a combination of the considerable guest size and 2- charge minimises entry into the βCD annulus [136].

Many structures of CD complexes have been determined by X-ray crystallography [24,27], and often there appears to be a substantial similarity between these solid state structures and those determined in solution and computed. However, the obvious difference between the solution and solid states is the relatively high CD and guest mobility and the competing interactions between CD, guest and solvent which influence complex stoichiometry and stereochemistry in solution. Accordingly, it is unsurprising when structural differences appear to exist between CD complexes in the two states. This is exemplified by the solution structures of the βCD.pyrene and βCD$_2$.pyrene complexes shown schematically as **1.5** and **1.6**, in which spectroscopic studies are consistent with a substantial portion of the pyrene guest being inside the βCD annulus and aligned parallel to its axis [137-143]. However, X-ray diffraction studies show that in crystalline βCD.1.5 octanol.pyrene.14.5 H$_2$O, pyrene sits parallel to the secondary faces of two βCDs which are hydrogen bonded to each other (**1.7**) [144]. A second example of such differences in guest orientation is provided by the αCD.4-fluorophenol complex where X-ray crystallographic studies show the flourine substituent to protrude from the secondary face of αCD (**1.8** [145]) while ROESY ^{1}H NMR studies show the guest orientation to be reversed in solution so that the hydroxy substituent protrudes from the secondary face of αCD (**1.9**) [146]. Thus, some caution is appropriate in extrapolating from solid state observations of CD complexes to the solution state.

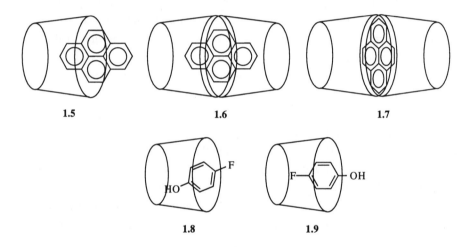

1.5 **1.6** **1.7**

1.8 **1.9**

1.6. Kinetics of Cyclodextrin Complexation

The complexation of a guest species by a CD involves the partial or complete dehydration of the guest as it enters the CD annulus, displacement of water from the annulus and adjustment of the guest to the thermodynamically most favourable orientation within the annulus. This involves a sequence of many small steps as the guest moves towards its optimum orientation in the complex, and both the guest and the CD adjust their hydration to that of the final complex. In principle, depending on the energy barrier associated with each complexation step, some of these steps should produce a spectroscopic or other change at a rate within the timescale of a kinetic technique. In practice, usually only one, and sometimes two, of these steps are observed. The latter sequence for the CD complexation of a guest molecule (G) to form a 1:1 complex may be represented as in Eq. (1.10) where $k_1/k_{-1} = K_1$ and $k_2/k_{-2} = K_2$ which are related to the stability constant for the CD.G complex discussed in section 1.3 by $K_{11} = k_1/k_{-1} + (k_1k_2)/(k_{-1}k_{-2})$.

$$CD + G \underset{k_{-1}}{\overset{k_1}{\rightleftharpoons}} CD.G' \underset{k_{-2}}{\overset{k_2}{\rightleftharpoons}} CD.G \qquad (1.10)$$

In many cases, simple inspection does not reveal which end of the guest first enters the CD annulus. This problem is minimised when one end of the guest is too large to enter the CD annulus as is the case of the naphthalene moiety in **1.10** in the formation of 1:1 complexes with αCD [147]. In this temperature-jump spectrophotometric study only a single relaxation is found consistent with the k_1/k_{-1} equilibrium processes (Eq. (1.10)) being too fast to be kinetically characterised so that the slower k_2/k_{-2} processes are quantified alone. When R^1 is varied from H to Me to Et, and R^2 is either OH or O$^-$, both k_2 and k_{-2} decrease greatly consistent with increasing steric hindrance similarly slowing both rate processes, and as a consequence K_{11} is not greatly affected by this change in **1.10**.

1.10

When both ends of the guest are of similar size the identification of the end entering the CD annulus through the secondary face first is less obvious. This difficulty may be tackled through a systematic comparison of complexation rates with structural variations in the guest as exemplified by a temperature-jump and stopped-flow spectrophotometric study of αCD complexation of the azo-dye guests **1.11a-1.13** [148]. While the k_1/k_{-1} ($= K_1$) equilibrium processes (Eq. (1.10)) are again too fast to kinetically characterise so that only the slower k_2 and k_{-2} processes are kinetically quantified, they still strongly affect the kinetics of the approach to equilibrium. Thus, the variation of the observed first order rate constant for the approach to equilibrium, k_{obs}, with change in the total concentration of αCD ($[\alpha CD]_t$), where $[\alpha CD]_t$ is in great excess over that of the guest dye, facilitates the derivation of K_1 values through Eq. (1.11).

$$k_{obs} = \frac{K_1 k_2 [\alpha CD]_t}{1 + K_1 [\alpha CD]_t} + k_{-2} \tag{1.11}$$

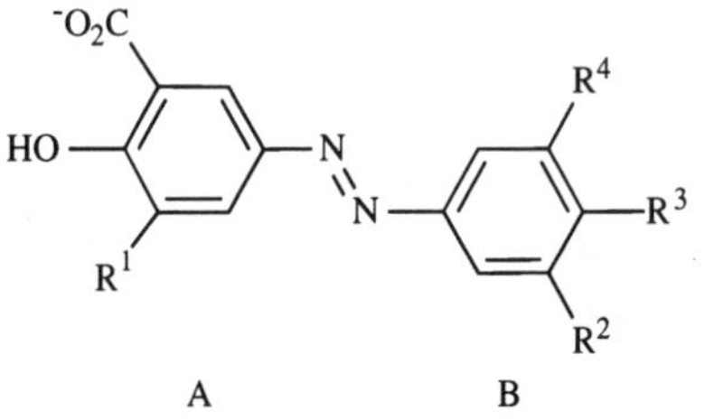

	R$_1$	R$_2$	R$_3$	R$_4$
1.11a	Me	H	SO$_3^-$	H
1.11b	H	H	SO$_3^-$	H
1.11c	H	H	NH$_2$	H
1.11d	H	Me	H	Me
1.11e	H	H	Cl	H
1.11f	H	H	OEt	H
1.11g	H	H	Me	H
1.11h	Me	Me	H	Me
1.11i	H	H	Pr	H

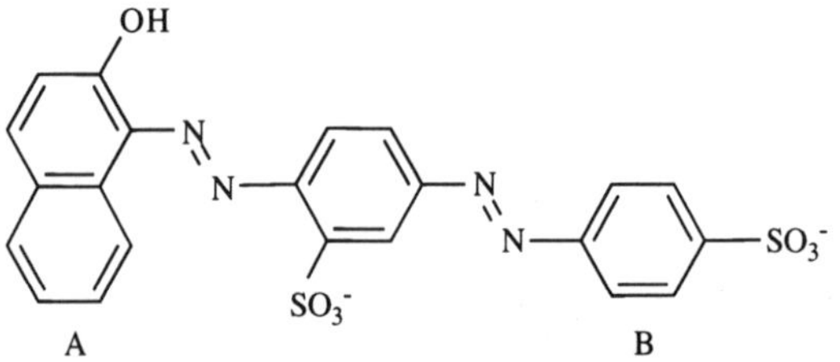

1.12

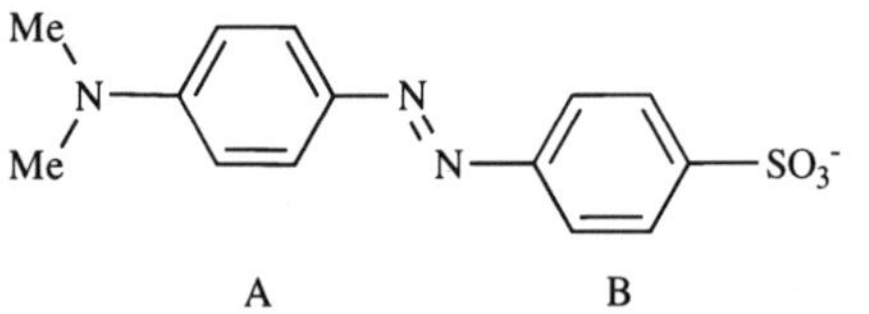

1.13

By varying the substituents on the A and B rings of the azo-dyes, charge and steric effects are found to substantially influence the magnitude of K_1, k_2 and k_{-2} (Table 1.5). It is assumed that for guests with the same entering group, the quantified rates of entry (k_2) are comparable, irrespective of the structure of the rest of the molecule, and that the leaving rate should show a greater dependence on the whole structure of the guest. This assumption infers that the rate determining step for entry into the CD annulus involves a transition state close to the final ground state of the complex. As **1.11h** is not complexed by αCD, it is deduced that the 3,5-dimethylphenyl and 3-methylsalicylate moieties are too large to enter the αCD annulus. Thus, the phenylsulfonate and salicylate moieties are the entering ends for **1.11a** and **1.11d**.

Table 1.5. Parameters for the Complexation of Aza-Dyes by αCD in Aqueous Solution at 298.2 K.

Guest	K_1 $dm^3\ mol^{-1}$	k_2 s^{-1}	k_{-2} s^{-1}	End entering first
1.11a	4.3×10^2	1.6	0.28	B
1.11b	6.2×10^2	2.9×10^3	6.0×10^2	A
1.11c	2.0×10^3	1.1×10^4	2.7×10^3	B
1.11d	4.8×10^2	4.4×10^3	7.0×10^2	A
1.11e	1.1×10^3	5.7×10^4	2.2×10^4	B
1.11f	1.5×10^3	1.9×10^4	9.8×10^2	B
1.11g	7.4×10^2	3.1×10^4	3.0×10^4	B
1.11h	no complexation			none
1.11i	2.0×10^3	5.4×10^3	2.6×10^3	B
1.12	1.1×10^3	0.6	0.5	B
1.13	1.6×10^2	1.2×10^4	3.2×10^2	A

Within the magnitude range of the parameters in Table 1.5, those for **1.11a** and **1.12** are similar and confirm phenylsulfonate as the entering moiety for both as the naphthalene moiety of **1.12** is too big to enter the αCD annulus. The similarity of parameters for **1.11b** and **1.11d** confirms that the salicylate is the entering moiety for both. The similar k_{-2} values for **1.11b** and **1.11d** are consistent with the other end of these guests having a minor influence, as is also the case for **1.11a** and **1.12**. For **1.11c, 1.11e, 1.11f, 1.11g** and **1.11i** both k_2 and k_{-2} are greater than those of **1.11d** consistent with the salicylate moiety not being the entering end for these guests. For Methyl Orange, **1.13**, k_2 is much larger than for **1.11a** and **1.12** were the phenylsulfonate moiety is thought to enter the annulus and accordingly it appears that the *N,N*-dimethylaniline moiety enters first for this guest.

The complexation of the alkyl-substituted hydroxyphenylazosulfanilic acid derivatives in their HA^- and A^{2-} forms, **1.14a-1.14i**, by αCD exhibits both of the k_1/k_{-1} and k_2/k_{-2} steps of Eq. (1.10), with the exception of **1.14a**, **1.14b** and **1.14h** [113,149,150]. The rate constants k_1, k_{-1}, k_2 and k_{-2}, determined by stopped-flow spectrophotometry, show significant variations which can be rationalised in terms of the steric crowding caused by the changing size of R^1 and R^2 (Table 1.6) On steric grounds, it is considered that either end of **1.14a-1.14c** may enter through the secondary face of αCD while the smaller size of the phenyl sulfonate ends of **1.14d-1.14i** favours entry of this end predominantly. This is supported by the marked decrease in the size of k_1 and k_{-1} from **1.14a** to **1.14c** and the much smaller variation in k_1 and k_{-1} from **1.14c** to **1.14f** and for **1.14h**. Only a single rate

The monoanions, HA^- are as shown and the dianions, A^{2-}, have deprotonated hydroxy groups.

	R^1	R^2
1.14a	H	H
1.14b	Me	H
1.14c	Et	H
1.14d	Pr	H
1.14e	*i*-Pr	H
1.14f	*s*-Bu	H
1.14g	*t*-Bu	H
1.14h	Me	Me
1.14i	*i*-Pr	*i*-Pr

process is observed for **1.14a** and **1.14b** which may indicate that these guests are sufficiently slender to approach their optimum complexation site after passing over one major energy barrier. In contrast, the well established two-step complexation of **1.14d-1.14f** and **1.14h** is consistent with passage over two major energy barriers where the appearance of the second suggests a steric interaction between the R^1 substituents and αCD as these guests move deeper into the annulus in the second step. The disubstituted **1.14h** shows similar kinetic characteristics to those of **1.14e** and **1.14f** which are singly substituted with larger substituents. The slowness of complexation of **1.14g** and **1.14i** and the observation of a single complexation step is attributed to the bulkiness of their R^1 and R^2 substituents limiting the depth of entry of these guests into the αCD annulus.

Table 1.6. Rate Constants for the Complexation of Alkyl-Substituted Hydroxyphenylazosulfanilic Acid Derivatives by αCD.

Guest	HA$^-$				A^{2-}			
	k_1 $\mathrm{mol^{-1}\,dm^3\,s^{-1}}$	k_{-1} $\mathrm{s^{-1}}$	k_2 $\mathrm{s^{-1}}$	k_{-2} $\mathrm{s^{-1}}$	k_1' $\mathrm{mol^{-1}\,dm^3\,s^{-1}}$	k_{-1}' $\mathrm{s^{-1}}$	k_2' $\mathrm{s^{-1}}$	k_{-2}' $\mathrm{s^{-1}}$
1.14a	$>10^7$	$>10^3$	b	b	$>10^7$	$>10^3$	b	b
1.14b	7.0×10^5	77	b	b	9.3×10^3	2.0	b	b
1.14c	2.1×10^4	6.5	c	c	8.1×10^3	2.1	c	c
1.14d	2.0×10^4	6.0	0.87	0.55	6.9×10^3	3.6	0.25	0.08
1.14e	1.2×10^4	9.4	0.58	0.26	9.0×10^3	15.0	1.47	0.10
1.14f,	1.1×10^4	14.0	0.8	0.16	7.0×10^3	20.0	1.67	0.08
1.14g	4.6×10^2	0.55	c	c	3.4×10^2	0.41	c	c
1.14h,	1.1×10^4	4.0	0.47	0.28	9.63×10^3	10	0.64	0.14
1.14i	5.4×10^2	0.7	c	c	3.5×10^2	0.5	c	c

[a]In aqueous solution at 298.2 K and $I = 0.1$ mol dm^{-3} (NaCl). [b]Undetected by stopped-flow spectrophotometry [c]Absorbance change too small for rate constant determination.

An interpretation of the sequence of events accompanying the complexation of **1.14d** and the associated activation parameters is shown in Fig. 1.3 [113]. Here, the hydration of **1.14d** is shown in **1.15a** as two components, the hydrogen bonding interactions between water and the hydrophilic hydroxy and sulfonate groups, and the self-hydrogen bonding water clustering around the hydrophobic regions of **1.14d**. In the first complexation step (k_1) to form the transition state, $\mathbf{1.15b^{\ddagger}}$, $\Delta S_1^{\ddagger}$ makes a major contribution to $\Delta G_1^{\ddagger}$. This is consistent with the sulfonate losing much of its hydration to displace some of the water in the αCD annulus to produce a large negative $\Delta S_1^{\ddagger}$ mainly through a loss of motional freedom of both participants in the first transition state. (This loss is expected to contribute -209 to -251 J mol^{-1} K^{-1} to $\Delta S_1^{\ddagger}$, with positive contributions from dehydration of the sulfonate. The entropy increase from the partial collapse of the water cluster around the hydrophobic region of

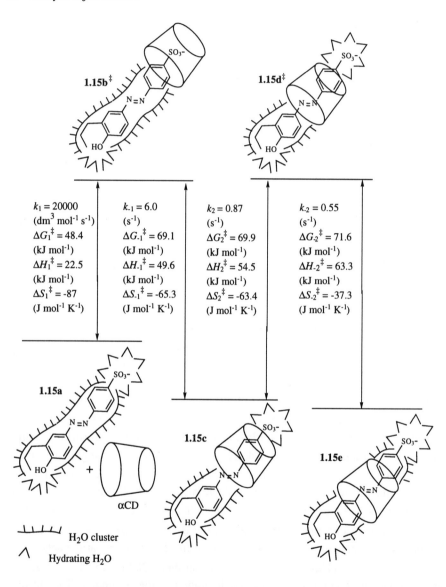

Fig. 1.3. The two-step mechanism proposed for the formation of the αCD complex (**1.15e**) of **1.15a** (**1.14d** shown with its hydration shell) through the intermediate complex **1.15c**. Two transition states, **1.15b‡** and **1.15d‡**, are shown although the latter does not appear in the original reference [113] and is shown here as midway between the structures of **1.15c** and **1.15e**.

the guest and the expulsion of water from the αCD annulus apparently sum to about 0.6 of $\Delta S_1^{\ddagger}$.) The reverse process from the intermediate species (**1.15c**) to **1.15b**‡ is also characterised by a negative but smaller $\Delta S_{-1}^{\ddagger}$, and now $\Delta G_{-1}^{\ddagger}$ is dominated by the larger $\Delta H_{-1}^{\ddagger}$ contribution. The dominance of $\Delta G_2^{\ddagger}$ and $\Delta G_{-2}^{\ddagger}$ by their large $\Delta H_2^{\ddagger}$ and $\Delta H_{-2}^{\ddagger}$ contributions accounts for the relative slowness of the k_2 and k_{-2} processes with the negative $\Delta S_2^{\ddagger}$ and $\Delta S_{-2}^{\ddagger}$ making a lesser contribution to this slowness.

Based on the kinetic data, the formation of the intermediate complex, **1.15c**, is characterised by $K_1 = 3.3 \times 10^3$ dm^3 mol^{-1}, $\Delta G_1^o = -20.7$ kJ mol^{-1}, the formation of the final complex, **1.15e**, from **1.15c** by $K_2 = 1.6$ dm^3 mol^{-1}, $\Delta G_2^o = -1.7$ kJ mol^{-1}, and the formation of **1.15e** from **1.15a** and αCD by $K_{11} = K_1 + K_1 K_2 = 8.6 \times 10^3$ dm^3 mol^{-1}. This compares with $K_{11} = 8.3 \times 10^3$ dm^3 mol^{-1}, determined by equilibrium spectrophotometry, $\Delta G_{11}^o = -22.5$ kJ mol^{-1}, $\Delta H_{11}^o = -33.3$ kJ mol^{-1} and $\Delta S_{11}^o = -32.0$ J K^{-1} mol^{-1}.

With the exception of **1.14a**, k_1 for HA$^-$ is larger than k_1' for A^{2-} and this is attributed to the increased hydration of the guests as the hydroxy group deprotonates (Table 1.6). While the difference between k_1 and k_1' for **1.14b** is large and consistent with entry by the phenolic end being dominant, the differences for **1.14c**-**1.14i** are modest consistent with entry from the sulfonate end. The changes in k_2 and k_{-2} are also quite small and consistent with this interpretation. Studies of the analogues of **1.14h** where SO$_3^-$ is replaced by either CO$_2^-$, SO$_2$NH$_2$ or AsO$_3$H$^-$, show that the magnitudes of k_1 and k_{-1} decrease in the sequence CO$_2^-$ > SO$_2$NH$_2$ > > AsO$_3$H$^-$ as the substituent size increases, and one-step complexations are observed for the first and third systems while a two-step complexation applies to the second [150]. The complexation of the analogues of **1.14a**-**1.14g** and **1.14i**, where the phenylsulfonate moieties are replaced by 2-naphthylsulfonate moieties, exhibit one-step and two-step sequences depending on the steric nature and the charge of the guest [151].

Having considered the kinetic aspects of their complexation, it is appropriate to explore some of the factors affecting the stabilities of the CD complexes of some of the above guests. The effect of CD and guest size on complexation is shown by the variation of K_{11} for the complexes of **1.14a**-**1.14g** and **1.14i** formed with αCD and βCD which vary in the sequence: 6.3×10^3 (2.0×10^3), 8.3×10^3 (2.0×10^3), 1.0×10^4 (4.8×10^3), 8.3×10^3 (5.6×10^3), 4.5×10^3 (2.5×10^3), 5.6×10^3 (4.2

$\times 10^3$), 1.9×10^3 (3.4×10^3) and 9.1×10^2 (1.3×10^3) dm^3 mol^{-1} (K_{11} for βCD is shown in brackets), respectively, in aqueous solution at 298.2 K and $I = 0.1$ mol dm^{-3} (NaCl) [152]. The complex stability increases to a maximum for αCD with **1.14c** and thereafter decreases. A similar pattern is seen for βCD which forms its most stable complex with **1.14d**. However, the αCD complexes are the more stable, except with the larger **1.14g** and **1.14i**. This pattern of stability variation is consistent with complex stability being dominated by the optimisation of fit of the guest to the CD annulus for this series of guests.

When a natural CD is modified it is to be expected that its complexation characteristics will also be modified. This expectation is fulfilled in the complexation of **1.14d**, **1.14e** and **1.14g** by hexakis(2,6-di-*O*-methyl)-αCD (DMαCD), hepta-kis(2,6-di-*O*-methyl)-βCD (DMβCD) and heptakis(2,3,6-tri-*O*-methyl)-βCD (TMβCD) where methylation at O(2) and O(6) in the first two cases and O(2), O(3) and O(6) in the third case greatly extends the hydrophobic region of the annuli by 8-11 Å beyond that in the natural CDs [24,153]. The K_{11} for the αCD complexes of **1.14d**, **1.14e** and **1.14g** are 8.3×10^3, 4.5×10^3 and 1.9×10^3 dm^3 mol^{-1} while those for their DMαCD analogues are increased to 4.9×10^4, 2.7×10^4 and 7.4×10^3 dm^3 mol^{-1} probably because the increased hydrophobicity of the DMαCD annulus enhances the interaction with the hydrophobic moieties of the guests in the complexes [154]. A similar stability increase is seen for the βCD-based systems where K_{11} for the βCD complexes of **1.14d**, **1.14e** and **1.14g** are 5.6×10^3, 2.5×10^3 and 3.4×10^3 dm^3 mol^{-1} and those of their DMβCD analogues are increased to 2.0×10^4, 2.2×10^4 and 2.2×10^4 dm^3 mol^{-1}, and is similarly explained. However, $K_{11} = 5.5 \times 10^3$, 3.6×10^3 and 4.7×10^3 dm^3 mol^{-1} found for the TMβCD complexes show only modest changes from those for the βCD complexes of **1.14d**, **1.14e** and **1.14g** possibly because the methylation at O(3) inhibits entry through the secondary face of the annulus. (X-Ray crystallographic studies show distortion of the CD ring of TMβCD and steric crowding of the -O(2)-CH$_3$ and -O(3)-CH$_3$ groups [153], and similar steric crowding and distortions occur in TMαCD [155]. By comparison both DMαCD [156] and DMβCD [157] are much less sterically crowded.) The αCD complexes of the deprotonated dianionic **1.14d** and **1.14e** (A^{2-}) show small decreases in stability by comparison with their HA$^-$ analogues, and that of **1.14g** shows a substantial decrease. With βCD, DMαCD and DMβCD, dianionic **1.14d**, **1.14e** and **1.14g** form significantly less stable complexes than their monoanionic (HA$^-$) analogues.

This is probably because the decreased hydrophobicity of these guests lessens their interactions with the hydrophobic annuli of the CDs.

The modification of αCD to DMαCD has the effect of rendering distinguishable three steps of the DMαCD complexation of **1.14d**, **1.14e** and **1.14g** and their dianionic analogues so that the three-step Eq. (1.12) applies [154]:

$$DM\alpha CD + G \xrightleftharpoons[k_{-1}]{k_1} DM\alpha CD.G'' \xrightleftharpoons[k_{-2}]{k_2}$$

$$DM\alpha CD.G' \xrightleftharpoons[k_{-3}]{k_3} CD \cdot G \qquad (1.12)$$

The rapidity of the complexation steps decreases from the first to the third such that the first step is only measurable as the ratio $k_1/k_{-1} = K_1$. Thus, when G = **1.14g**, stopped-flow spectrophotometric measurements yield $K_1 = 2.6 \times 10^2$ dm^3 mol^{-1}, $k_2 = 1.3 \times 10^2$ s^{-1}, $k_{-2} = 6.0$ s^{-1}, $k_3 \sim 0.7$ s^{-1} and $k_{-3} \sim 0.3$ s^{-1} in aqueous solution at $I = 0.1$ (NaCl) and 298.2 K. It is envisaged that the sequence of complexation steps is similar to that in Fig. 1.3, with the addition of a third detectable step as a result of the extension of the DMαCD annulus through methylation.

1.16a   1.16b

The volumes of activation and reaction, $\Delta V^{\ddagger}$ and ΔV^{o}, respectively, give interesting insights into the two-step complexation of the dyes Ethyl Orange (**1.16a**) and Mordant Yellow (**1.16b**) by αCD [158-159]. Their complexations are envisaged as being similar to that shown in Fig. 1.3 with the phenylsulfonate end entering the αCD first. The parameters for both complexations are given in Table 1.7, and $\Delta V^{\ddagger}$ and ΔV^{o} for the complexation of Mordant Yellow are shown in the volume profile of Fig. 1.4. The substantial negative value for $\Delta V_1^{\ddagger}$ is attributed to dehydration of the sulfonate group and its entry into the annulus and its interaction with water therein in the first transition state, $\alpha CD.\mathbf{1.16b}^{\ddagger\prime}$, which represents a contraction in volume compared with that of the ground state $\alpha CD + \mathbf{1.16b}$. In the intermediate $\alpha CD.\mathbf{1.16b}'$ the water has been expelled from the annulus and the sulfonate has become rehydrated as it protrudes from the primary end of αCD, and ΔV_1° is only slightly negative. The very negative $\Delta V_2^\ddagger$ is attributed to the formation of hydrogen bonds between the hydroxy and carboxylate groups of $\mathbf{1.16b}$ and αCD during the formation of the second transition state, $\alpha CD.\mathbf{1.16b}^\ddagger$, where the αCD may be distorted as a consequence of rotation of one or more of the glucopyranose rings about the glycosidic linkages [42]. Subsequently, this distortion may relax to produce an

Table 1.7. Parameters for αCD Complexation of Ethyl Orange, **1.16a**, and Mordant Yellow, **1.16b**, in Water.

Parameters	Guest		Parameters	Guest	
	Ethyl Orange	Mordant Yellow		Ethyl Orange	Mordant Yellow
$k_1/dm^3\ mol^{-1}\ s^{-1\ a}$	12200	15200	$k_2/s^{-1\ a}$	0.199	1.83
$k_{-1}/s^{-1\ a}$	1.84	25.4	$k_{-2}/s^{-1\ a}$	0.093	0.17
$\Delta H_1^\ddagger/kJ\ mol^{-1}$	20.1	27.3	$\Delta H_2^\ddagger/kJ\ mol^{-1}$	72.0	50.8
$\Delta H_{-1}^\ddagger/kJ\ mol^{-1}$	54.8	38.6	$\Delta H_{-2}^\ddagger/kJ\ mol^{-1}$	43.8	39.5
$\Delta S_1^\ddagger/J\ K^{-1}\ mol^{-1}$	-99.3	-73.3	$\Delta S_2^\ddagger/J\ K^{-1}\ mol^{-1}$	-16.8	-69.4
$\Delta S_{-1}^\ddagger/J\ K^{-1}\ mol^{-1}$	-55.9	-88.6	$\Delta S_{-2}^\ddagger/J\ K^{-1}\ mol^{-1}$	-117.7	-127.0
$\Delta V_1^\ddagger/cm^3\ mol^{-1}$	-22.1[b]	-20.9[c]	$\Delta V_2^\ddagger/cm^3\ mol^{-1}$	-1.8[b]	-15.8[c]
$\Delta V_{-1}^\ddagger/cm^3\ mol^{-1}$	-15.8[b]	-17.3[c]	$\Delta V_{-2}^\ddagger/cm^3\ mol^{-1}$	-18.8[b]	-21.8[c]
$\log(K_{11}/dm^3\ mol^{-1})^{a,d}$	4.3	3.8			

[a] At 298 K. [b] At 308 K. [c] At 288 K. [d] $K_{11} = k_1/k_{-1} + (k_1 k_2)/(k_{-1} k_{-2})$

increase in volume as the final ground state, $\alpha CD.\mathbf{1.16b}$, forms with $\Delta V_2^\circ = +6.1$ $cm^3\ mol^{-1}$. Interestingly, the overall volume change from $\alpha CD + \mathbf{1.16b}$ to $\alpha CD.\mathbf{1.16b}$ is +2.5 $cm^3\ mol^{-1}$ which corresponds to the sum of hydration changes and conformational changes in αCD. A similar volume profile pertains to Ethyl Orange with a larger volume increase on going from the starting reactants to the final complex of +10.7 $cm^3\ mol^{-1}$.

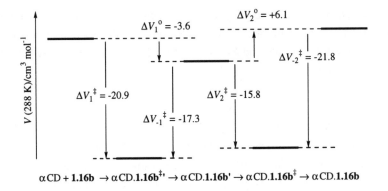

$$\alpha CD + \textbf{1.16b} \rightarrow \alpha CD.\textbf{1.16b}^{\ddagger\prime} \rightarrow \alpha CD.\textbf{1.16b}' \rightarrow \alpha CD.\textbf{1.16b}^{\ddagger} \rightarrow \alpha CD.\textbf{1.16b}$$

Fig. 1.4. Volume profile for the complexation of Mordant Yellow by αCD.

The kinetics of the formation of 1:2 $CD.G_2$ complexes have also been studied (Eq. (1.13)). When G is Tropaeolin 000 No. 2 (**1.17**) and CD is either TMβCD or γCD, temperature-jump spectrophotometric studies show that k_2 and $k_{-2} = 1.68 \times 10^8$ and 2.27×10^9 dm^3 mol^{-1} s^{-1}, and 3.48×10^3 and 1.38×10^3 s^{-1}, respectively, at 298.2 K in aqueous 0.200 mol dm^{-3} K_2SO_4, and k_1 and k_{-1} are too large to measure [110]. (In contrast to the earlier discussed examples, only a single step is detected in

$$CD + 2G \underset{k_{-1}}{\overset{k_1}{\rightleftharpoons}} CD.G + G \underset{k_{-2}}{\overset{k_2}{\rightleftharpoons}} CD.G_2 \qquad (1.13)$$

the formation of CD.G, and also for $CD.G_2$). The corresponding k_1/k_{-1} and k_2/k_{-2} are 1.24×10^2 and 4.82×10^4, and 4.18×10^2 and 1.68×10^6 dm^3 mol^{-1}. The dimerisation of Tropaeolin in the absence of CD is characterised by $K_d = 9.1 \times 10^2$ dm^3 mol^{-1} which by comparison k_2/k_{-2}, shows that the complexed dimer is substantially stabilised in $CD.G_2$.

1.17

1.18

1.19

The formation of a 2:2 $\beta CD_2.G_2$ complex is detected when G is 1-(2-naphthyl)-ethanol (**1.18**) in the equilibria shown in Eq. (1.14) where $k_1 = 2.9 \times 10^8$ dm^3 mol^{-1} s^{-1}, $k_{-1} = 1.8 \times 10^5$ s^{-1} and $k_d = (0.2\text{-}2.5) \times 10^3$ s^{-1} determined by laser flash photolysis at 293 K in aqueous solution [160]. The excitation of guests from their electronic ground states to their triplet states may change their complexing characteristics as exemplified by the complexing of xanthone, **1.19** [161]. Thus, for βCD and γCD the ground state $K_{11} = 1.1 \times 10^3$ dm^3 mol^{-1} and 2.2×10^2 dm^3 mol^{-1} in water at 293 K, respectively, while for the triplet excited state $K_{11} = 4.8 \times 10^2$ dm^3 mol^{-1} and < 4 dm^3 mol^{-1}. It is thought that triplet state xanthone has a greater dipole moment than its ground state so that water hydrates the triplet state more strongly [162]. For the βCD and γCD triplet state complexes, $k_1 = 4 \times 10^8$ dm^3 mol^{-1} s^{-1} and $< 3 \times 10^7$ dm^3 mol^{-1} s^{-1}, respectively, and $k_{-1} = 8.4 \times 10^6$ s^{-1} and 7.3×10^6 s^{-1}.

$$2\beta CD + 2G \; \underset{k_{-1}}{\overset{k_1}{\rightleftharpoons}} \; 2\beta CD.G \; \underset{k_{-d}}{\overset{k_d}{\rightleftharpoons}} \; \beta CD_2.G_2 \qquad (1.14)$$

1.7. Chiral Discrimination in Cyclodextrin Complexes

The homochirality of CDs gives them the potential to discriminate between guests in forming complexes according to their chirality and this has been a major area of study for both natural and modified CDs [21,163,164]. Since early studies of the enantioselectivity displayed by natural CDs in the formation of diastereomeric complexes [165] the quantification of the extent of chiral discrimination and interpretation of its origins have proceeded in parallel for both the natural CDs and the modified CDs synthesised in an effort to enhance this thermodynamic chiral discrimination. A model for such thermodynamic discrimination, derived by Armstrong, has formed the basis of much of the interpretation of observed chiral discrimination by CDs and the design of modified CDs to enhance such discrimination [164]. It is postulated that i) the guest must be complexed within the CD annulus, ii) the fit of the guest into the annulus must be tight, iii) the stereogenic centre of the guest should form one strong interaction with the hydroxy groups at the annulus entrance and iv) the secondary hydroxy groups of the CD are particularly important in chiral discrimination. With the increasing sophistication of molecular modelling techniques, this model has been subjected to some refinement [55,166], but it still forms a sound basis for the understanding of chiral discrimination in natural CD complexes.

Generally, thermodynamic discrimination displayed by αCD, βCD and γCD in complexing chiral guests in aqueous solution is modest [21,86,90,164,167], but markedly different NMR chemical shifts are often displayed by the enantiomers and diastereomers of complexed guests [45,86,168-171]. This is exemplified by a particularly detailed ^{1}H and ^{13}C NMR and molecular dynamics simulation study of the complexation of tryptophan by αCD [166]. Chemical shift, coupling constant and relaxation time changes caused by complexation are greater for (R)- than for (S)-tryptophan consistent with the (R)-enantiomer interacting more strongly with αCD than does the (S)-enantiomer. Intermolecular NOE interactions between αCD and both (R)- and (S)-tryptophan are consistent with the indole ring being in the vicinity of the αCD secondary hydroxy groups and the benzene ring protruding into the annulus consistent with similar complexing modes in both diastereomeric complexes. In both cases molecular dynamic modelling shows that this tryptophan orientation is about 4 kJ mol^{-1} more stable than the reversed orientation. This modelling also shows that (R)- and (S)-tryptophan are off-centre and differently oriented in the αCD annulus with (R)-tryptophan forming twice as many hydrogen bonds as (S)-tryptophan. This is attributed to (R)-tryptophan carboxylate oxygens and the indole NH readily hydrogen bonding with αCD secondary hydroxy groups, while the opposite chirality of (S)-tryptophan renders it less able to form hydrogen bonds. In neither case does the NH$_3^+$ group of the tryptophan zwitterion appear to hydrogen bond with αCD as it projects too far out from the annulus. Overall, hydrogen bonding stabilises the αCD (R)-tryptophan complex by 12.6 kJ mol^{-1} more than is the case for the αCD (S)-tryptophan complex. A ^{1}H NMR and molecular mechanics study of the complexation of (R,S)-3-(2-thienyl)piperizin-2-one (Tenilsetam, CAS-997 - an anti-amnesic drug) shows guest complexing by the thiophene ring through the secondary face of αCD and βCD, with deeper penetration into the annulus in the latter case, while the piperizinone ring protrudes from the annulus [172]. Hydrogen bonding with αCD and βCD provides specific sites for differentiation between the enantiomeric guests in the diastereomeric complexes. No complexation by γCD is detected. Similar modelling studies have been reported for chiral discrimination by DMβCD [173,174]. Generally, molecular modelling studies indicate that the forces determining the extent of chiral discrimination are small by comparison with the total guest binding force in the complex.

Substantial differences between the diastereomeric complexes formed by (*R,S*)-2-(3-phenoxy)phenylpropionic acid (Fenoprofen, a non-steroidal anti-inflammatory drug) occur in the solid state [175]. X-ray crystallography shows that in the βCD complexes of (*R*)- and (*S*)-Fenoprofen βCD is present as a head-to-head dimer as a result of extensive intermolecular hydrogen bonding between secondary hydroxy groups of adjacent βCDs with one molecule of Fenoprofen present in each βCD annulus. However, while βCD is isomorphous in the two complexes, the (*R*)-Fenoprofen guest is oriented with its propionic acid group protruding from the primary face of βCD and its phenoxy group protruding from the secondary face, whereas (*S*)-Fenoprofen shows this orientation and its reverse in a repeating pairwise arrangement. (*S*)-Fenoprofen is the pharmaceutically active enantiomer. Due to the homochiral nature of the CD annulus, chiral interactions are ubiquitous in CD chemistry and accordingly examples of chiral discrimination appear throughout the succeeding chapters of this book.

1.8. References

1.   A. Villiers, *Compt. Rend.* **112** (1891) 536.
2.   a) F. Z. Schardinger, *Unters. Nahr. u. genussm.* **6** (1903) 865. b) F. Z. Schardinger, *Wien. Klin. Wochenscht.* **17** (1904) 207. c) F. Z. Schardinger, *Zentralbl. Bakteriol. Parasintenk. Abt. 2* **14** (1905) 772. d) F. Z. Schardinger, *Zentralbl. Bakteriol. Parasintenk. Abt. 2* **29** (1911) 118.
3.   J. Szejtli, *Chem. Rev.* **98** (1998) 1743.
4.   J. Szejtli, *J. Mater. Chem.* **7** (1997) 575.
5.   *Cyclodextrins and Their Industrial Uses*, ed. D. Duchêne (Editions Santé, Paris, 1987).
6.   J. Szejtli, *Cyclodextrin Technology* (Kluwer, Dordrecht, 1988).
7.   S. M. Han and D. W. Armstrong, in *Chiral Separations by HPLC*, ed. A. M. Krstalovic (John Wiley and Sons, New York, 1989), p. 208.
8.   D. Duchêne, *New Trends in Cyclodextrins and Derivatives* (Editions Santé, Paris, 1991).

9. S. Li and W. C. Purdy, *Chem. Rev.* **92** (1992) 1457.

10. K. H. Frömming and J. Szejtli, in *Cyclodextrins in Pharmacy* (Kluwer, Dordrecht, 1994).

11. B. Chankvetadze, G. Endresz and G. Blaschke, *Chem. Soc. Rev.* **25** (1996) 141.

12. *Comprehensive Supramolecular Chemistry*, eds. J. L. Atwood, J. E. D. Davies, D. D. MacNicol, F. Vögtle and J.-M. Lehn, Vol. 3, eds. J. Szejtli and T. Osa (Pergamon, Oxford, 1996).

13. A. R. Hedges, *Chem. Rev.* **98** (1998) 2035.

14. K. Uekama, F. Hirayama and T. Irie, *Chem. Rev.* **98** (1998) 2045.

15. M. L. Bender and M. Komiyama, *Cyclodextrin Chemistry* (Springer-Verlag, New York, 1978).

16. W. Saenger, *Angew. Chem., Int. Ed. Engl.* **19** (1980) 344.

17. W. Saenger, in *Inclusion Compounds*, eds. J. L. Atwood, J. E. D. Davies and D. D. MacNicol (Academic Press, London, 1984) Vol. 2, p. 231.

18. R. J. Clarke, J. H. Coates and S. F. Lincoln, *Adv. Carbohyd. Chem. Biochem.* **45** (1989) 205.

19. K. Harata, in *Inclusion Compounds*, eds. J. L. Atwood, J. E. D. Davies and D. D. MacNicol (Academic Press, London, 1991), Vol. 5, p. 311.

20. W. Saenger, C. Niemann, R. Herbst, W. Hinrichs and T. Steiner, *Pure Appl. Chem.* **65** (1993) 809.

21. C. J. Easton and S. F. Lincoln, *Chem. Soc. Rev.* **35** (1996) 163.

22. J. Szejtli, *Inclusion of Guest Molecules, Selectivity and Molecular Recognition by Cyclodextrins*, in *Comprehensive Supramolecular Chemistry*, eds. J. L. Atwood, J. E. D. Davies, D. D. MacNicol, F. Vögtle and J.-M. Lehn, Vol. 3, eds. J. Szejtli and T. Osa (Pergamon, Oxford, 1996), p. 189.

23. K. A. Connors, *Measurement of Cyclodextrin Complex Stability Constants*, in *Comprehensive Supramolecular Chemistry*, eds. J. L. Atwood, J. E. D. Davies, D. D. MacNicol, F. Vögtle and J.-M. Lehn, Vol. 3, eds. J. Szejtli and T. Osa (Pergamon, Oxford, 1996), p. 205.

24. K. Harata, *Crystallographic Studies*, in *Comprehensive Supramolecular Chemistry*, eds. J. L. Atwood, J. E. D. Davies, D. D. MacNicol, F. Vögtle and J.-M. Lehn, Vol. 3, eds. J. Szejtli and T. Osa (Pergamon, Oxford, 1996), p. 279.

25. K. A. Connors, *Chem. Rev.* **97** (1997) 1325.

26. W. Saenger, J. Jacob, K. Gessler, T. Steiner, D. Hoffmann, H. Sanbe, K. Koizumi, S. M. Smith and T. Takaha, *Chem. Rev.* **98** (1998) 1787.

27. K. Harata, *Chem. Rev.* **98** (1998) 1803.

28. M. V. Rekharsky and Y. Inoue, *Chem. Rev.* **98** (1998) 1875.

29. S. F. Lincoln and C. J. Easton, *Inclusion Complexes of the Cyclodextrins*, in *Polysaccharides: Structural Diversity and Functional Versatility*, ed. S. Dumitriu (Marcel Dekker, New York, 1998), p. 473.

30. T. Fujiwara, N. Tanaka and S. Kobayashi, *Chem. Lett.* (1990) 739.

31. H. Ueda, T. Endo, H. Nagase, S. Kobayashi and T. Nagai, *J. Inclusion Phenom. Mol. Recogn. Chem.* **25** (1996) 17.

32. T. Endo, H. Nagase, H. Ueda, S. Kobayashi and T. Nagai, *Chem. Pharm. Bull.* **45** (1997) 532.

33. J. Jacob, K. Gessler, D. Hoffman, H. Sanbe, K. Koizumi, S. M. Smith, T. Takaha and W. Saenger, *Angew. Chem., Int. Ed. Engl.* **37** (1998) 606.

34. T. Takaha, M. Yanase, S. Takata, S. Okada and S. M. Smith, *J. Biol. Chem.* **271** (1996) 2902.

35. K. Gessler, I. Uson, T. Takaha, N. Krauss, S. M. Smith, S. Okada, G. M. Sheldick and W. Saenger, quoted in reference 26.

36. M. J. Jozwiakowski and K. A. Connors, *Carbohydr. Res.* **143** (1985) 51.

37. K. Miyajima, M. Sawada and M. Nakagaki, *Bull. Chem. Soc. Jpn.* **56** (1983) 3556.

38. a) R. I. Gelb, L. M. Schwartz, J. J. Bradshaw and D. A. Laufer, *Bioorg. Chem.* **9** (1980) 299. b) R. I. Gelb, L. M. Schwartz and D. A. Laufer, *Bioorg. Chem.* **11** (1982) 274.

39. P. C. Manor and W. Saenger, *J. Am. Chem. Soc.* **96** (1974) 3630.

40. K. Lindner and W. Saenger, *Acta Crystallogr.* **B38** (1982) 203.

41. K. K. Chacko and W. Saenger, *J. Am. Chem. Soc.* **103** (1981) 1708.

42. W. Saenger, M. Noltemeyer, P. C. Manor, B. Hingerty and B. Klar, *Bioorg. Chem.* **5** (1976) 187.

43. M. J. Gidley and S. Bociek, *Carbohydr. Res.* **183** (1988) 126.

44. M. Suzuki, J. Szejtli and L. Szente, *Carbohydr. Res.* **192** (1989) 61.

45. H.-J. Schneider, F. Hacket, V. Rüdiger and H. Ikeda, *Chem. Rev.* **98** (1998) 1755.

46. A. F. Bell, L. Hecht and L. D. Barron, *Chem. Eur. J.* **3** (1997) 1292.

47. J. E. H. Koehler, W. Saenger and W. F. van Gunsteren, *J. Mol. Biol.* **203** (1988) 241.

48. K. B. Lipkowitz, *J. Org. Chem.* **56** (1991) 6357.

49. S. P. van Helden, B. P. van Eijck and L. H. M. Janssen, *J. Biol. Struct. Dyn.* **9** (1992) 1269.

50. K. B. Lipkowitz, K. Green and J.-A. Yang, *Chirality* **4** (1992) 205.

51. D. A. Wertz, C.-X. Shi and C. A. Vernanzi, *J. Comput. Chem.* **13** (1992) 41.

52. H. Dodziuk and K. Nowinski, *J. Mol. Struct. (Theochem.)* **304** (1994) 61.

53. G. Marconi, S. Monti, B. Mayer and G. Köhler, *J. Phys. Chem.* **99** (1995) 3943.

54. B. Mayer and G. Kohler, *J. Mol. Struct. (Theochem.)* **363** (1996) 217.

55. K. B. Lipkowitz, *Chem. Rev.* **98** (1998) 1829.

56. K. Lindner and W. Saenger, *Angew. Chem., Int. Ed. Engl.* **17** (1978) 694.

57. K. Harata, *Bull. Chem. Soc. Jpn.* **60** (1987) 2763.

58. W. Linert, L.-F. Han and I. Likovits, *Chem. Phys.* **139** (1989) 441.

59. Y. Inoue, T. Hakushi, Y. Liu, L.-H. Tong, B.-J. Shen and D.-S. Jin, *J. Am. Chem. Soc.* **115** (1993) 475.

60. Y. Inoue, Y. Liu, L.-H. Tong, B.-J. Shen and D.-S. Jin, *J. Am. Chem. Soc.* **115** (1993) 10637.

61. W. P. Jencks, *Catalysis in Chemistry and Enzymology* (McGraw-Hill, New York, 1969) p. 427.

62. D. L. Vander Jagt, F. L. Killian and M. L. Bender, *J. Am. Chem. Soc.* **92** (1970) 1016.

63. O. Exner, *Nature* **227** (1970) 366.

64. O. Exner, *Prog. Phys. Org. Chem.* **10** (1973) 411.

65. E. Fenyvesi, L. Szente, N. R. Russell and M. McNamara, *Specific Guest Types*, in *Comprehensive Supramolecular Chemistry*, eds. J. L. Atwood, J. E. D. Davies, D. D. MacNicol, F. Vögtle and J.-M. Lehn, Vol. 3, eds. J. Szejtli and T. Osa (Pergamon, Oxford, 1996), p. 305.

66. Y. Matsui and K. Mochida, *Bull. Chem. Soc. Jpn.* **52** (1979) 2808.

67. I. M. Brereton, T. M. Spotswood, S. F. Lincoln and E. H. Williams, *J. Chem. Soc., Faraday Trans. 1* **80** (1984) 3147.

68.   R. I. Gelb and L. M. Schwartz, *J. Inclusion Phenom. Mol. Recogn. Chem.* **7** (1989) 465.

69.   D. Hallén, A. Schön, I. Shehatta and I. Wadsö, *J. Chem. Soc, Faraday Trans.* **88** (1992) 2859.

70.   M. V. Rekharsky, F. P. Schwarz, Y. B. Tewari and R. N. Goldberg, *J. Phys. Chem.* **98** (1994) 10282.

71.   M. V. Rekharsky, F. P. Schwarz, Y. B. Tewari, R. N. Goldberg, M. Tanaka and Y. Yamashoji, *J. Phys. Chem.* **98** (1994) 4098.

72.   M. V. Rekharsky, M. P. Mayhew, R. N. Goldberg, P. D. Ross, Y. Yamashoji and Y. Inoue, *J. Phys. Chem. B* **101** (1997) 87.

73.   A. Gadre, V. Rüdiger, H.-J. Schneider and K. A. Connors, *J. Pharm. Sci.* **86** (1997) 236.

74.   V. Rüdiger, A. Eliseev, S. Simova, H.-J. Schneider, M. J. Blandamer, P. M. Cullis and A. J. Meyer, *J. Chem. Soc., Perkin Trans. 2* (1996) 2119.

75.   Y. Inoue, K. Yamamoto, T. Wada, S. Everitt, X.-M. Gao, Z.-J. Hou, L.-H. Tong, S.-K. Jiang and H.-M. Wu, *J. Chem. Soc, Perkin Trans. 2* (1998) 1807.

76.   S.-F. Lin and K. A. Connors, *J. Pharm. Sci.* **72** (1983) 1333.

77.   G. L. Bertrand, J. R. Faulkner, S. M. Han and D. W. Armstrong, *J. Phys. Chem.* **93** (1989) 6863.

78.   R. Dhillon, C. J. Easton, S. F. Lincoln and J. Papageorgiou, *Aust. J. Chem.* **48** (1995) 1117.

79.   K. Hendrickson, C. J. Easton and S. F. Lincoln, *Aust. J. Chem.* **48** (1995) 1125.

80.   P. D. Ross and M. V. Rekharsky, *Biophys. J.* **71** (1996) 2144.

81.   Y. Liu, Y.-M. Zhang, A.-D. Qi, R.-T. Chen, K. Yamamoto, T. Wada and Y. Inoue, *J. Org. Chem.* **62** (1997) 1826.

82.   D. M. Davies, M. E. Deary and D. I. Wealleans, *J. Chem. Soc., Perkin Trans. 2* (1998) 193.

83.   W. H. Tan, N. Niino, T. Ishikura, A. Maruta, T. Yamamoto and Y. Matsui, *Bull. Chem. Soc. Jpn.* **71** (1998) 1285.

84.   J. M. Madrid, F. Mediculi and W. L. Mattice, *J. Phys. Chem. B* **102** (1998) 2037.

85.   N. J. Smith, T. M. Spotswood and S. F. Lincoln, *Carbohydr. Res.* **192** (1989) 9.

86. S. E. Brown, J. H. Coates, S. F. Lincoln, D. R. Coghlan and C. J. Easton, *J. Chem. Soc.*, *Faraday Trans.* **87** (1991) 2699.

87. S. E. Brown, J. H. Coates, C. J. Easton, S. J. van Eyk, S. F. Lincoln, B. L. May, M. A. Stile, C. B. Whalland and M. L. Williams, *J. Chem. Soc.*, *Chem. Commun.* (1994) 47.

88. S. E. Brown, C. A. Haskard, C. J. Easton and S. F. Lincoln, *J. Chem. Soc.*, *Faraday Trans.* **91** (1995) 1013, 4335.

89. S. E. Brown, C. J. Easton and S. F. Lincoln, *J. Chem. Research* (1995) 2.

90. S. E. Brown, C. J. Easton and S. F. Lincoln, *Aust. J. Chem.* **48** (1995) 505.

91. C. J. Easton, S. F. Lincoln, J. Papageorgiou and D. M. Schliebs, *J. Chem. Res.* (1995) 381.

92. G. Castronuovo, V. Elia, D. Fessas, A. Giordano and F. Velleca, *Carbohydr. Res.* **272** (1995) 31.

93. Y. Liu, Y.-M. Zhang, A.-D. Qi., R.-T. Chen., K. Yamamoto, T. Wada and Y. Inoue, *J. Org. Chem.* **62** (1997) 1826.

94. Y. Liu, B.-H. Han, B. Li, Y.-M. Zhang, P. Zhao, Y.-T. Chen, T. Wada and Y. Inoue, *J. Org. Chem.* **63** (1998) 1444.

95. D. L. Pisaniello, S. F. Lincoln and J. H. Coates, *J. Chem. Soc.*, *Faraday Trans. 1* **81** (1985) 1247.

96. S. F. Lincoln, J. H. Coates, B. G. Doddridge and A. M. Hounslow, *J. Inclusion Phenom. Mol. Recogn. Chem.* **5** (1987) 49.

97. S. F. Lincoln, A. M. Hounslow, J. H. Coates and B. G. Doddridge, *J. Chem. Soc.*, *Faraday Trans. 1* **83** (1987) 2697.

98. S. F. Lincoln, A. M. Hounslow, J. H. Coates, R. P. Villani and R. L. Schiller, *J. Inclusion Phenom. Mol. Recogn. Chem.* **6** (1988) 183.

99. S. E. Brown, J. H. Coates, C. J. Easton, S. F. Lincoln, Y. Luo and A. K. W. Stephens, *Aust. J. Chem.* **44** (1991) 855.

100. I. Orienti, A. Fini, V. Bertasi and V. Zecchi, *Eur. J. Pharm. Biopharm.* **37** (1991) 110.

101. A. F. Danil de Namor, D. A. P. Tanaka, L. N. Regueira and I. G. Orellana, *J. Chem. Soc, Faraday Trans.* **88** (1992) 1665.

102. A. Ashnagar, P. T. Culnane, C. J. Easton, J. B. Harper and S. F. Lincoln, *Aust. J. Chem.* **50** (1997) 447.

103. E. Junquera, M. Martin-Pastor and E. Aicart, *J. Org. Chem.* **63** (1998) 4349.

104. N. Schaschke, S. Fiori, E. Weyher, C. Escrieut, D. Fourmy, G. Müller and L Moroder, *J. Am. Chem. Soc.* **120** (1998) 7030.

105. R. J. Clarke, J. H. Coates and S. F. Lincoln, *Carbohydr. Research* **127** (1984) 181.

106. R. J. Clarke, J. H. Coates and S. F. Lincoln, *J. Chem. Soc., Faraday Trans. 1* **82** (1986) 2333. ("Annular radii" should read "annular diameters" in this paper.)

107. R. L. Schiller, S. F. Lincoln and J. H. Coates, *J. Chem. Soc., Faraday Trans. 1* **82** (1986) 2123. ("Internal radii" should read "internal diameters" in this paper.)

108. R. L. Schiller, S. F. Lincoln and J. H. Coates, *J. Inclusion Phenom. Mol. Recogn. Chem.* **5** (1987) 59.

109. S. F. Lincoln, J. H. Coates and R. L. Schiller, *J. Inclusion Phenom. Mol. Recogn. Chem.* **5** (1987) 709.

110. R. P. Villani, S. F. Lincoln and J. H. Coates, *J. Chem. Soc., Faraday Trans. 1* **83** (1987) 2751.

111. R. L. Schiller, S. F. Lincoln and J. H. Coates, *J. Chem. Soc., Faraday Trans. 1* **83** (1987) 3237.

112. A. F. Danil de Namor, R. Traboulssi and D. F. V. Lewis, *J. Am. Chem. Soc.* **112** (1990) 8442.

113. N. Yoshida, A. Seiyama and M. Fujimoto, *J. Phys. Chem.* **94** (1990) 4246.

114. A. F. Danil de Namor, P. M. Blackett, M. C. Cabaleiro and J. M. A. Al Rawi, *J. Chem. Soc, Faraday Trans.* **90** (1994) 845.

115. R. P. Rohrbach. L. J. Rodriguez, E. M. Eyring and J. F. Wojcik, *J. Phys. Chem.* **81** (1977) 944.

116. P. Mu, T. Okada, N. Iwami and Y. Matsui, *Bull. Chem. Soc. Jpn.* **66** (1993) 1924.

117. Y. Yamashoji, M. Fujiwara, T. Matsushita and M. Tanaka, *Chem. Lett.* (1993) 1029.

118. Y. Matsui, M. Ono and S. Tokunaga, *Bull. Chem. Soc. Jpn.* **70** (1997) 535.

119. P. S. Bates, R. Kataky and D. Parker, *J. Chem. Soc., Chem. Commun.* (1993) 691.

120. F. Cramer and F. M. Henglein, *Angew. Chem.* **68** (1956) 649.

121. D.-D. Zhang, J.-W. Chen, Y. Yang, R.-F. Cai, X.-L. Shen and S.-H. Wu, *J. Inclusion Phenom. Mol. Recogn. Chem.* **16** (1993) 245.

122. Z. Yoshida, H. Takekuma, S. Takekuma and Y. Matsubara, *Angew. Chem., Int. Ed. Engl.* **33** (1994) 1597.

123. H.-S. Kim and S.-J. Jeon, *J. Chem. Soc., Chem. Commun.* (1996) 817.

124. R. J. Bergeron, M. A. Channing, G. J. Gibeily and D. M. Pillor, *J. Am. Chem. Soc.* **99** (1977) 5146.

125. M. Kitagawa, H. Hoshi, M. Sakurai, Y. Inoue and R. Chûjô, *Carbohydr. Res.* **163** (1987) c1.

126. M. Sakurai, M. Kitagawa, H. Hoshi, Y. Inoue and R. Chûjô, *Chem. Lett.* (1988) 895.

127. M. Sakurai, M. Kitagawa, H. Hoshi, Y. Inoue and R. Chûjô, *Carbohydr. Res.* **198** (1990) 181.

128. Y. Yamamoto, M. Onda, M. Kitagawa, Y. Inoue and R. Chûjô, *Carbohydr. Res.* **167** (1987) c11.

129. K. Harata, *Bull. Chem. Soc. Jpn.* **50** (1977) 1416.

130. I. Bakó and L. Jicsinszky, *J. Inclusion Phenom. Mol. Recogn. Chem.* **18** (1994) 275.

131. A. Botsi, K. Yannakopolou, E. Hadjoudis and J. Waite, *Carbohydr. Res.* **283** (1996) 1.

132. Z. Yang and R. Breslow, *Tetrahedron Lett.* **38** (1997) 6171.

133. M. R. Eftink, M. L. Andy, K. Bystrom, H. D. Perlmutter and D. S. Kristol, *J. Am. Chem. Soc.* **111** (1989) 6765.

134. R. Palepu and V. C. Reinsborough, *Aust. J. Chem.* **43** (1990) 2119.

135. K. Kano, K. Yoshiyasu, H. Yasuoka, S. Hata and S. Hashimoto, *J. Chem. Soc., Perkin Trans. 2* (1992) 1265.

136. K. Kano, S. Negi, H. Kamo, T. Kitae, M. Yamaguchi, H. Okubo and M. Hirama, *Chem. Lett.* (1998) 151.

137. K. Kano, I. Takenoshita and T. Ogawa, *Chem. Lett.* (1982) 321.

138. N. Kobayashi, R. Saito, H. Hino, Y. Hino, A. Ueno and T. Osa, *J. Chem. Soc., Perkin Trans. 2* (1983) 1031.

139. G. Patonay, A. Shapira, P. Diamond and I. M. Warner, *J. Phys. Chem.* **90** (1986) 1963.

140. G. Nelson, G. Patonay and I. M. Warner, *Anal. Chem.* **60** (1988) 274.

141. S. Hamai, *J. Phys. Chem.* **99** (1989) 2074.

142. W. Xu, J. N. Demas, B. A. DeGraff and M. Whaley, *J. Phys. Chem.* **97** (1993) 6546.

143. S. De Feyter, J. van Stam, N. Boens and F. C. De Schryver, *Chem. Phys. Lett.* **249** (1996) 46.

144 K. A. Udachin and J. A. Ripmeester, *J. Am. Chem. Soc.* **120** (1998) 1080.

145. M. Shibakami and A. Sekiya, *J. Chem. Soc., Chem. Commun.* (1992) 1742.

146. J. L. Alderfer and A. V. Eliseev, *J. Org. Chem.* **62** (1997) 8225.

147. F. Cramer, W. Saenger and H.-C. Spatz, *J. Am. Chem. Soc.* **89** (1967) 14.

148. A. Hersey and B. H. Robinson, *J. Chem. Soc., Faraday Trans. 1* **80** (1984) 2039.

149. A. Seiyama, N. Yoshida and M. Fujimoto, *Chem. Lett.* (1985) 1013.

150. N. Yoshida and K. Hayashi, *J. Chem. Soc., Perkin Trans. 2* (1994) 1285.

151. N. Yoshida, *J. Chem. Soc., Perkin Trans. 2* (1995) 2249.

152. N. Yoshida, A. Seiyama and M. Fujimoto, *J. Phys. Chem.* **94** (1990) 4254.

153. a) K. Harata, K. Uekama, M. Otagiri and F. Hirayama, *Bull. Chem. Soc. Jpn.* **56** (1983) 1732. b) K. Harata, *J. Chem. Soc., Chem. Commun.* (1988) 928. c) M. R. Caira, V. J. Griffith, L. R. Nassimbeni and B. van Oudtshoorn, *J. Chem. Soc., Perkin Trans. 2* (1994) 2071. d) M. K. Dowd, A. D. French and P. J. Reilly, *Carbohydr. Res.* 264 (1994) 1.

154. N. Yoshida and Y. Fujita, *J. Phys. Chem.* **99** (1995) 3671.

155. a) K. Harata, K. Uekama, M. Otagiri and F. Hirayama, *Bull. Chem. Soc. Jpn.* **55** (1982) 407, 3904. b) K. Harata, K. Uekama, M. Otagiri, F. Hirayama and Y. Sugiyama, *Bull. Chem. Soc. Jpn.* **55** (1982) 3386. c) K. Harata, K. Uekama, M. Otagiri and F. Hirayama, *J. Inclusion Phenom.* **1** (1984) 279. d) K. Harata, K. Uekama, M. Odagiri and F. Hirayama, *Bull. Chem. Soc. Jpn.* **60** (1987) 497. e) K. Harata, *J. Chem. Soc., Perkin Trans. 2* (1990) 799. f) T. Steiner and W. Saenger, *Carbohydr. Res.* **282** (1996) 53.

156. a) K. Harata, *Carbohydr. Res.* **192** (1989) 33. b) K. Harata, *Bull. Chem. Soc. Jpn.* **63** (1990) 2481. c) K. Harata, *Supramol. Chem.* **5**. (1995) 231.

157. a) K. Harata, *Bull. Chem. Soc. Jpn.* **61** (1988) 1939. b) T. Steiner and W. Saenger, *Carbohydr. Res.* **275** (1995) 73.

158. P. Bugnon, P. G. Lye, A. Abou-Hamdan and A. E. Merbach, *J. Chem. Soc., Chem. Commun.* (1996) 2787.

159. P. Bugnon, P. G. Lye, A. Abou-Hamdan and A. E. Merbach, in *High Pressure Research in the Biosciences and Biotechnology*, ed. K. Heremans (Leuven University Press, Leuven, Belgium, 1997) p. 519.

160. T. C. Barros, K. Stefaniak, J. F. Holzwarth and C. Bohne, *J. Phys. Chem. A* **102** (1998) 5639.

161. Y. Liao, J. Frank, J. F. Holzwarth and C. Bohne, *J. Chem. Soc., Chem. Commun.* (1995) 199.

162. M. Barra, C. Bohne and J. C. Scaino, *J. Am. Chem. Soc.* **112** (1990) 8075.

163. W. L. Hinze, T. E. Riehl, D. W. Armstrong, W. DeMond, A. Alak and T. Ward, *Anal Chem.* **57** (1985) 237.

164. D. W. Armstrong, T. J. Ward, R. D. Armstrong and T. E. Beesley, *Science* **232** (1986) 1132.

165. F. Cramer and W. Dietsche, *Chem. Ber.* **92** (1959) 378.

166. K. B. Lipkowitz, S. Raghothama and J.-A. Yang, *J. Am. Chem. Soc.* **114** (1992) 1554.

167. A. Cooper and D. D. MacNicol, *J. Chem. Soc., Perkin Trans.* 2 (1978) 760.

168. D. D. MacNicol and D. S. Ryecroft, *Tetrahedron Lett.* (1977) 2173.

169. D. Greatbanks and R. Pickford, *Mag. Reson. Chem.* **25** (1987) 208.

170. A. F. Casy and A. D. Mercer, *Mag. Reson. Chem.* **26** (1988) 765.

171. A. F. Casy, *Mag. Reson. Chem.* **31** (1993) 416.

172. M. E. Amato, F. Djedaïni-Pilard, B. Perly and G. Scarlata, *J. Chem. Soc., Perkin Trans.* 2 (1992) 2065.

173. R. Reinhardt, M. Richter, P. Mager and W. Engewald, *Chromatographia* **43** (1996) 187.

174. K. B. Lipkowitz, R. Coner and M. A. Peterson, *J. Am. Chem. Soc.* **119** (1997) 11269.

175. J. A. Hamilton and L. Chen, *J. Am. Chem. Soc.* **110** (1988) 4379.

CHAPTER 2
MODIFIED CYCLODEXTRINS
STRATEGIES FOR SYNTHESIS

2.1. Introduction

While the natural cyclodextrins (CDs) are themselves of interest as molecular hosts, much of their utility in supramolecular chemistry derives from their modification. This involves altering the shape, size and other physical properties of the CD annuli and surfaces, and introducing new functional groups. The geometry of the modified CDs is well defined, reflecting that of the parent CDs. Consequently, CDs can be carefully tailored to match particular guests and meet specific requirements in their host-guest complexes. Reasons for the modifications include the need to introduce photochemically active and pH sensitive groups in order to develop chemical sensors, the necessity to introduce coordination sites for the construction of metalloCDs, the desire to alter the physical properties of external surfaces of CDs in order to construct monolayers and micellar structures, and the aspiration to build enzyme mimics and related catalysts with predetermined alignment of binding sites and catalytically active groups. Objectives such as these are discussed in more detail in subsequent sections of this book while this chapter covers the underlying strategies available for covalent modification of the natural CDs in a closely controlled manner. Related reviews [1-4] have surveyed other aspects of this field from different perspectives and provide complementary assessments of the enormous amount of work published in this area.

The most straightforward way to elaborate CDs is through reaction of their hydroxy groups. This survey is concerned only with such modifications of αCD, βCD and γCD, but it is worth noting that access to larger CD homologues [5-12] broadens even further the scope for the synthesis of novel molecular hosts. Each glucopyranose residue of a CD has a primary hydroxy group at C(6) and secondary hydroxy groups at C(2) and C(3), and αCD, βCD and γCD comprise six, seven and eight glucopyranose residues, respectively (see Chapter 1, Fig. 1.1, p.3). Thus, each CD contains numerous sites for elaboration. The difficulty is that reactions of the hydroxy groups tend to occur in a random fashion, to afford complex mixtures of products. Such mixtures have proved useful in a number of applications, including the chromatographic separation of chemicals and the administration of pharmaceuticals (see Chapter 1), however, they are generally unsuitable for use in the contexts discussed in this book. Instead, these require individual modified CDs which are uniquely defined and well characterised, and are only obtained when a strategic approach is used to control the regioselectivity and limit the extent of reaction, to substitution of either one or a specific combination of the hydroxy groups.

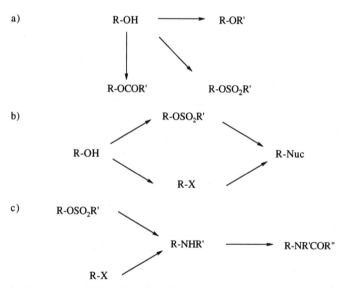

Scheme 2.1. Methods for modifying CD hydroxy groups.

The methods that have been developed for the controlled modification of CDs exploit the different reactivity of the C(2), C(3) and C(6) hydroxy groups, protecting groups to mask reactive centres, the complexation of reagents in the CD annuli, and a variety of other factors. Required substituents may be introduced directly, for example through alkylation, acylation and sulfonation (Scheme 1.1a). Alternatively, sulfonates, halides and related species may be prepared as intermediates, for the subsequent introduction of substituents through nucleophilic displacement (Scheme 1.1b). In turn, amino groups incorporated in this manner provide nucleophilic sites for further elaboration (Scheme 1.1c). These multi-step methods comprise initially making one or more sites of a CD chemically distinct from the others, in a controlled manner, such that further reactions occur without competing involvement of unreacted hydroxy groups. Since this generally involves the synthesis of CD sulfonates and halides, procedures to accomplish such transformations have attracted a great deal of attention and are a main focus of this review.

2.2a. Modifications Involving Reaction of a Single C(6) Primary Hydroxy Group

The primary hydroxy groups of the CDs are the most basic. They are also the most nucleophilic, as a consequence of which they may be selectively modified through their reactions with electrophilic species. For example, αCD and βCD react with *p*-toluenesulfonyl chloride in pyridine, to give mixtures of products arising from sulfonation of these groups [13,14]. The corresponding C(6) monotosylates **2.1** and **2.2** may then be obtained by separation from the mixtures, through chromatography in the case of the αCD **2.1**, and through recrystallisation from water for the βCD **2.2**. In this manner, the sulfonates **2.1** and **2.2** have been prepared in yields of 20 and 31%, respectively. The importance of the sulfonates **2.1** and **2.2** as intermediates for the synthesis of other modified CDs is such that several alternative procedures for their preparation have also been reported [15-17]. In a recent example, treatment of βCD with *p*-toluenesulfonic anhydride, under basic aqueous conditions, afforded the tosylate **2.2** in 61% yield [17]. The high yield of the monosubstitution product **2.2** is attributed to complexation of the anhydride in the CD annulus prior to reaction.

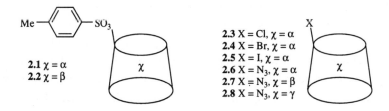

Me —⟨ ⟩— SO₃

2.1 χ = α
2.2 χ = β

2.3 X = Cl, χ = α
2.4 X = Br, χ = α
2.5 X = I, χ = α
2.6 X = N₃, χ = α
2.7 X = N₃, χ = β
2.8 X = N₃, χ = γ

Through reactions with nucleophiles, the tosylates **2.1** and **2.2** are suitable intermediates for the introduction of a variety of other substituents. Thus, reaction of the tosylate **2.1** with sodium chloride, bromide, iodide and azide gives the halides **2.3-2.5** and the azide **2.6**, respectively [13]. Alternatively, the azide **2.6** and the corresponding βCD and γCD derivatives **2.7** and **2.8** have been prepared directly, in yields of 18, 15 and 22%, from αCD, βCD and γCD, respectively, through treatment with lithium azide, triphenylphosphine and carbon tetrabromide [18]. As with the preparation of the tosylates **2.1** and **2.2**, the synthesis of the azides **2.6-2.8** demonstrates selective reaction of the C(6) hydroxy groups of the natural CDs.

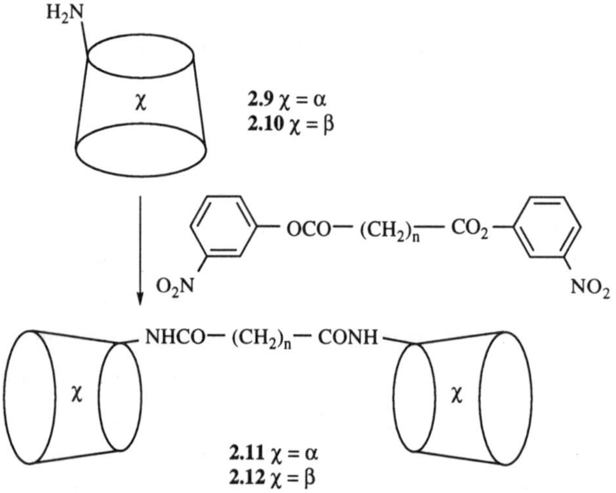

Scheme 2.2. Selective elaboration of the amino group of the CDs **2.9** and **2.10** in the synthesis of the dimers **2.11** and **2.12**, respectively.

Catalytic hydrogenation of the azides **2.6** and **2.7** is used to prepare the corresponding 6^A-amino-6^A-deoxyCDs **2.9** and **2.10** [13]. The monoamines **2.9** and **2.10** have also been prepared more directly, by treatment of the corresponding tosylates **2.1** and **2.2** with ammonia [14]. Analogous reactions of primary and secondary amines have also been reported, as a method for attaching substituted amino groups at C(6) of a CD [19]. These are incorporated as basic residues, which become ionic when protonated, and metal coordinating ligands (see Chapter 5). In addition, they can be exploited for further modification of the CDs, since the amino groups are substantially more nucleophilic than the CD hydroxy groups. This is demonstrated in reactions of the amines **2.9** and **2.10**, with complexed esters to give good yields of the amides **2.13** and **2.14**, respectively [20], and with diesters to give the corresponding specifically linked CD dimers **2.11** and **2.12** (Scheme 2.2) [21,22].

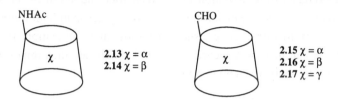

The βCD tosylate **2.2** has been converted to the corresponding aldehyde **2.16**, through oxidation using dimethyl sulfoxide [23-25]. Alternatively, the aldehydes **2.15-2.17** may be prepared directly from αCD, βCD and γCD, respectively, in yields ranging from 85-100%, by oxidation with Dess-Martin periodinane [26]. The efficiency of this monofunctionalisation is attributed to formation of a covalent intermediate between the oxidising agent and each of the natural CDs, which limits access of more reagent to the other C(6) hydroxy groups. The aldehydes **2.15-2.17** are suitable precursors for the synthesis of the corresponding carboxylic acids, and for attaching other substituents to CDs through the formation of imines.

From the above examples it is clear that a sulfonate can be introduced in place of a single C(6) hydroxy group of a CD, then replaced through nucleophilic substitution. Alkaline bases are unsuitable for use in these substitution reactions, however, because they react instead by deprotonation of the C(3) hydroxy group of the modified glucopyranose residue and intramolecular displacement of the sulfonate, to give the corresponding 3,6-anhydroCDs [27]. Direct *O*-alkylation of a C(6) hydroxy group using the corresponding CD alkoxide is also not feasible, as the C(2) and C(3) hydroxy groups are more acidic than those at C(6) [28-30], and they react preferentially. It is possible to prepare a mixture of modified CDs where the primary and secondary hydroxy groups have been *O*-alkylated in a random fashion, and then separate the C(6) monosubstituted CD chromatographically. Better alternative methods for the C(6) *O*-alkylation of CDs involve the use of protecting groups. For example, silylation of the primary hydroxy groups of αCD, then peracetylation of the secondary hydroxy groups and desilylation, gave a product which was treated with a glucopyranosyl bromide in the presence of tetraethylammonium bromide, before base hydrolysis to remove the ester protecting groups, in order to prepare a C(6) monopyranosyl CD [31]. In another example [32], which is illustrated in Scheme 2.3, reaction of βCD with sodium hydride and *tert*-butyldimethylsilyl chloride gives a mixture of products from silylation of between four and seven of the secondary hydroxy groups. This degree of substitution proves sufficient to limit further reaction of the secondary hydroxy groups, such that subsequent alkylation of a primary hydroxy group can then be achieved. The silyl groups are then removed through treatment with *tert*-butylammonium fluoride.

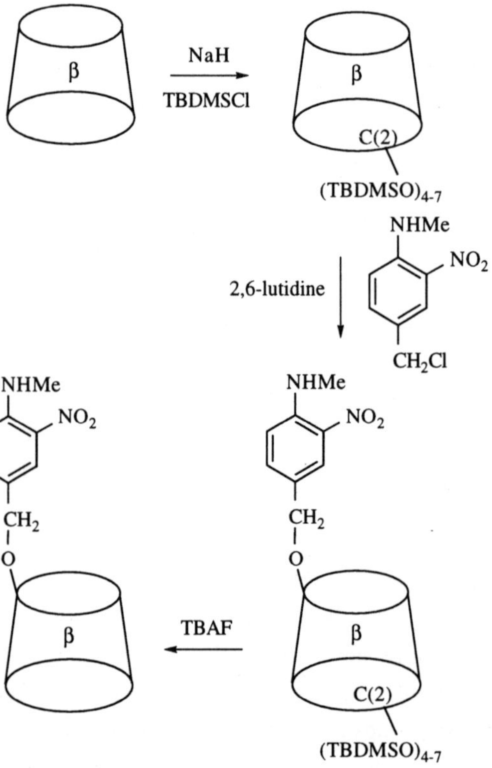

Scheme 2.3. Use of protecting groups for the alkylation of a βCD C(6) hydroxy group.

2.2b. Facial Modifications Involving Reaction of All the C(6) Primary Hydroxy Groups

The methods that have been used to modify each of the primary hydroxy groups of the CDs, without competing reaction of the secondary alcohols, are similar to those described above for reaction of a single C(6) hydroxy group. They rely on the greater nucleophilicity of the primary hydroxy groups but, whereas the main difficulty with modification of a single residue is to avoid competing reactions of the other primary groups, the limitation to elaboration of each of the primary hydroxy groups is that competing reactions of the secondary hydroxy groups become more likely as the degree of reaction increases. Thus, purification of a CD in which each of the primary hydroxy groups has been modified often involves its separation from mixtures with other CDs in which all but one of those groups has reacted or in which some reaction of the secondary hydroxy groups has also occurred. In this manner, the *p*-toluenesulfonates **2.18-2.20** are obtained by treatment of αCD, βCD, and γCD, respectively, with excess *p*-toluenesulfonyl chloride in pyridine, followed by chromatography of the product mixtures [33-41].

2.18 χ = α, n = 6
2.19 χ = β, n = 7
2.20 χ = γ, n = 8

The polytosylates **2.18-2.20** undergo nucleophilic substitution reactions, which are analogous to those of the monotosylates **2.1** and **2.2** described above. Accordingly, the βCD **2.19** reacted with methylamine and ethylamine, to give the polyamines **2.21** and **2.22**, respectively [37,42]. CD polyhalides have also been prepared by treatment of the corresponding polytosylates with sodium halides [35,36]. Alternatively, the polyhalides can be prepared more directly. Treatment of αCD, βCD and γCD with methanesulfonyl bromide and *N,N*-dimethylformamide, followed by sodium methoxide, gives the corresponding bromides **2.23-2.25** [43]. Reactions of αCD and βCD with triphenylphosphine and iodine are used to prepare the iodides **2.26** and **2.27**, in 80 and 88% yield, respectively, and βCD gives the bromide **2.24** in 93% yield on treatment with triphenylphosphine and bromine [44-46]. The chloride **2.28** has been obtained in 90% yield through treatment of βCD with methanesulfonyl chloride and imidazole in *N,N*-dimethylformamide [47]. Halomethylenemorpholinium halides have also been used as reagents for these transformations [48].

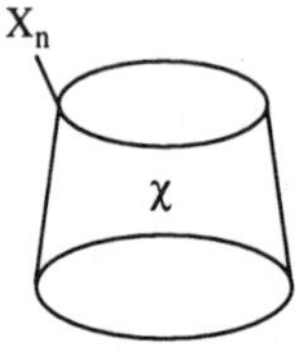

2.21 χ = β, X = NHMe, n = 7
2.22 χ = β, X = NHEt, n = 7
2.23 χ = α, X = Br, n = 6
2.24 χ = β, X = Br, n = 7
2.25 χ = γ, X = Br, n = 8
2.26 χ = α, X = I, n = 6
2.27 χ = β, X = I, n = 7
2.28 χ = β, X = Cl, n = 7

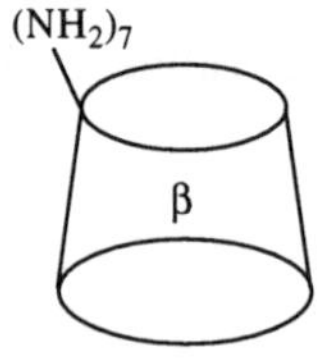

2.29

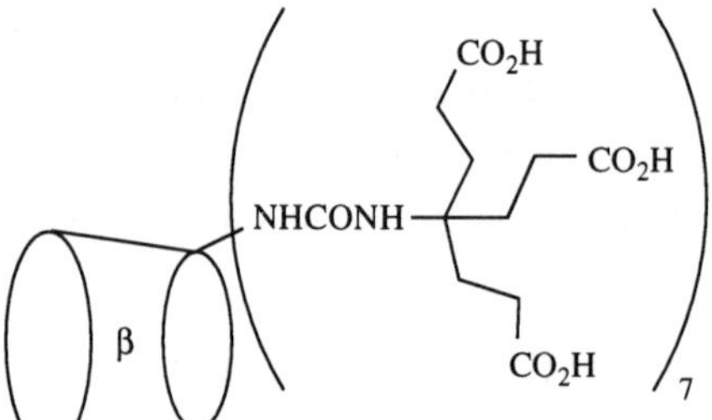

2.30

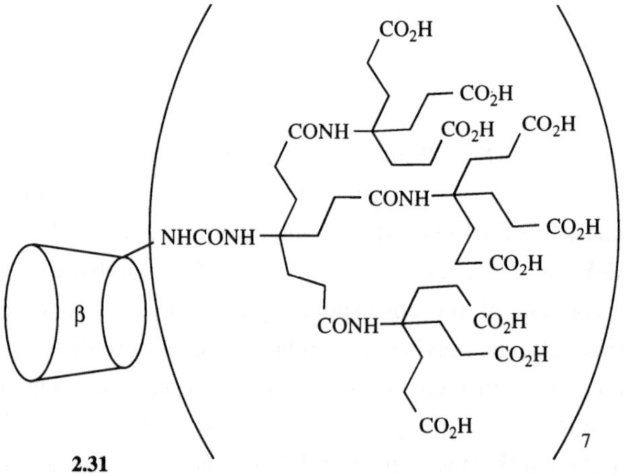

2.31

Nucleophilic substitution of the polyiodide **2.27** with sodium azide, followed by reduction, affords the polyamine **2.29** [45]. Direct substitution of halides such as **2.23**-**2.27** with alkylamines gives the corresponding per-6-alkylamino-substituted CDs [49,50]. While these modifications render the face of the C-6 carbons of the CDs basic, and therefore ionic as a result of protonation under acidic conditions, reduction of per-6-haloCDs with sodium borohydride affords the corresponding per-6-deoxyCDs [51], which have a hydrophobic face. Such amino-substituted and reduced CDs are particularly useful for the production of CD monolayers on polar and non-polar surfaces, respectively (Chapter 8). The polyamine **2.29** has also been used as the template for the synthesis of the first and second generation dendrimers **2.30** and **2.31** [52].

The reactions of the polytosylates **2.18**-**2.20** with alkaline bases such as sodium hydroxide are analogous to those of the monotosylates **2.1** and **2.2**, and afford the corresponding per-3,6-anhydroCDs **2.32**-**2.34** [38,39,41,53-55]. The anhydrides **2.32** and **2.33** are also obtained by treatment of the corresponding periodides **2.26** and **2.27** with hydroxide in dimethyl sulfoxide [44,56]. The CDs **2.32**-**2.34** are substantially distorted, compared with αCD, βCD and γCD, and show quite different behaviour as molecular hosts. Their cation complexing ability is particularly noteworthy.

2.32 n = 6
2.33 n = 7
2.34 n = 8

O-Alkylation of each of the primary hydroxy groups of a CD is best achieved through prior protection of the secondary residues, as described above for the *O*-alkylation of a single C(6) hydroxy group. The general approaches which have been reported involve either esterification of the secondary hydroxy groups, alkylation of the primary residues, and then hydrolysis of the esters [57], or silylation of the primary alcohols, either esterification or benzylation of the secondary residues, desilylation and alkylation of the primary hydroxy groups, and then hydrolysis of the esters or debenzylation [31,58,59]. The latter approach again exploits the greater nucleophilicity of the primary hydroxy residues in their silylation, and all such residues of αCD, βCD and γCD can be protected through reaction with *tert*-butyldimethylsilyl chloride [31,45,60-63].

2.2c. Modifications Involving Reaction of Two, Three, or Four C(6) Primary Hydroxy Groups

In addition to being able to modify either one or all of the primary hydroxy groups of a CD, it is also feasible to modify an intermediate number of these residues. Where the extent and regioselectivity of these reactions can be controlled, it is possible to build a CD derivative with a predetermined geometry and orientation of the substituents, both with respect to each other and with respect to the CD annulus. Such modified CDs are particularly well suited for application in the construction of enzyme mimics and other catalysts (Chapter 4).

One way to access individual disubstituted CDs is through their separation from mixtures obtained through random modification, but this generally involves tedious chromatography, and individual products are obtained only in low yields. A preferable alternative is to use a bifunctional reagent, where the geometry of that reagent determines the relative orientation of the CD residues which react. Arenedisulfonyl chlorides are commonly used for this purpose [64-71]. For example, the $6^A,6^D$-disulfonated CDs **2.35** and **2.36** are obtained as the predominant products, in yields of 20 and 18%, by treatment of βCD with *trans*-stilbene-4,4'-disulfonyl chloride and biphenyl-4,4'-disulfonyl chloride, respectively [68,69].   Using benzophenone-3,3'-disulfonyl chloride as the reagent gives the $6^A,6^C$-disubstituted βCD **2.37** in 40% yield [68,69]. The disulfonated CDs **2.38** [70] and **2.39** [71] have been prepared in 40 and 12% yields, by treatment of βCD with 1,3-benzenedisulfonyl chloride and 4,6-dimethoxybenzene-1,3-disulfonyl chloride, respectively. Although formed in lower yield, the latter of these $6^A,6^B$-capped CDs **2.38** and **2.39** is the more stable and less susceptible to hydrolysis.

Regiospecifically trisubstituted CDs are more difficult to obtain. Nevertheless each of the five possible isomers of tri-6-tosylated βCD has been isolated, by chromatography of a mixture obtained through random reaction of the parent CD [72]. Reactions on adjacent glucopyranose residues may be restricted using bulky reagents, as a method to control the regioselectivity of modification of CDs. This has been exploited in the reaction of αCD with trityl chloride, which gives a 23% yield of the $6^A,6^C,6^E$-trityl species **2.40** [73]. An alternative procedure for the controlled synthesis of $6^A,6^C,6^E$-trisubstituted αCDs is through enzyme-catalysed cyclotrimerisation of the corresponding monosubstituted maltoses (Scheme 2.4) [74,75]. Cyclooligomerisation of other disaccharides is also used to prepare symmetrically substituted αCDs and γCDs, as well as related cyclic oligosaccharides and their higher homologues [76-79].

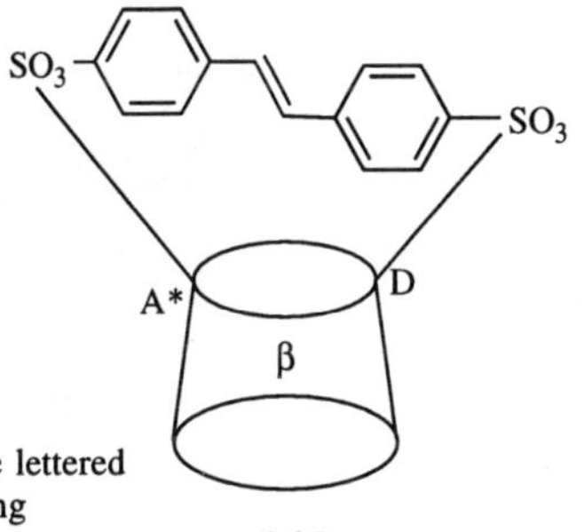

*denotes the substituted glucopyranoses which are lettered sequentially around the ring

2.35   **2.36**

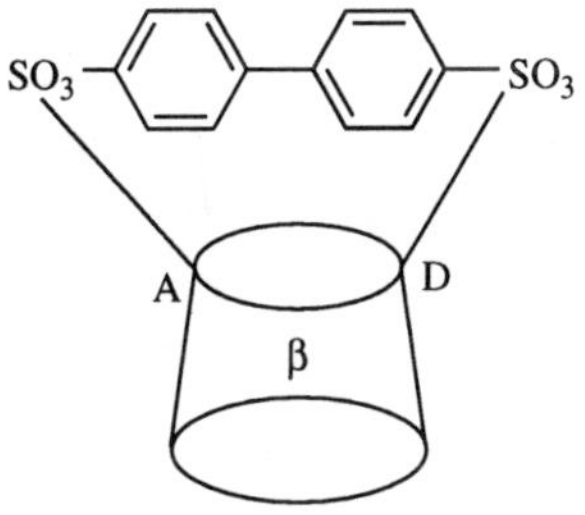

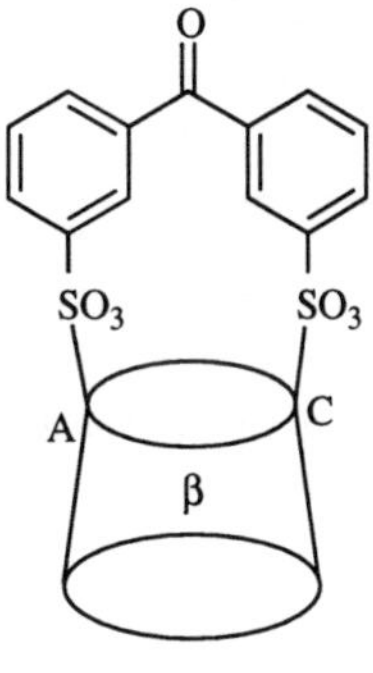

2.37 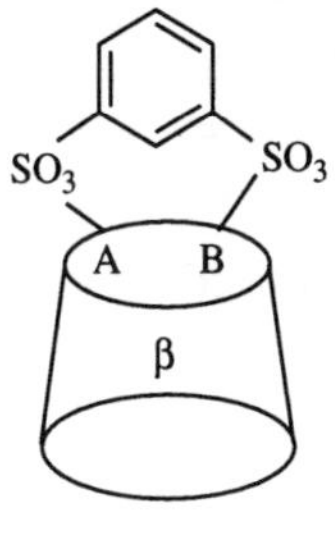   **2.38** 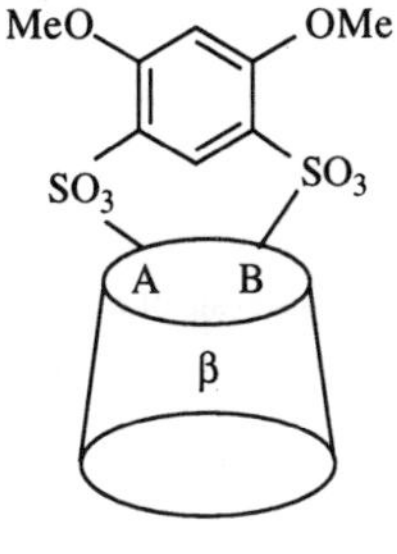   **2.39**

The regiospecifically $6^A,6^C,6^D,6^F$-tetrasubstituted CD **2.41** has also been prepared and characterised [80,81]. It is obtained in 35% yield, by treatment of βCD with benzophenone-3,3'-disulfonyl chloride. As outlined above, the initial reaction of βCD affords the disulfonate **2.37**, which then reacts with a second equivalent of the bifunctional reagent. The sulfonates **2.35-2.39** and **2.41**, and related di-, tri- and tetra-functionalised CDs, undergo reactions which are analogous to those of the tosylates **2.1**, **2.2**, and **2.18**-**2.20**, outlined above. These include intermolecular nucleophilic substitution reactions with halides and amines, and intramolecular nucleophilic displacement on treatment with hydroxide to give the corresponding 3,6-anhydroCDs. Thus, it is possible to construct a range of regiospecifically C(6) modified CDs in a controlled manner.

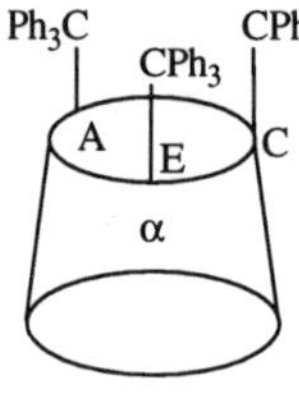

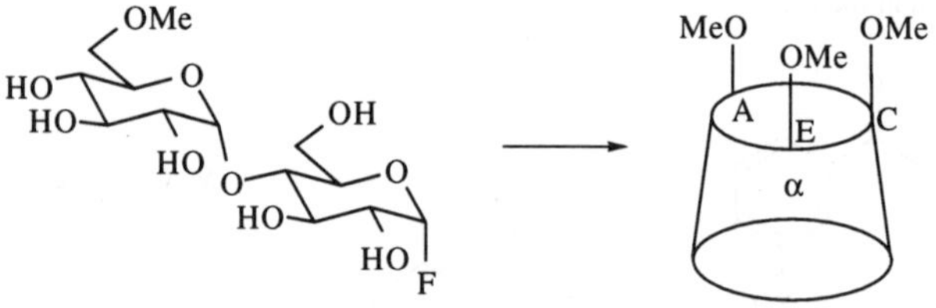

Scheme 2.4. Synthesis of a symmetrically trisubstituted αCD through cyclooligomerisation of a modified maltose.

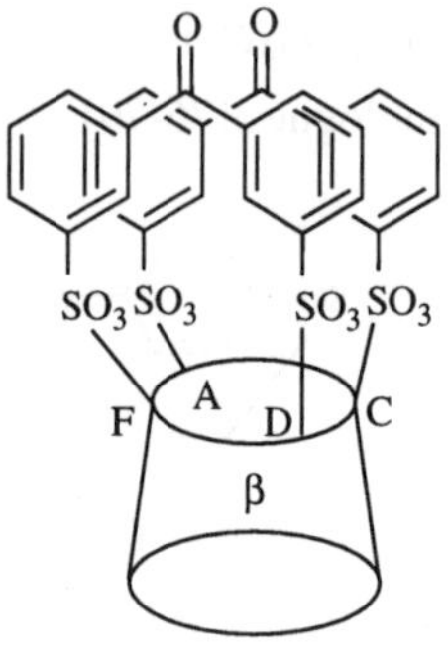

2.3a. Modifications Involving Reaction of a Single C(2) Secondary Hydroxy Group

The secondary hydroxy groups of the CDs are the most acidic, with pK_a values near 12.2 [28-30]. The C(2) groups are also exposed at the wider end of the CD annuli, such that when they deprotonate under basic conditions, the resultant alkoxides often react readily and selectively as nucleophiles. That is, whereas the primary hydroxy groups are the most nucleophilic under neutral and acidic conditions, above pH 10-11 it is the C(2) secondary hydroxy groups which are generally selectively modified through treatment with electrophilic reagents. The inclusion of reagents in the annuli of the CDs and the relative orientations of reactive groups in the complexes also affect the regioselectivity of reactions of the CDs. As a consequence of these effects, reactions of phenyl esters with αCD and βCD at pH 10 or higher result in acylation of the C(2) hydroxy groups [28,82]. The CD esters thus formed are base labile and hydrolyse under the reaction conditions, however, so this is not an efficient method for their preparation.

The orientation of reagent complexation and selective deprotonation of the CD C(2) hydroxy groups also account for the production of the C(2) tosylate **2.43**, through reaction of βCD with *m*-nitrophenyl tosylate **2.46**, in aqueous buffer at pH 10 [83]. The reagent **2.46** is thought to complex in the CD cavity as shown in Fig. 2.1. A similar rationale accounts for the reaction of αCD with *m*-nitrobenzenesulfonyl chloride at pH 12 to give the C(2) sulfonate **2.45** [84]. The sulfonate **2.43** is also accessible by treatment of βCD with *p*-toluenesulfonyl chloride and sodium hydride in *N,N*-dimethylformamide [85]. The corresponding αCD tosylate **2.42** has been prepared using *p*-toluenesulfonyl chloride in aqueous alkaline solution [15,86]. By contrast, the reaction of βCD with *p*-toluenesulfonyl chloride in aqueous alkaline solution, under the same conditions, gives the C(6) tosylate **2**, instead of the regioisomer **2.43** [15]. This illustrates the effect that the size and complexation properties of the CDs can have on the regioselectivity of their reactions.

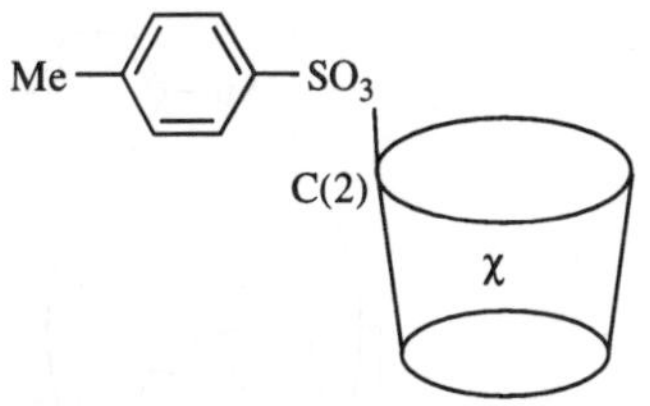

2.42 χ = α
2.43 χ = β
2.44 χ = γ

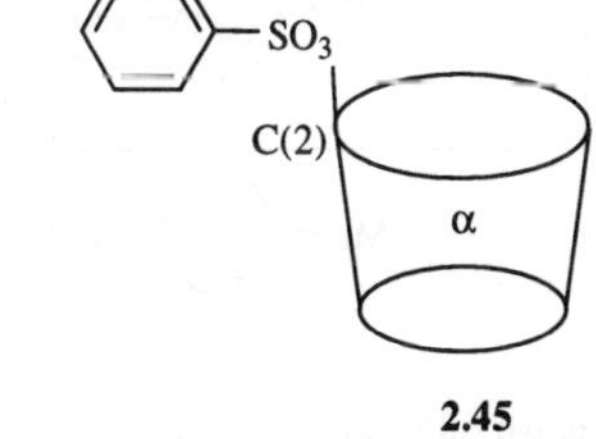

2.45

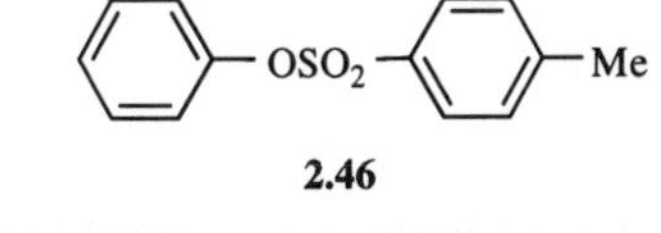

2.46

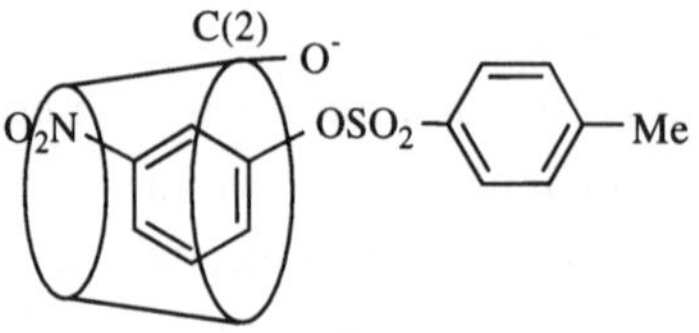

Fig. 2.1. Inclusion of the sulfonate **2.46** in the annulus of βCD determines the regioselectivity of reaction.

An alternative method for synthesis of the monotosylates **2.42** and **2.43**, which is also suitable for preparation of the corresponding γCD derivative **2.44**, is illustrated in Scheme 2.5 [87]. It exploits the reaction of 1,2-diols with dibutyltin oxide, to activate the C(2) hydroxy group of a CD glucopyranose residue. Another variation on the synthesis of the tosylate **2.43** is to first silylate the primary hydroxy groups of βCD, so that the product of reaction with *p*-toluenesulfonyl chloride is more easily purified through chromatography [88].

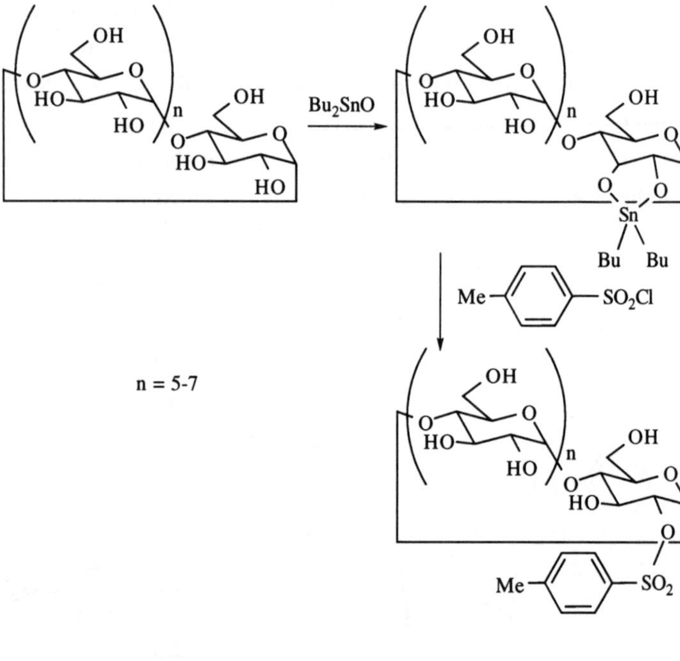

Scheme 2.5. Use of dibutyltin oxide to control the regioselective sulfonation of the CDs.

Displacement reactions of CD C(2) sulfonates generally result in the formation of *manno*-2,3-epoxyCDs (Scheme 2.6) [84,86,89,90]. Independent investigators [91] were unable to substantiate the only report to the contrary [92]. Consequently this approach is generally regarded as unsuitable for the direct introduction of substituents at C(2). Nucleophilic ring-opening reactions of the *manno*-epoxyCDs afford C(3) substituted products, with overall inversion of stereochemistry at both C(2) and C(3) of the modified glucopyranose residue. These reactions are discussed in more detail below, as a route for the synthesis of C(3) monosubstituted CDs.

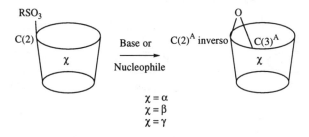

Scheme 2.6. Displacement reactions of CD C(2) sulfonates afford *manno*-2,3-epoxyCDs.

CD C(3) sulfonates undergo displacement reactions to afford *allo*-2,3-epoxyCDs [84,93-95]. Reactions of these epoxides with nucleophiles afford, as the major products, C(3) substituted CDs, with overall retention of stereochemistry at C(2) and C(3). They also afford minor amounts of the C(2) modified analogues, with stereochemical inversion at C(2) and C(3). Accordingly, the *allo*-epoxide, prepared from βCD, reacts with imidazole to give the C(2) substituted CD **2.47**, in 18% yield [91].

Nevertheless, the elaboration of the corresponding sulfonates is inefficient as a method for the synthesis of C(2) substituted CDs, so other approaches have been investigated. The C(2) amino-substituted βCD **2.48** has been prepared by insertion of a glucosamine residue into αCD [96,97]. The amine **2.48** has also been prepared from heptakis(2,3,6-tri-*O*-benzoyl)-βCD, by hydrolysis of one of the C(2) esters, then elaboration of the unmasked hydroxy group, before hydrolysis of the remaining ester moieties [98]. Alternatively, a substituent may be introduced directly at C(2) of a CD by *O*-alkylation [85,99,100]. For example, treatment of βCD with sodium hydride and *N*-methyl-4-chloromethyl-2-nitroaniline gives the ether **2.49**, in 35% yield

[85,99]. This type of chemistry has also been carried out using CDs in which the C(6) hydroxy groups have first been silylated, to aid chromatographic separation of the products [88].

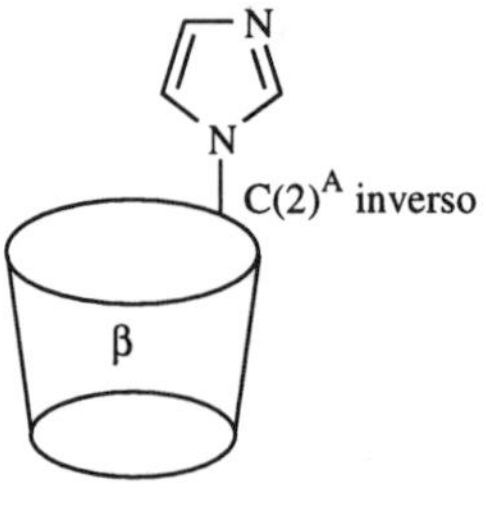

| 2.47 | 2.48 | 2.49 |

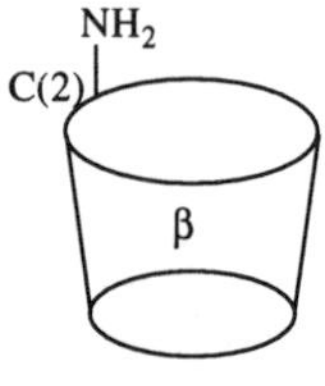

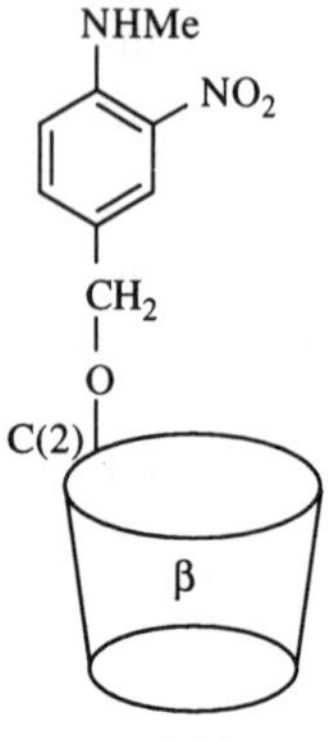

2.3b. Modifications Involving Reaction of Multiple C(2) Secondary Hydroxy Groups

The methods outlined above for modification of a single C(2) hydroxy group have also been used for multiple substitutions of the CDs. For example, the reaction of αCD with *m*-nitrobenzenesulfonyl chloride at pH 12 has been used to prepare a mixture of the corresponding $2^A,2^B$-, $2^A,2^C$-, and $2^A,2^D$-disubstituted CDs, which have been separated through chromatography [84]. The procedure outlined in Scheme 2.5, exploiting dibutyltin oxide to activate the C(2) hydroxy group of the CD glucopyranose residues, toward reaction with *p*-toluenesulfonyl chloride, has been used to obtain a mixture of the isomeric C(2) ditosylated βCDs, from which the individual components have been isolated [101].

Complete sulfonation of each of the C(2) hydroxy groups can only be accomplished if the primary hydroxy groups are first protected, to prevent their involvement in competing reactions. Thus, silylation of the primary hydroxy groups of βCD, followed by sulfonation of the C(2) alcohols through reaction with *p*-toluenesulfonyl chloride, and then desilylation with boron trifluoride etherate, affords the CD per-2-tosylate **2.50** [102,103]. Similar methodology has been used to prepare the corresponding per-2-benzenesulfonate [104].

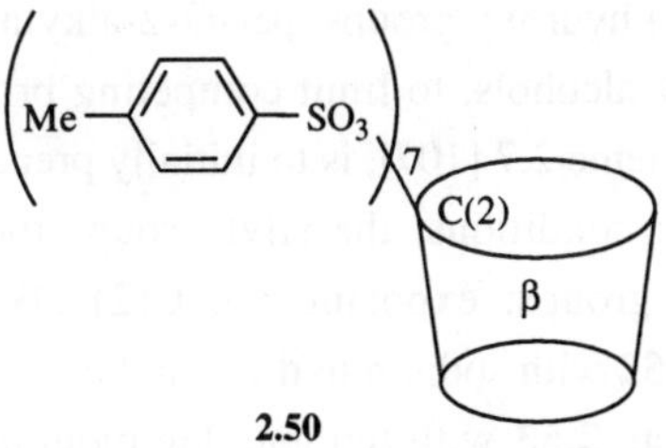

2.50 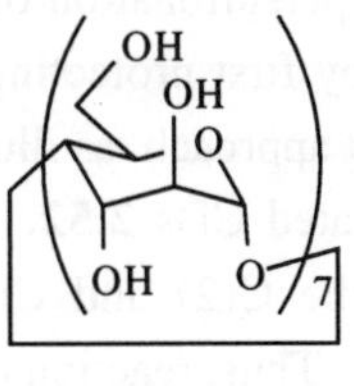   **2.51**

On treatment with base, the CD di- and per-2-sulfonates described above each react by intramolecular displacement to give the corresponding di- and per-*manno*-2,3-epoxyCDs [84,101,103-105]. Little has been reported on the reactions of these compounds, but per-*manno*-2,3-epoxy-βCD has been treated with water to give β-cycloaltrin **2.51**, in 73% yield [106], opening the way to a new class of cyclic oligosaccharides.

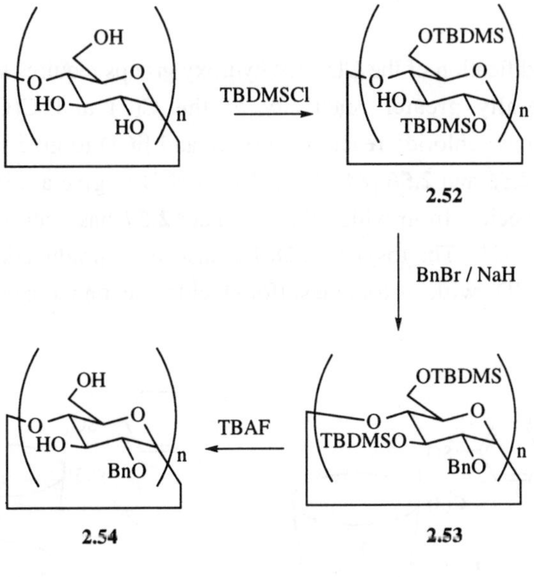

Scheme 2.7. Use of protecting groups for the synthesis of per-*O*-2-alkylated CDs.

As for persulfonation of the CD C(2) hydroxy groups, per-*O*-2-alkylation is best achieved by first protecting the primary alcohols, to limit competing processes. A convenient approach, as illustrated in Scheme 2.7 [107], is to initially prepare the per-2,6-disilylated CDs **2.52**. Under basic conditions, the silyl groups then migrate between the C(2) and C(3) hydroxy groups, exposing the C(2) alkoxides for alkylation. Thus, reaction of the CDs **2.52** with sodium hydride and benzyl bromide, followed by desilylation of the products **2.53** with tetrabutylammonium fluoride, gives the corresponding per-*O*-2-benzylCDs **2.54**. Heptakis(2-*O*-methyl)-βCD has also been prepared using similar methodology [107]. Alternatively, silylation of only the primary hydroxy groups followed by alkylation of all the C(2) hydroxy groups has also been reported [59].

2.4a.   Modifications Involving Reaction of a Single C(3) Secondary Hydroxy Group

Selective modification of the CD C(3) hydroxy groups requires special reagents to avoid the generally greater reactivity of the C(2) and C(6) alcohols. β-Naphthalenesulfonyl chloride reacts with αCD and βCD to give the corresponding monosulfonates **2.55** and **2.56** [84,94], and with γCD to give a mixture of C(2) and C(3) modified species, from which the sulfonate **2.57** has been separated through chromatography [95]. The tosylate **2.58** has also been produced, in low yield, by treatment of αCD with *p*-toluenesulfonyl chloride under carefully controlled conditions [93].

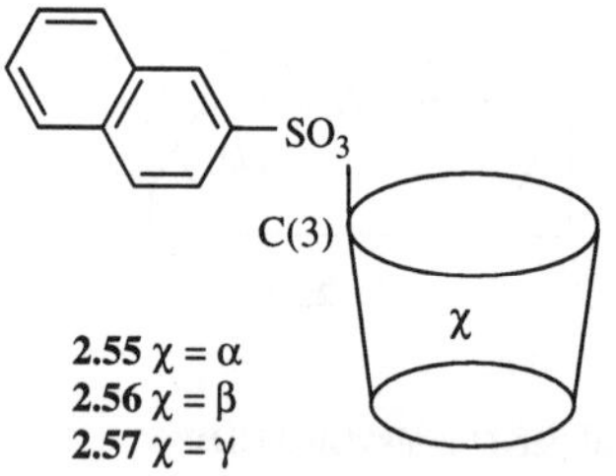

2.55 χ = α
2.56 χ = β
2.57 χ = γ

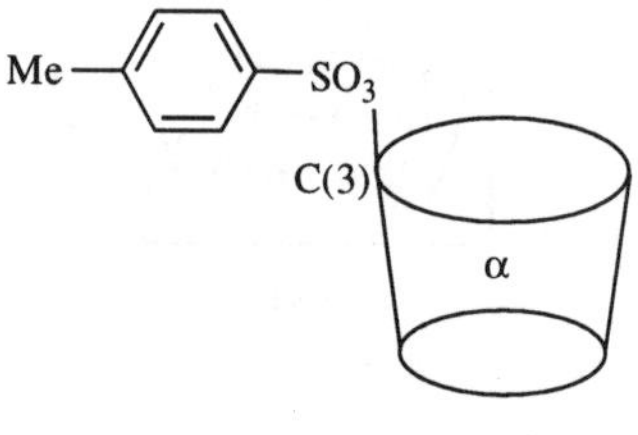

2.58

Treatment of the sulfonates **2.55-2.58** with bases gives the corresponding *allo*-2,3-epoxyCDs [84,93-95]. These react with nucleophiles to give mainly C(3) substituted

CDs, with overall retention of stereochemistry at C(2) and C(3), together with minor amounts of the C(2) substituted analogues, through stereochemical inversion at C(2) and C(3) [91]. Cyclodextrin *manno*-2,3-epoxides, obtained as described above by treatment of C(2) sulfonates with base [84,86,89,90], react with nucleophiles to give C(3) substituted CDs, also with stereochemical inversion at C(2) and C(3) of the modified glucopyranose [86]. In this way a pyridoxamine moiety has been attached at C(3) of βCD [89,90], and the amines **2.59** and **2.60** have been prepared by treatment of the corresponding αCD and βCD *manno*-epoxides with ammonia [86,108]. The amino group of the CDs **2.59** and **2.60** reacts selectively as a nucleophile and therefore provides a site for further elaboration, as illustrated in the synthesis of the acetamide **2.61** [108] and the CD dimers **2.62** [21,22] through acylation of the amine **2.60**.

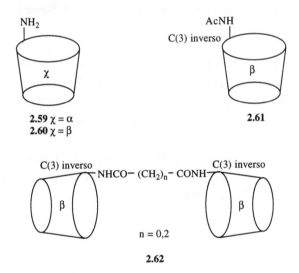

Selective alkylation of CD C(3) hydroxy groups has been accomplished by exploiting insertion reactions of aromatic diazo compounds. For example, the reaction of βCD with phenyldiazomethane affords mainly the ether **2.63** [109]. Selective modification of a C(3) hydroxy group of heptakis(6-*O-tert*-butyldimethyl-silyl)-βCD is also achieved using sodium hydride and *N*-methyl-4-chloromethyl-2-nitroaniline, in order to prepare the ether **2.64** [110].

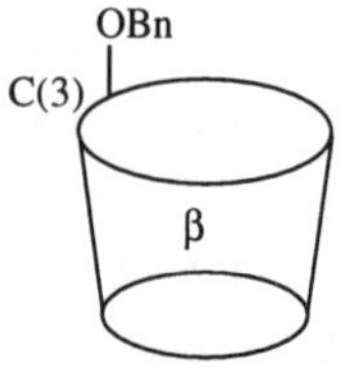

2.63

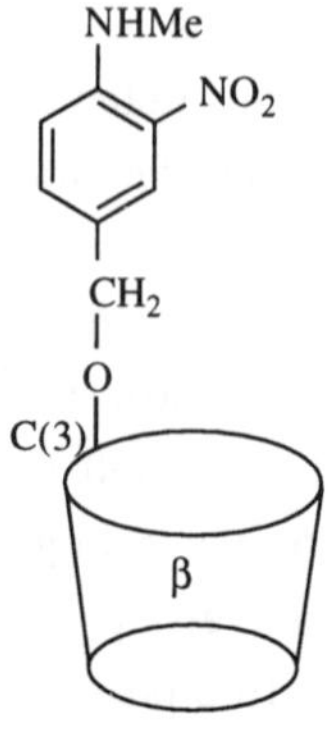

2.64

2.4b. Modifications Involving Reaction of Multiple C(3) Secondary Hydroxy Groups

Multiply C(3) substituted CDs have also been reported. The disulfonates **2.65** and **2.66** are obtained by treatment of βCD with β-naphthalenesulfonyl chloride [94]. In a similar fashion, $3^A,3^C,3^E$-tri-*O*-(β-naphthylsulfonyl)-βCD **2.67** has been prepared in 18% yield [111]. The sulfonates **2.65-2.67** are converted to the corresponding di- and tri-*allo*-epoxides [94,111], which in turn react with imidazole to give the regiospecifically modified CDs **2.68-2.70**, respectively [112]. Thus, the behaviour of the di- and tri-sulfonates **2.65-2.67** is analogous to that of the corresponding monosulfonates **2.55-2.58**, described above.

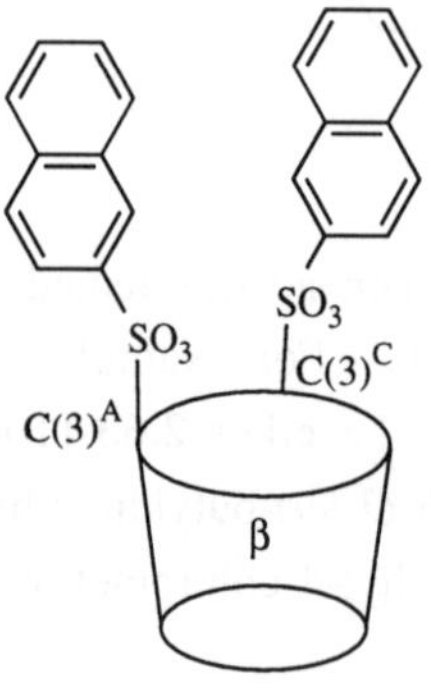

2.65

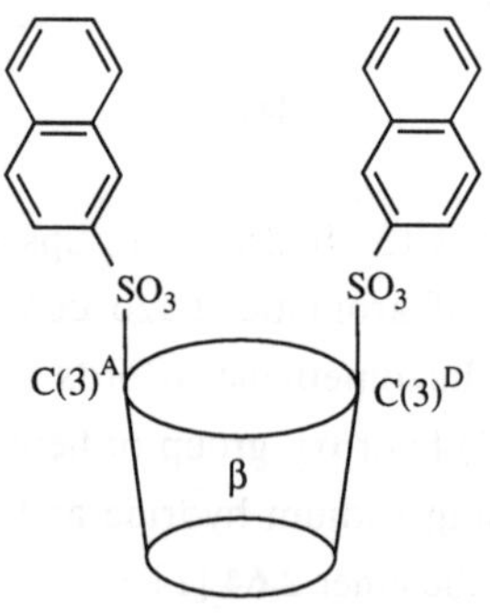

2.66

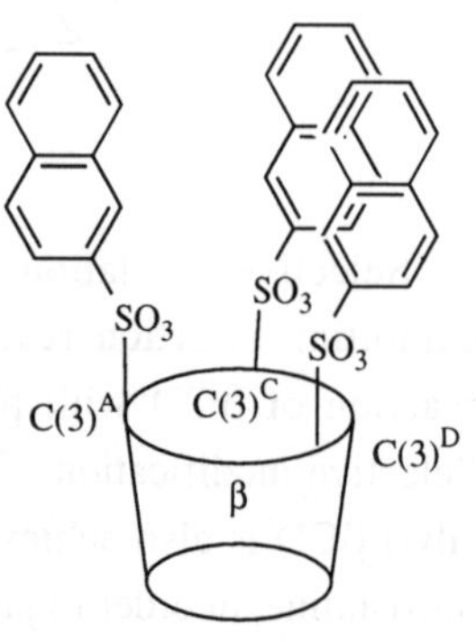

2.67

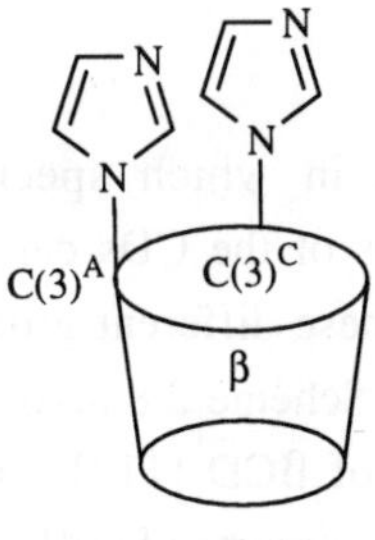

2.68* 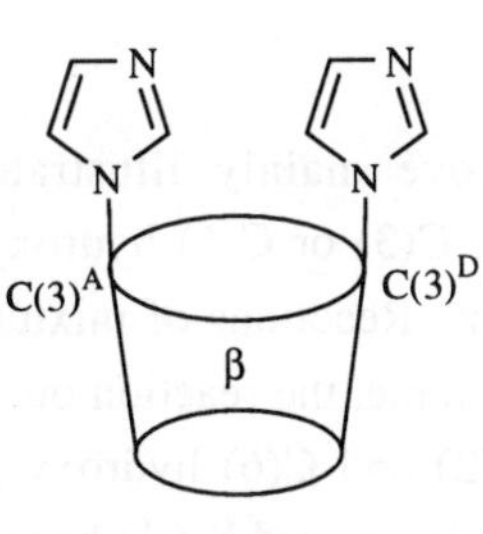   2.69* 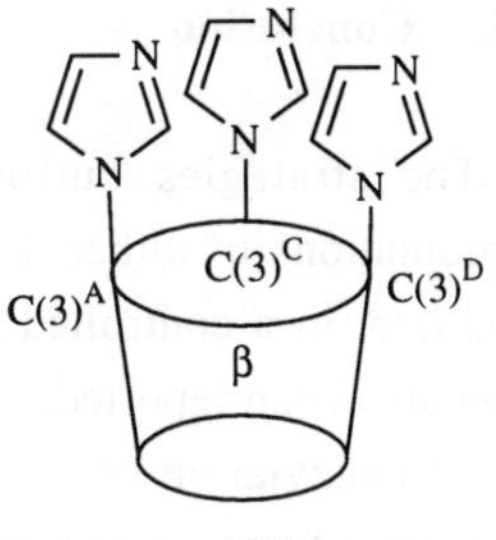   2.70*

*The stereochemistry of the CDs **2.68-2.70** is inverted at C(2) and C(3) of the modified glucopyranose residues.

To prevent modification of the C(2) and C(6) hydroxy groups in the synthesis of heptakis(3-O-methyl)-βCD from the parent oligosaccharide, protecting groups are employed, as illustrated in Scheme 2.8 [113]. This illustrates an alternative approach for the modification of multiple CD C(3) hydroxy groups.

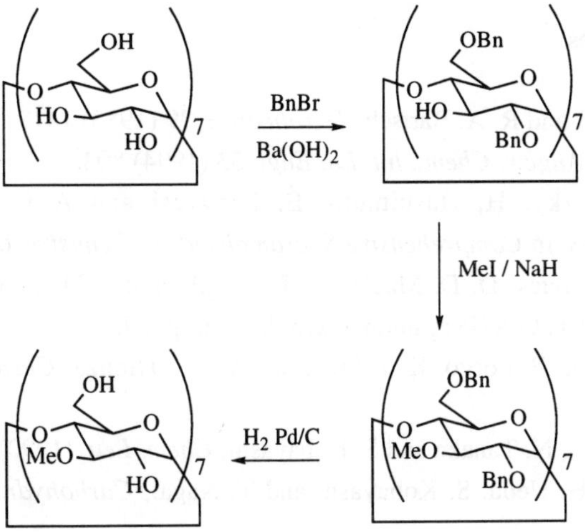

Scheme 2.8. Synthesis of heptakis(3-O-methyl)-βCD.

2.5.   Conclusion

The strategies outlined above mainly illustrate ways in which specific combinations of either the C(2), C(3) or C(6) hydroxy groups of the CDs can be modified, in a controlled manner. Reactions of mixtures of these different groups have also been reported. For example, the reaction outlined in Scheme 2.8 involves initial benzylation of all the C(2) and C(6) hydroxy groups of βCD [113]. The synthesis of per-2,6-*O*-trimethylsilyl-α- and β-CD has also been reported [114], and procedures for modification of combinations of the C(2) and C(3) hydroxy groups have also been described [62,63,115-118]. Generally these modifications exploit the same properties of the different types of hydroxy groups which have already been discussed. Consequently the methods that have been developed for modification of the natural CDs provide enormous scope for the construction of molecular hosts, which may be tailored to meet quite specific requirements in supramolecular chemistry. Applications of these are discussed in the following chapters of this book.

2.6.   References

1.   A. P. Croft and R. A. Bartsch, *Tetrahedron* **39** (1983) 1417.
2.   G. Wenz, *Angew. Chem., Int. Ed. Engl.* **33** (1994) 803.
3.   L. Jicsinszky, H, Hashimoto, É. Fenyvesi and A Ueno, *Cyclodextrin Derivatives*, in *Comprehensive Supramolecular Chemistry*; eds. J. L. Atwood, J. E. D. Davies, D. D. MacNicol, F. Vögtle and J.-M. Lehn, Vol. 3, eds. J. Szejtli and T. Osa (Pergamon, Oxford, 1996) p. 57.
4.   A. R. Khan, P. Forgo, K. J. Stine and V. T. D'Souza, *Chem. Rev.* **98** (1998) 1977.
5.   T. Fujiwara, N. Tanaka and S. Kobayashi, *Chem. Lett.* (1990) 739.
6.   T. Endo, H. Ueda, S. Kobayashi and T. Nagai, *Carbohydr. Res.* **269** (1995) 369.
7.   I. Miyazawa, H. Ueda, H. Nagase, T. Endo, S. Kobayashi and T. Nagai, *Eur. J. Pharm. Sc.* **3** (1995) 153.
8.   H. Ueda, T. Endo, H. Nagase, S. Kobayashi and T. Nagai, *J. Inclusion Phenom. Mol. Recogn. Chem.* **25**, (1996) 17.

9.   T. Endo, H. Nagase, H. Ueda, S. Kobayashi and T. Nagai, *Chem. Pharm. Bull.* **45** (1997) 532.

10.   T. Endo, H. Nagase, H. Ueda, A. Shigihara, S. Kobayashi and T. Nagai, *Chem. Pharm. Bull.* **45** (1997) 1856.

11.   J. Jacob, K. Geßler, D. Hoffmann, H. Sanbe, K. Koizumi, S. M. Smith, T. Takaha and W. Saenger, *Angew. Chem.*, *Int. Ed. Engl.* **37** (1998) 606.

12.   W. Saenger, J. Jacob, K. Gessler, T. Steiner, D. Hoffmann, H. Sanbe, K. Koizumi, S. M. Smith and T. Takaha, *Chem. Rev.* **98** (1998) 1787.

13.   L. D. Melton and K. N. Slessor, *Carbohydr. Res.* **18** (1971) 29.

14.   S. E. Brown, J. H. Coates, D. R. Coghlan, C. J. Easton, S. J. van Eyk, W. Janowski, A. Lepore, S. F. Lincoln, Y. Luo, B. L. May, D. S. Schiesser, P. Wang and M. L. Williams, *Aust. J. Chem.* **46** (1993) 953.

15.   K. Takahashi, K. Hattori and F. Toda, *Tetrahedron Lett.* **25** (1984) 3331.

16.   R. C. Petter, J. S. Salek, C. T. Sikorski, G. Kumaravel and F.-T. Lin, *J. Am. Chem. Soc.* **112** (1990) 3860.

17.   N. Zhong, H.-S. Byun and R. Bittman, *Tetrahedron Lett.* **39** (1998) 2919.

18.   S. Hanessian, A. Benalil and C. Laferrière, *J. Org. Chem.* **60** (1995) 4786.

19.   B. L. May, S. D. Kean, C. J. Easton and S. F. Lincoln, *J. Chem. Soc.*, *Perkin Trans. 1* (1997) 3157.

20.   C. J. Easton, S. Kassara, S. F. Lincoln and B. L. May, *Aust. J. Chem.* **48** (1995) 269.

21.   J. H. Coates, C. J. Easton, S. J. van Eyk, S. F. Lincoln, B. L. May, C. B. Whalland and M. L. Williams, *J. Chem. Soc.*, *Perkin Trans. 1* (1990) 2619.

22.   C. J. Easton, S. J. van Eyk, S. F. Lincoln, B. L. May, J. Papageorgiou and M. L. Williams, *Aust. J. Chem.* **50** (1997) 9.

23.   J. B. Huff and C. Bieniarz, *J. Org. Chem.* **59** (1994) 7511.

24.   K. A. Martin and A. W. Czarnik, *Tetrahedron Lett.* **35** (1994) 6781.

25.   J. Yoon, S. Hong, K. A. Martin and A. W. Czarnik, *J. Org. Chem.* **60** (1995) 2792.

26.   M. J. Cornwell, J. B. Huff and C. Bieniarz, *Tetrahedron Lett.* **36** (1995) 8371.

27.   K. Fujita, H. Yamamura, T. Imoto and I. Tabushi, *Chem. Lett.* (1988) 543.

28.   R. L. VanEtten, G. A. Clowes, J. F. Sebastian and M. L. Bender, *J. Am. Chem. Soc.* **89** (1967) 3253.

29.   R. I. Gelb, L. M. Schwartz, J. J. Bradshaw and D. A. Laufer, *Bioorg. Chem.* **9** (1980) 299.

30.   R. I. Gelb, L. M. Schwartz and D. A. Laufer, *Bioorg. Chem.* **11** (1982) 274.

31.   K. Takeo, K. Uemura and H. Mitoh, *J. Carbohydr. Chem.* **7** (1988) 293.

32.   S. Tian and V. T. D'Souza, *Tetrahedron Lett.* **35** (1994) 9339.

33.   W. Lautsch, R. Wiechert and H. Lehmann, *Kolloid-Z.* **135** (1954) 134.

34.   S. Umezawa and K. Tatsuta, *Bull. Chem. Soc. Jpn.* **41** (1968) 464.

35.   F. Cramer, G. Mackensen and K. Sensse, *Chem. Ber.* **102** (1969) 494.

36.   F. Cramer and G. Mackensen, *Chem. Ber.* **103** (1970) 2138.

37.   R. Breslow, M. F. Czarniecki, J. Emert and H. Hamaguchi, *J. Am. Chem. Soc.* **102** (1980) 762.

38.   H. Yamamura and K. Fujita, *Chem. Pharm. Bull.* **39** (1991) 2505.

39.   P. R. Ashton, P. Ellwood, I. Staton and J. F. Stoddart, *J. Org. Chem.* **56** (1991) 7274.

40.   H. Yamamura, Y. Kawase, M. Kawai and Y. Butsugan, *Bull. Chem. Soc. Jpn.* **66** (1993) 585.

41.   H. Yamamura, T. Ezuka, Y. Kawase, M. Kawai, Y. Butsugan and K. Fujita, *J. Chem. Soc., Chem. Commun.* (1993) 636.

42.   J. Emert and R. Breslow, *J. Am. Chem. Soc.* **97** (1975) 670.

43.   K. Takeo, T. Sumimoto and T. Kuge, *Staerke* **24** (1974) 111.

44.   A. Gadelle and J. Defaye, *Angew. Chem., Int. Ed. Engl.* **30** (1991) 78.

45.   P. R. Ashton, R. Königer, J. F. Stoddart, D. Alker and V. D. Harding, *J. Org. Chem.* **61** (1996) 903.

46.   B. I. Gorin, R. J. Riopelle and G. R. J. Thatcher, *Tetrahedron Lett.* **37** (1996) 4647.

47.   A. R. Khan and V. T. D'Souza, *J. Org. Chem.* **59** (1994) 7492.

48.   K. Chmurski and J. Defaye, *Tetrahedron Lett.* **38** (1997) 7365.

49.   D. P. Parazak, A. R. Khan, V. T. D'Souza and K. J. Stine, *Langmuir* **12** (1996) 4046.

50.   D. Vizitiu, C. S. Walkinshaw, B. I. Gorin and G. R. J. Thatcher, *J. Org. Chem.* **62** (1997) 8760.

51.   H. H. Baer, A. Vargas Berenguel, Y. Y. Shu, J. Defaye, A. Gadelle and F. Santoyo González, *Carbohydr. Res.* **228** (1992) 307.

52. G. R. Newkome, L. A. Godínez and C. N. Moorefield, *J. Chem. Soc., Chem. Commun.* (1998) 1821.

53. P. R. Ashton, P. Ellwood, I. Staton and J. F. Stoddart, *Angew. Chem., Int. Ed. Engl.* **30** (1991) 80.

54. H. Yamamura, H. Masuda, Y. Kawase, M. Kawai, Y. Butsugan and H. Einaga, *J. Chem. Soc., Chem. Commun.* (1996) 1069.

55. H. Yamamura, H. Kawai, T. Yotsuya, T. Higuchi, Y. Butsugan, S. Araki, M. Kawai and K. Fujita, *Chem. Lett.* (1996) 799.

56. P. R. Ashton, G. Gattuso, R. Königer, J. F. Stoddart and D. J. Williams, *J. Org. Chem.* **61** (1996) 9553.

57. J. Boger, R. J. Corcoran and J.-M. Lehn, *Helv. Chim. Acta* **61** (1978) 2190.

58. K. Takeo, H. Mitoh and K. Uemura, *Carbohydr. Res.* **187** (1989) 203.

59. P. S. Bansal, C. L. Francis, N. K. Hart, S. A. Henderson, D. Oakenfull, A. D. Robertson and G. W. Simpson, *Aust. J. Chem.* **51** (1998) 915.

60. P. Fügedi, *Carbohydr. Res.* **192** (1989) 366.

61. M. J. Pregel and E. Buncel, *Can. J. Chem.* **69** (1991) 130.

62. P. Zhang, C.-C. Ling, A. W. Coleman, H. Parrot-Lopez and H. Galons, *Tetrahedron Lett.* **32** (1991) 2769.

63. D. Alker, P. R. Ashton, V. D. Harding, R. Königer, J. F. Stoddart, A. J. P. White and D. J. Williams, *Tetrahedron Lett.* **35** (1994) 9091.

64. I. Tabushi, K. Shimokawa, N. Shimizu, H. Shirakata and K. Fujita, *J. Am. Chem. Soc.* **98** (1976) 7855.

65. I. Tabushi, K. Shimokawa and K. Fujita, *Tetrahedron Lett.* (1977) 1527.

66. R. Breslow, J. B. Doherty, G. Guillot and C. Lipsey, *J. Am. Chem. Soc.* **100** (1978) 3227.

67. R. Breslow, P. Bovy and C. Lipsey Hersh, *J. Am. Chem. Soc.* **102** (1980) 2115.

68. I. Tabushi, Y. Kuroda, K. Yokota and L. C. Yuan, *J. Am. Chem. Soc.* **103** (1981) 711.

69. I. Tabushi, K. Yamamura and T. Nabeshima, *J. Am. Chem. Soc.* **106** (1984) 5267.

70. I. Tabushi, T. Nabeshima, K. Fujita, A. Matsunaga and T. Imoto, *J. Org. Chem.* **50** (1985) 2638.

71. R. Breslow, J. W. Canary, M. Varney, S. T. Waddell and D. Yang, *J. Am. Chem. Soc.* **112** (1990) 5212.

72.   K. Fujita, T. Tahara and T. Koga, *Chem. Lett.* (1989) 821.

73.   J. Boger, D. G. Brenner and J. R. Knowles, *J. Am. Chem. Soc.* **101** (1979) 7630.

74.   S. Cottaz and H. Driguez, *J. Chem. Soc., Chem. Commun.* (1989) 1088.

75.   S. Cottaz, C. Apparu and H. Driguez, *J. Chem. Soc., Perkin Trans. 1* (1991) 2235.

76.   L. Bornaghi, J.-P. Utille, D. Penninga, A. K. Schmidt, L. Dijkhuizen, G. E. Schulz and H. Driguez, *J. Chem. Soc., Chem. Commun.* (1996) 2541.

77.   P. R. Ashton, C. L. Brown, S. Menzer, S. A. Nepogodiev, J. F. Stoddart and D. J. Williams, *Chem. Eur. J.* **2** (1996) 580.

78.   P. R. Ashton, S. J. Cantrill, G. Gattuso, S. Menzer, S. A. Nepogodiev, A. N. Shipway, J. F. Stoddart and D. J. Williams, *Chem. Eur. J.* **3** (1997) 1299.

79.   G. Gattuso, S. A. Nepogodiev and J. F. Stoddart, *Chem. Rev.* **98** (1998) 1919.

80.   I. Tabushi, L. C. Yuan, K. Shimokawa, K. Yokota, T. Mizutani and Y. Kuroda, *Tetrahedron Lett.* **22** (1981) 2273.

81.   I. Tabushi, Y. Kuroda and K. Yokota, *Tetrahedron Lett.* **23** (1982) 4601.

82.   R. L. VanEtten, J. F. Sebastian, G. A. Clowes and M. L. Bender, *J. Am. Chem. Soc.* **89** (1967) 3242.

83.   A. Ueno and R. Breslow, *Tetrahedron Lett.* **23** (1982) 3451.

84.   K. Fujita, S. Nagamura, T. Imoto, T. Tahara and T. Koga, *J. Am. Chem. Soc.* **107** (1985) 3233.

85.   D. Rong and V. T. D'Souza, *Tetrahedron Lett.* **31** (1990) 4275.

86.   H. Ikeda, Y. Nagano, Y. Du, T. Ikeda and F. Toda, *Tetrahedron Lett.* **31** (1990) 5045.

87.   T. Murakami, K. Harata and S. Morimoto, *Tetrahedron Lett.* **28** (1987) 321.

88.   E. van Dienst, B. H. M. Snellink, I. von Peikartz, M. H. B. Grote Gansey, F. Venema, M. C. Feiters, R. J. M. Nolte, J. F. J. Engbersen and D. N. Reinhoudt, *J. Org. Chem.* **60** (1995) 6537.

89.   R. Breslow and A. W. Czarnik, *J. Am. Chem. Soc.* **105** (1983) 1390.

90.   R. Breslow, A. W. Czarnik, M. Lauer, R. Leppkes, J. Winkler and S. Zimmerman, *J. Am. Chem. Soc.* **108** (1986) 1969.

91.   D.-Q. Yuan, K. Ohta and K. Fujita, *J. Chem. Soc., Chem. Commun.* (1996) 821.

92. K. Rama Rao, T. N. Srinivasan, N. Bhanumathi and P. B. Sattur, *J. Chem. Soc., Chem. Commun.* (1990) 10.

93. K. Fujita, S. Nagamura and T. Imoto, *Tetrahedron Lett.* **25** (1984) 5673.

94. K. Fujita, T. Tahara, T. Imoto and T. Koga, *J. Am. Chem. Soc.* **108** (1986) 2030.

95. T. Tahara, K. Fujita and T. Koga, *Bull. Chem. Soc. Jpn.* **63** (1990) 1409.

96. N. Sakairi, L.-X. Wang and H. Kuzuhara, *J. Chem. Soc., Chem. Commun.* (1991) 289.

97. N. Sakairi, L.-X. Wang and H. Kuzuhara, *J. Chem. Soc., Perkin Trans. 1* (1995) 437.

98. N. Sakairi, H. Kuzuhara, T. Okamoto and M. Yajima, *Bioorg. Med. Chem.* **4** (1996) 2187.

99. D. Rong, H. Ye, T. R. Boehlow and V. T. D'Souza, *J. Org. Chem.* **57** (1992) 163.

100. J. Jindrich, J. Pitha, B. Lindberg, P. Seffers and K. Harata, *Carbohydr. Res.* **266** (1995) 75.

101. K. Fujita, T. Ishizu, K. Oshiro and K. Obe, *Bull. Chem. Soc. Jpn.* **62** (1989) 2960.

102. A. W. Coleman, P. Zhang, H. Parrot-Lopez, C.-C. Ling, M. Miocque and L. Mascrier, *Tetrahedron Lett.* **32** (1991) 3997.

103. A. W. Coleman, P. Shang, C. C. Ling, J. Manhutea, H. Parrot-Lopez and M. Miocque, *Supramolecular Chem.* **1** (1992) 11.

104. A. R. Khan, L. Barton and V. T. D'Souza, *J. Chem. Soc., Chem. Commun.* (1992) 1112.

105. A. R. Khan, L. Barton and V. T. D'Souza, *J. Org. Chem.* **61** (1996) 8301.

106. K. Fujita, H. Shimada, K. Ohta, Y. Nogami, K. Nasu and T. Koga, *Angew. Chem., Int. Ed. Engl.* **34** (1995) 1621.

107. P. R. Ashton, S. E. Boyd, G. Gattuso, E. Y. Hartwell, R. Königer, N. Spencer and J. F. Stoddart, *J. Org. Chem.* **60** (1995) 3898.

108. T. Murakami, K. Harata and S. Morimoto, *Chem. Lett.* (1988) 553.

109. S. H. Smith, S. M. Forrest, D. C. Williams, M. F. Cabell, M. F. Acquavella and C. J. Abelt, *Carbohydr. Res.* **230** (1992) 289.

110. S. Tian, P. Forgo and V. T. D'Souza, *Tetrahedron Lett.* **37** (1996) 8309.

111. K. Fujita, T. Tahara, H. Yamamura, T. Imoto, T. Koga, T. Fujioka and K. Mihashi, *J. Org. Chem.* **55** (1990) 877.

112. W.-H. Chen, D.-Q. Yuan and K. Fujita, *Tetrahedron Lett.* **38** (1997) 4599.

113. J. Canceill, L. Jullien, L. Lacombe and J.-M. Lehn, *Helv. Chim. Acta* **75** (1992) 791.

114. M. Bukowska, M. Maciejewski and J. Prejzner, *Carbohydr. Res.* **308** (1998) 275.

115. K. Fujita, T. Tahara, H. Sasaki, Y. Egashira, T. Shingu, T. Imoto and T. Koga, *Chem. Lett.* (1989) 917.

116. N. Sakairi and H. Kuzuhara, *Chem. Lett.* (1993) 2077.

117. L. Jullien, J. Canceill, L. Lacombe and J.-M. Lehn, *J. Chem. Soc., Perkin Trans. 2* (1994) 989.

118. N. Sakairi, N. Nishi, S. Tokura and H. Kuzuhara, *Carbohydr. Res.* **291** (1996) 53.

CHAPTER 3
MODIFIED CYCLODEXTRINS

INTRAMOLECULAR COMPLEXATION, CHIRAL DISCRIMINATION AND PHOTOCHEMICAL DEVICES

3.1. Intramolecular Complexation Modes Displayed by Modified Cyclodextrins

A substituent of a modified CD may include within the CD annulus, to form an intramolecular complex. Many of the factors which contribute to this complexation are the same as those which lead to the formation of intermolecular complexes, as discussed in Chapter 1. The extent of intramolecular complexation depends on complementary relationships between the substituent and the CD annulus, the most fundamental of these being size. This is exemplified by the 6^A-(4-(dimethylamino)benzamido)-6^A-deoxy-substituted αCD, βCD, and γCD **3.1-3.3** [1-3]. The ^{1}H NMR and induced circular dichroism spectra of these compounds show that only the βCD **3.2** spontaneously forms an intramolecular complex in aqueous solution. Presumably this is an example of what has been referred to as the *Goldilocks effect*, where the αCD annulus is too small to accommodate the substituent, the γCD annulus too large, and the βCD annulus *just right* [4]. In aqueous solution a single hydrophobic substituent is unlikely to complex in a relatively large hydrophobic CD annulus, because accommodation of residual water molecules in the resultant smaller cavity is energetically unfavourable. However, in the presence of another guest which fits the smaller cavity, intramolecular complexation may take place. This occurs with the γCD derivative **3.3**, where the substituent acts as a spacer to narrow the large annulus and allow binding of small hydrophobic molecules (Scheme 3.1c). The substituent must be displaced from the annulus of the βCD **3.2** for competitive guest binding to occur (Scheme 3.1b) but,

71

since the substituent of the αCD **3.1** is not complexed, in this case the annulus remains available for intermolecular guest complexation (Scheme 3.1a).

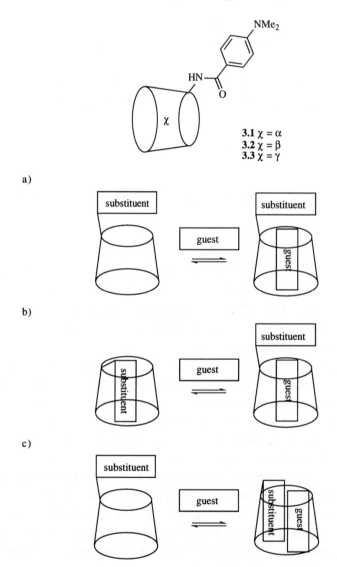

3.1 χ = α
3.2 χ = β
3.3 χ = γ

Scheme 3.1. Schematic representation of intra- and inter-molecular complexation modes for substituted CDs.

d)

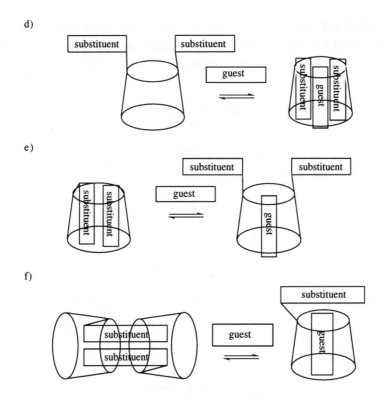

e)

f)

Scheme 3.1 cont. Schematic representation of intra- and inter-molecular complexation modes for substituted CDs.

The complexing characteristics of the CDs **3.2** and **3.3** in aqueous solution are similar to those of the dansyl-substituted βCD and γCD **3.4** and **3.5**, respectively [5,6]. The βCD **3.4** forms an intramolecular complex, which dissociates for competitive guest binding (Scheme 3.1b) [5]. The substituent of the γCD **3.5** may include within the annulus, to regulate the shape of the cavity for intermolecular inclusion of a guest (Scheme 3.1c) [6]. Alternatively, it sometimes acts as a cap, to increase the hydrophobicity of the CD annulus (Scheme 3.1a) [6]. The space-regulating effect of substituents of modified CDs (Scheme 3.1c) is an example of *induced-fit complexation*, and was originally observed with the naphthylacetate **3.6** [7-9]. Although the effect is most common with derivatives of γCD, where the annulus is relatively large, the phenomenon has also been seen with the anthranilatemodified βCD **3.7** [10]. A variant of the effect is observed with the $6^A,6^E$-disubstituted γCD **3.8**, where both substituents include in the CD annulus, to create a small hydrophobic pocket for guest complexation (Scheme 3.1d) [11].

3.4 χ = β
3.5 χ = γ

3.6

3.7

*denotes the substituted glucopyranoses which are lettered sequentially around the ring

3.8

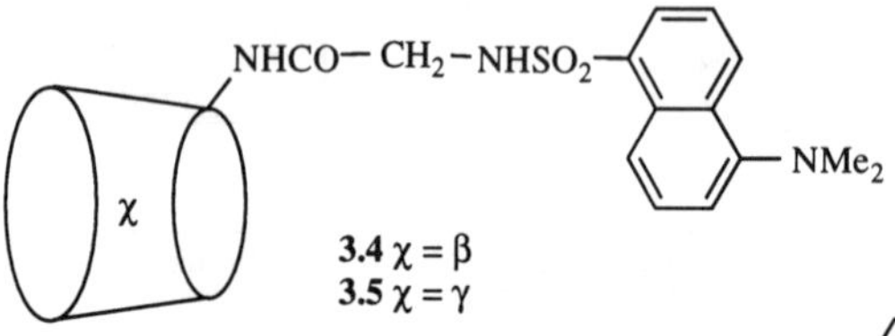

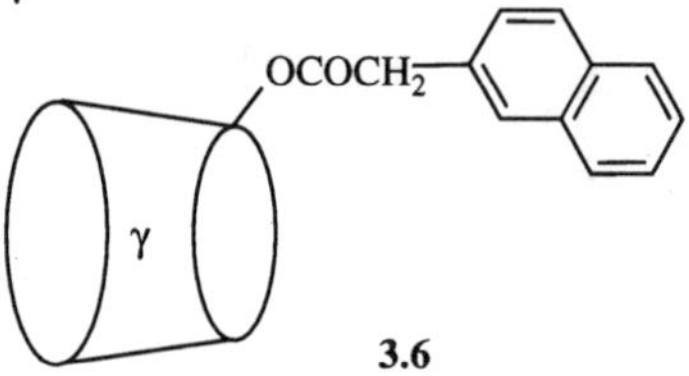

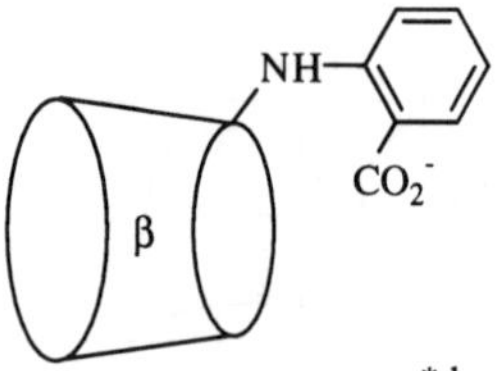

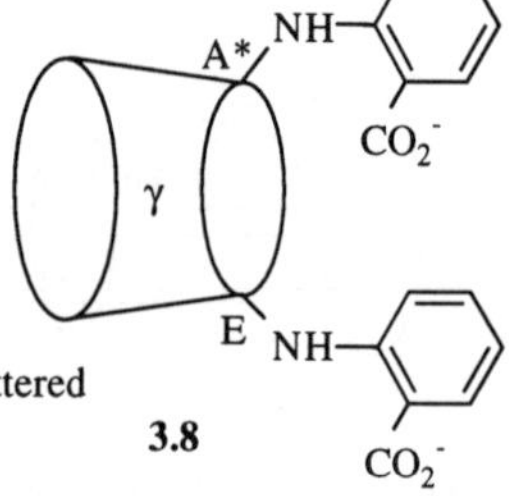

A variety of anthracene-appended CDs have been prepared and studied as molecular hosts [12,13]. In water, the mono-substituted βCD **3.9** and the doubly modified γCD **3.11** display competitive guest binding, according to Schemes 3.1b and 3.1e, respectively. The latter mode of guest complexation is also observed with a variety of dinaphthyl-substituted γCDs [14-19]. The ability of a γCD to accommodate two aryl substituents is reflected in the behaviour of the anthracene derivative **3.10** to form an association dimer, which undergoes guest-induced dissociation (Scheme 3.1f). Pyrene-substituted γCDs such as the amide **3.12** also form association dimers in aqueous solution, which dissociate on competitive guest complexation, according to Scheme 3.1f [20-24].

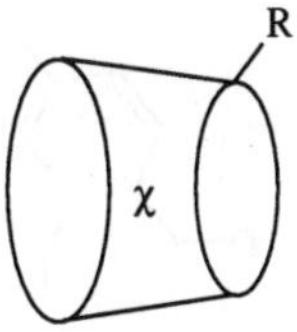

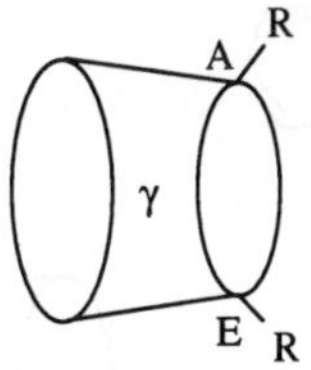

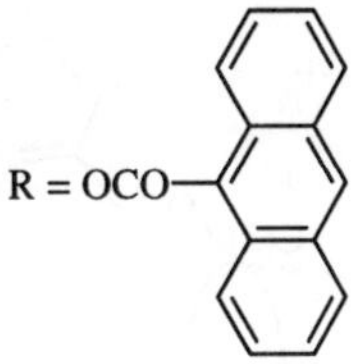

3.9 χ = β
3.10 χ = γ

3.11

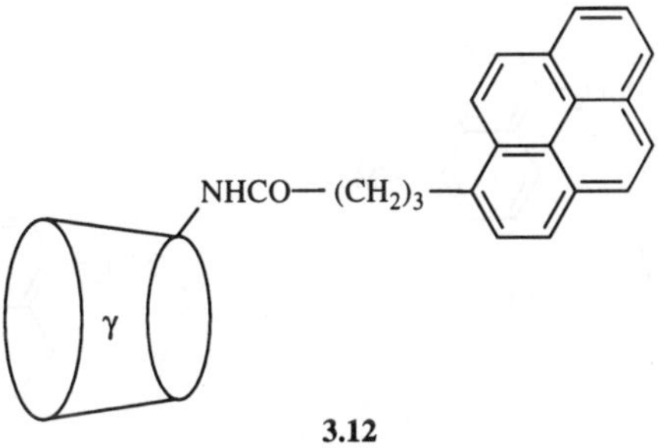

R = OCO—

NHCO— (CH₂)₃—

3.12

As discussed in Chapter 2, substituents may be readily attached to CDs through modification of the C(2), C(3) and C(6) hydroxy groups. The site of attachment affects the extent of intramolecular complexation [25-28], as illustrated with the regioisomeric βCDs **3.13** and **3.14** [27]. In aqueous solution, the bromonaphthyl moiety of the C(2) substituted CD **3.13** complexes in the annulus, from where it is displaced by dodecyltrimethylammonium bromide, acting as a competitive guest. The CD **3.14** does not form an intramolecular complex under the same conditions, presumably because the end of the annulus delineated by the primary hydroxy groups is too hindered. The dansylated βCDs **3.15-3.17** provide another example of the dependence of the intramolecular complexation of a substituent on its site of attachment to the CD nucleus. All three compounds form intramolecular complexes in water but, whereas the aromatic moiety of the C(6) and C(2) substituted derivatives **3.15** and **3.16** is readily displaced by other hydrophobic guests, the thermodynamic stability of the intramolecular complex of the C(3) substituted CD **3.17** is generally too high for competitive guest binding [25,26].

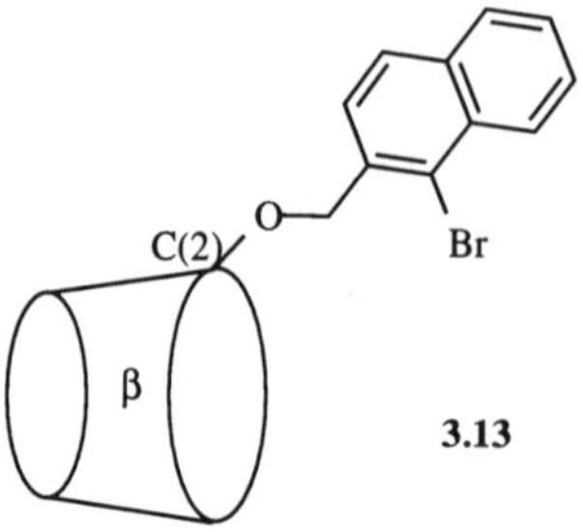

The extent of intramolecular complexation is also affected by other factors such as the nature of the link between the CD and the substituent [29,30], the chirality of the substituent [31-33], the temperature [34-36], and the extent of protonation of the substituent [30]. ^{1}H NMR studies in deuterium oxide have shown that of the CDs **3.18-3.20**, which vary in the way in which the substituent is tethered, the latter has the aromatic substituent most completely included [29]. The effect of the CD annulus to protect the ester moiety from hydrolysis reflects the extent of this complexation, with the pentane derivative **3.20** being the least reactive.

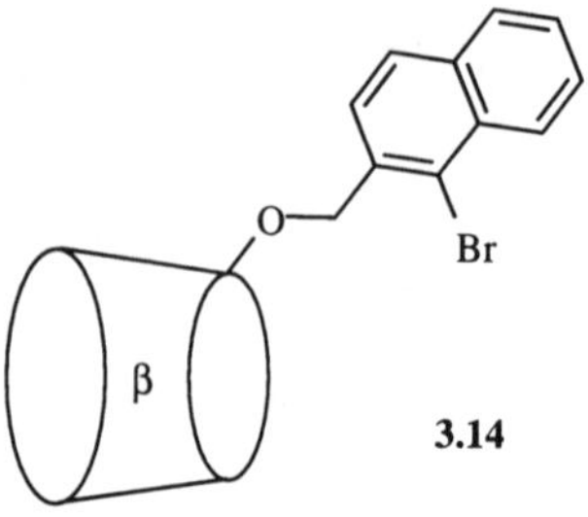

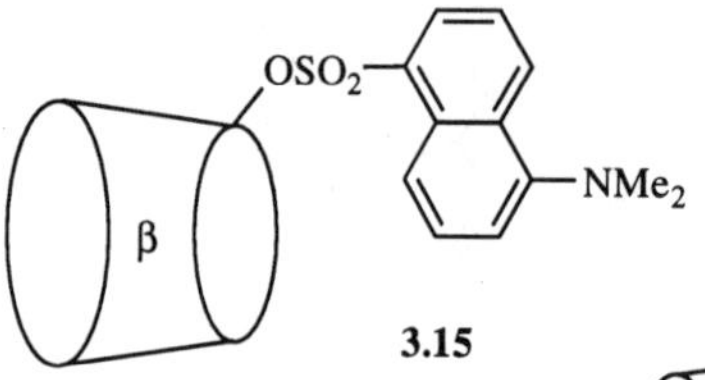

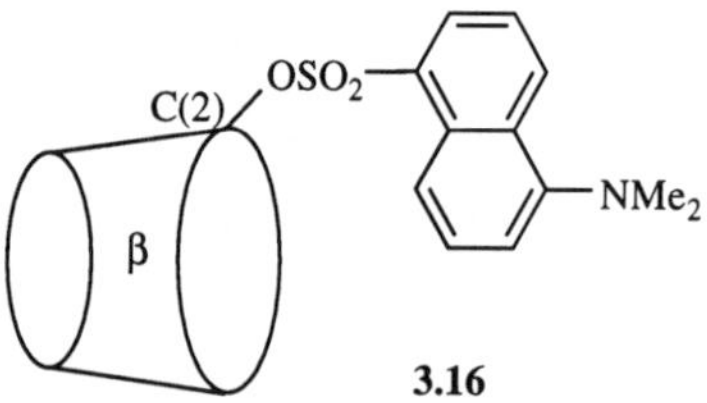

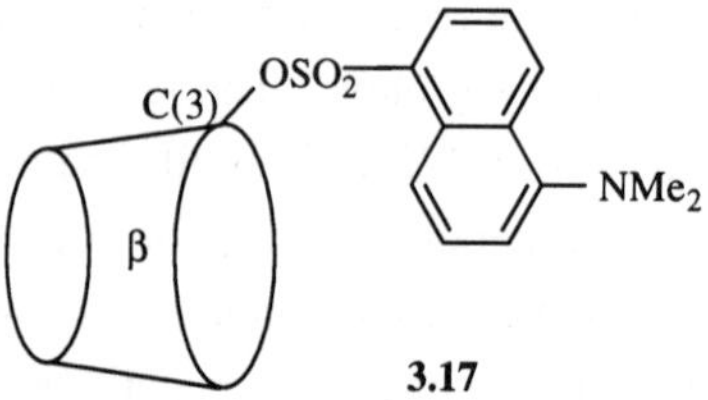

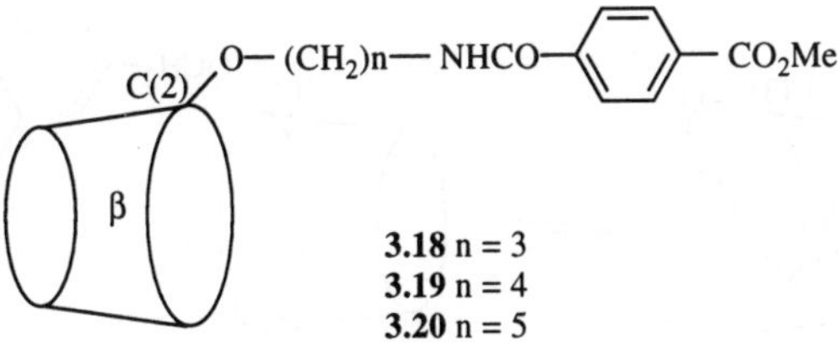

3.18 n = 3
3.19 n = 4
3.20 n = 5

Since the natural CDs each exist as a single enantiomer, their substitution with racemates produces diastereomers. The extent to which intramolecular complex formation with stereoisomers of this type can depend on the chirality of the substituent is illustrated with the *N*-dansylleucine-appended CDs **3.22** and **3.24**, where the dansyl portion of the (*S*)-leucine derivative **3.22** is more deeply included in its own annulus than is the case with the diastereomer **3.24** [33].

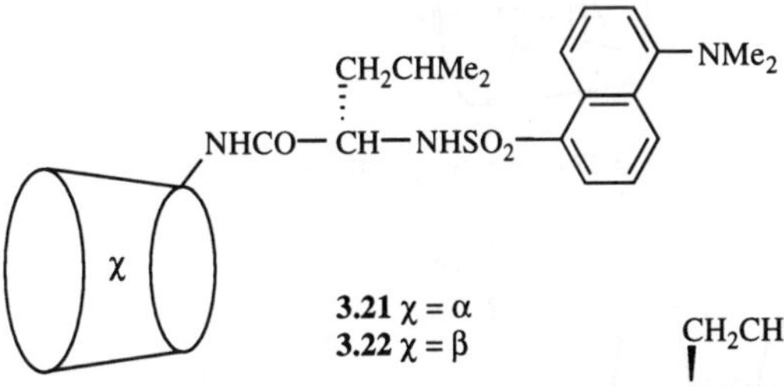

3.21 χ = α
3.22 χ = β

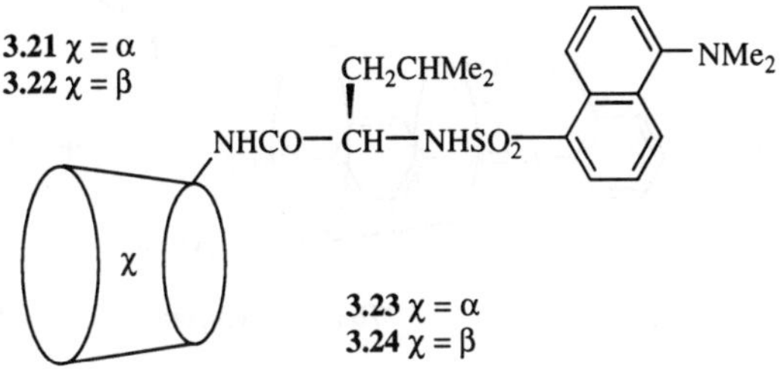

3.23 χ = α
3.24 χ = β

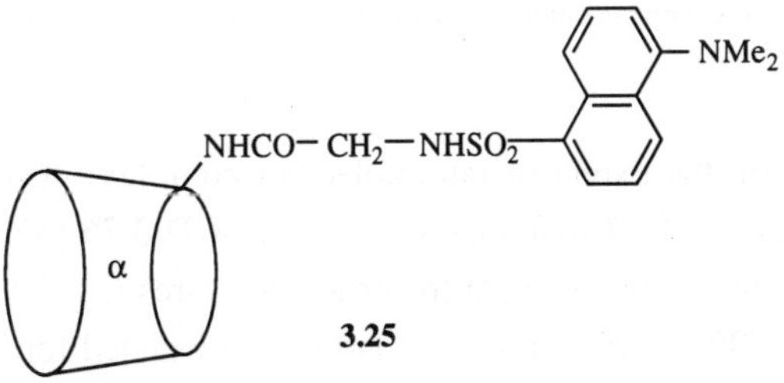

3.25

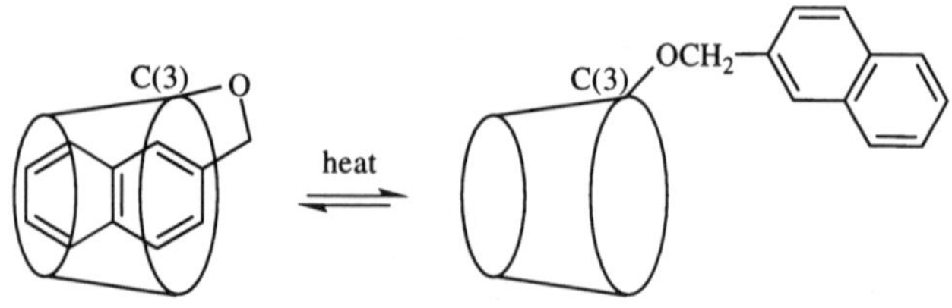

3.26a lower temperatures   **3.26b** higher temperatures

Scheme 3.2. Temperature-dependent inside-outside isomerism of the naphthyl-βCD **3.26**.

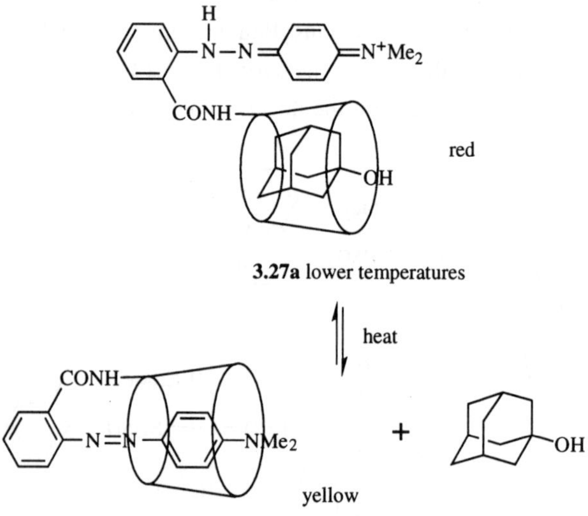

3.27a lower temperatures

heat

yellow

3.27b higher temperatures

Scheme 3.3. Heat-induced intramolecular complexation of the βCD **3.27** in the presence of adamantan-1-ol at pH 1.6.

The effect of temperature on the extent of intramolecular complex formation is clearly demonstrated in studies of 3^A-O-(naphth-2-ylmethyl)-βCD **3.26** [35,36]. In aqueous solution, the substituent is complexed at lower temperatures (*ca.* 20 °C) but not at higher temperatures (*ca.* 90 °C) (Scheme 3.2), and the dissociated form **3.26b** shows evidence of aggregation [37]. However, a lower temperature doesn't necessarily favour intramolecular complexation. In the presence of adamantan-1-ol, the intramolecular complex of the dye-substituted CD **3.27** increasingly dissociates with decreasing temperature over the range 65-25 °C (Scheme 3.3) [34].

In buffer at neutral pH, where the dansyl substituent is uncharged, the modified CDs **3.28** and **3.29** form thermodynamically stable intramolecular complexes, even in the presence of other potential guest molecules. The effect of protonation of the substituents on the stability of these complexes becomes evident as the pH is lowered, in that intramolecular complexation becomes less favourable and the tendency to form intermolecular complexes with other guests increases [30].

3.28 n = 1
3.29 n = 3

3.2. Molecular Recognition by Intramolecular Complexes of Substituted Cyclodextrins

Intramolecular complexation of the substituent of a modified CD affects the ease of competitive intermolecular guest complexation by such a host. In aqueous solution and in the absence of other guests, the substituents of the (*S*)- and (*R*)-phenylalanine-substituted CDs **3.30** and **3.31** are complexed within the CD annuli, from where they are displaced by the enantiomers of *N*-dansylalanine **3.32** and *N*-dansylphenylalanine **3.33**. This substituent displacement is reflected in the stability constants (K_{11}) of the complexes of βCD and the derivatives **3.30** and **3.31** with the enantiomers of the amino acid derivatives **3.32** and **3.33** (Table 3.1) [38-42], which are generally lower for the modified CDs **3.30** and **3.31**, due to competition between the intramolecular and intermolecular complexes in those cases. The range in thermodynamic stabilities of the complexes indicates the way in which CDs can be modified to control molecular recognition in guest binding. Other examples are given in Tables 3.2 and 3.3, for the complexation of cholic acid **3.34** and adamantan-1-ol by the dansylated αCDs **3.21**, **3.23**, and **3.25** [43], and the βCDs **3.2** and **3.22**

[2,3]. The data for the βCDs **3.2** and **3.22** indicate that the dimethylaminobenzoyl substituent of the former is more easily displaced from its annulus than is the dansyl group of the latter, since the stability constants of the intermolecular complexes of the former are much higher.

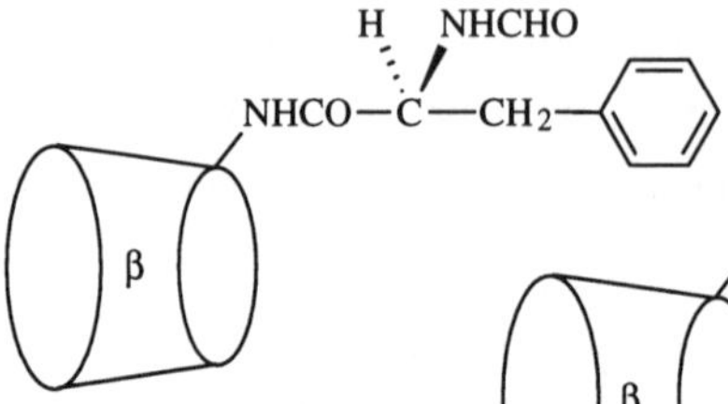

3.30

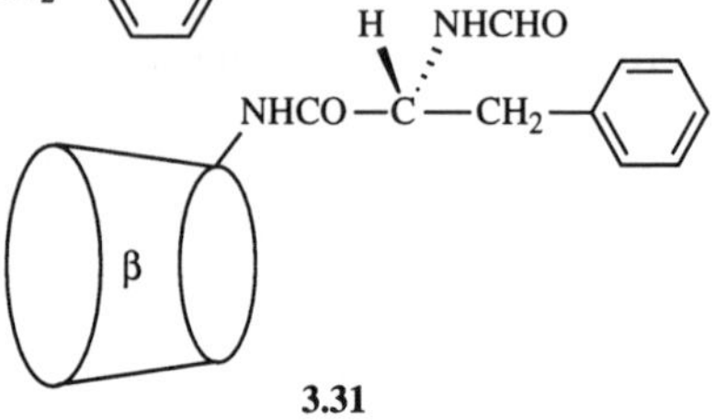

3.31

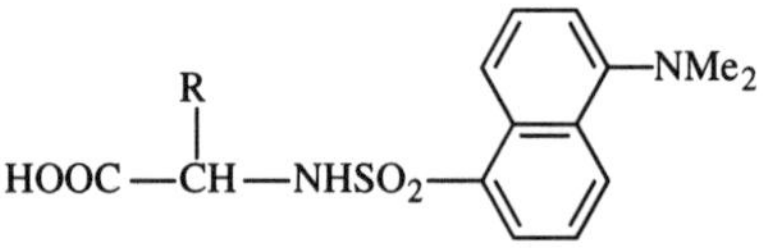

3.32 R = Me
3.33 R = CH₂Ph

Table 3.1. Host-Guest Stability Constants of the Complexes of βCD and the Derivatives **3.30** and **3.31** with the Enantiomers of the Amino Acid Derivatives **3.32** and **3.33**.[a]

Guest	K_{11} $dm^3\ mol^{-1}$		
	βCD	3.30	3.31
(R)-**3.32**	179	113	42
(S)-**3.32**	144	95	54
(R)-**3.33**	197	139	160
(S)-**3.33**	153	231	83

[a]At 298 K and pH 7.0 in 0.075 mol dm⁻³ phosphate buffer.

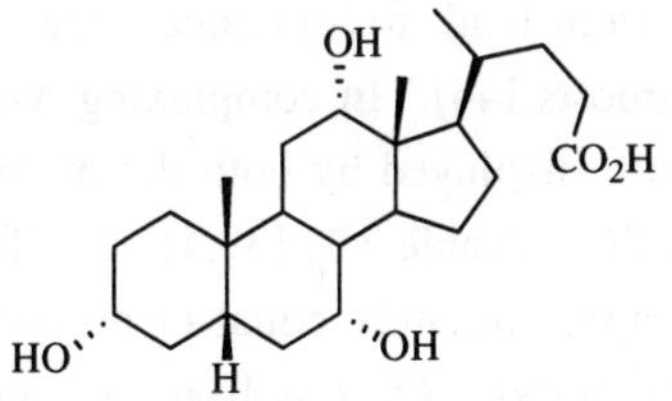

3.34

Table 3.2. Host-Guest Stability Constants of the Complexes of the Dansylated αCDs **3.21**, **3.23** and **3.25** with Cholic Acid **3.34** and Adamantan-1-ol.[a]

Guest	K_{11} $dm^3\ mol^{-1}$		
	3.21	**3.23**	**3.25**
3.34	1650	588	60.4
adamantan-1-ol	54200	19900	8910

[a]At 298 K and pH 7.0 in 0.01 mol dm^{-3} phosphate buffer.

Table 3.3. Host-Guest Stability Constants of the Complexes of the βCDs **3.2** and **3.22** with Cholic Acid **3.34** and Adamantan-1-ol.[a]

Guest	K_{11} $dm^3\ mol^{-1}$	
	3.2	**3.22**
3.34	27000	1320
adamantan-1-ol	128000	116

[a]At 298 K and pH 7.0 in 0.01 mol dm^{-3} phosphate buffer.

Chiral substituents often lead to enhanced stereoselectivity by CDs on complexation of racemic guests [44]. In complexing *N*-dansylphenylalanine **3.33**, clearly the enantioselectivity displayed by both the modified CDs **3.30** and **3.31** is greater than that shown by βCD (Table 3.1) [39,41,42]. The chiral discrimination by the CDs **3.30** and **3.31** is approximately equal in magnitude, although reversed in terms of absolute stereochemistry. On this basis it appears that the annuli of the modified CDs **3.30** and **3.31** serve mainly to bind the guests and contribute little toward the enantioselectivity. Instead, stereoselectivity probably results from interactions of chiral portions of the enantiomers of the guest **3.33** with the chiral substituents of the modified CDs **3.30** and **3.31**. A chiral substituent is not a prerequisite for enhanced enantioselectivity by a modified CD, however, as illustrated by the complexes of the toluidinyl-substituted CD **3.35** with amino acids [45]. Whereas βCD shows very little stereoselectivity in binding the enantiomers of alanine, serine, valine and leucine, the stability constants of the corresponding complexes of the modified CD **3.35** differ by factors of 3.0, 7.6, 3.6 and 33, respectively.

3.35

It is worth digressing at this point to briefly discuss some other examples of chiral discrimination by modified CDs. Thermodynamically, the chiral discrimination displayed by the natural CDs is usually quite small and of little practical utility in terms of separating guest enantiomers unless the CDs are incorporated into chromatographic materials. Even simple modifications to the CDs appear to increase their asymmetry and lead to greater diastereoselectivity of complexation [44]. Thus, the stability constants of the complexes of (*R*)- and (*S*)-2-phenylpropanoate with βCD, K_{11} = 63 and 52 dm^3 mol^{-1}, respectively, while those of the analogous complexes of protonated 6^A-amino-6^A-deoxy-βCD and 3^A-amino-3^A-deoxy-(2^{A}S,3^{A}S)-βCD are 36 and 13, and 51 and 32 dm^3 mol^{-1} [46,47]. It appears that unfavourable interactions between substituents of the modified CDs and the racemic guests destabilise the complexes and lead to greater enantioselectivity in these cases. Another example of chiral discrimination between amino acids by a simply modified CD is provided by 6^A-O-(ethoxyhydroxyphosporyl)-βCD complexes where K_{11} (dm^3 mol^{-1}) = 130 and 220, 1480 and 410, 1400 and 740, 1700 and 1400, and 510 and 330, respectively, when the guests are the (*R*)- and (*S*)-enantiomeric pairs of alanine, serine, valine, leucine and cysteine [48]. The difference in thermodynamic stability between such diastereomeric pairs of complexes gives little indication of the structural changes that may be involved. However, the differing NMR chemical shifts shown by enantiomers complexed by modified CDs do provide insight into complex structure and the origins of the chiral discrimination [49,50]. For example, NMR studies and complementary molecular modelling calculations of the enantioselective complexation of (*R*)- and (*S*)-atenolol with heptakis(2,3,6-tri-*O*-phenylcarbamate)-βCD show that the aromatic moiety of (*S*)-atenolol is complexed in the CD annulus, with the chiral centre outside the annulus (**3.36**), while the opposite is the case with the (*R*)-atenolol (**3.37**) [50]. Molecular dynamics studies show that for permethyl-βCD, the most often used CD chiral stationary phase in gas chromatography, the preferred guest complexation site is in the interior of the annulus and that the guest is quite mobile within the annulus [51,52].

3.36

3.37

The influence of CD and guest charge on chiral discrimination is illustrated by the 1:1 complexes formed between protonated 6^A-amino-6^A-deoxy-βCD (**3.38**) and heptakis(6-amino-6-deoxy)-βCD (**3.39**) and deprotonated *N*-acetylated tryptophan (**3.40**), phenylalanine (**3.41**), leucine (**3.42**) and valine (**3.43**) as shown in Table 3.4 [53,54]. The pK_as of **3.38** and **3.39** are 8.49 [47] and 6.9-8.5 [55],

Table 3.4. Stability Constants for Protonated Amino-βCD Complexes of Deprotonated Amino Acid Derivatives.[a]

Amino Acid Derivative	CD **3.38** K_{11}/dm^3 mol^{-1}	CD **3.39** K_{11}/dm^3 mol^{-1}
3.40	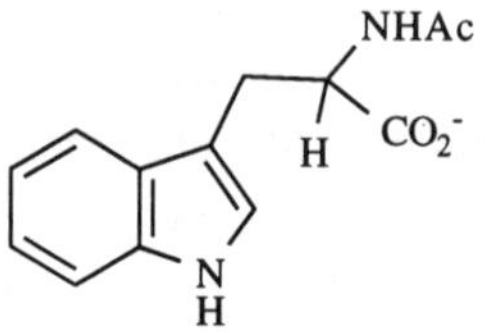 99 (*R*) 64 (*S*)	2310 (*R*) 1420 (*S*)
3.41	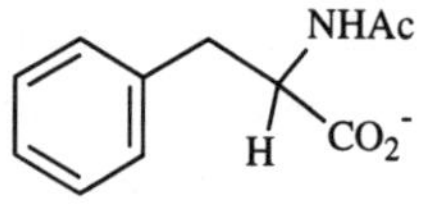 67 (*R*) 55 (*S*)	2180 (*R*) 2000 (*S*)
3.42	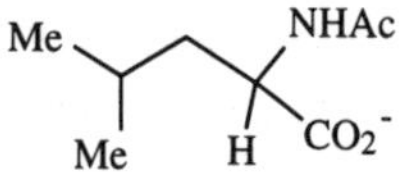 58 (*R*) 60 (*S*)	2480 (*R*) 2380 (*S*)
3.43	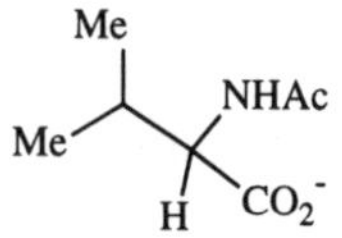	2090 (*R*) 1310 (*S*)

[a]Determined by ^{1}H NMR in D_2O at pD = 6.0 and 298.2 K.

respectively, and at the pD = 6.0 of the study, the charge of **3.39** is much greater than the unipositive charge of **3.38**. The enantioselectivity in favour of the (*R*)-enantiomers shown by the complexes is modest, but is greater than that shown by the analogous αCD and βCD complexes. It is clear that the greater electrostatic interactions between **3.39** and **3.40-3.42** greatly increase K_{11} over those characterising the complexes of **3.38** and **3.40-3.42** where the electrostatic interactions are weaker. This electrostatic contribution to complex stability also appears to be important in the complexation of nucleotides by **3.39** [56]. Molecular modelling shows the mutual electrostatic repulsion between the -NH_3^+ groups of

3.39 to force them apart, and attraction between them and the negatively charged amino acid function of **3.42** to occur in the complex. The modelled complex structure is in agreement with that deduced from ^{1}H NMR studies. Weaker enantioselectivity, also in favour of the (*R*)-enantiomers, occurs in the complexing of cationic protonated α-amino acid methyl esters by anionic deprotonated heptakis[(6-thioglycolic acid)-6-deoxy]-βCD [54].

Returning now to the discussion of intramolecular complexes of modified CDs, in addition to substituent chirality, factors such as the protonation state of substituents and temperature also affect the extent of intramolecular complexation. They can also be exploited to enhance molecular recognition through controlling intermolecular complexation. As already mentioned, the CD **3.27** complexes adamantan-1-ol only at lower temperatures, when formation of the intramolecular complex is disfavoured [34], and the CDs **3.28** and **3.29** only form intermolecular complexes at low pH, when the dansyl substituents are protonated and therefore readily displaced from the CD annuli [30]. Fluorescein-modified CDs provide other examples of the effect of substituent protonation on molecular recognition [57,58], as illustrated by the behaviour of the βCD **3.44**. In aqueous solution, the stability constants of its complexes with adamantan-1-ol are 670 dm^3 mol^{-1} at pH 9.30, 2800 dm^3 mol^{-1} at pH 4.00, and 18000 dm^3 mol^{-1} at pH 1.20, under which conditions the substituent is anionic **3.44a**, neutral **3.44b** and cationic **3.44c**, respectively. Under the same conditions, the stability constants of the complexes of adamantan-1-ol with βCD are 6900, 5900, and 5100 dm^3 mol^{-1}, respectively.

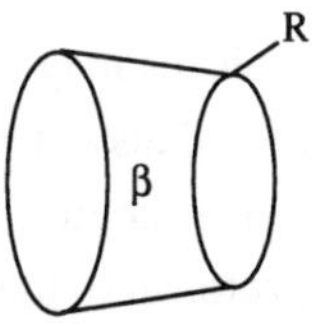

3.44a R = —OCO

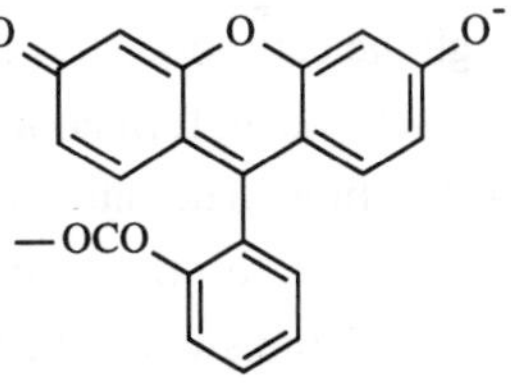

3.44b R = —OCO

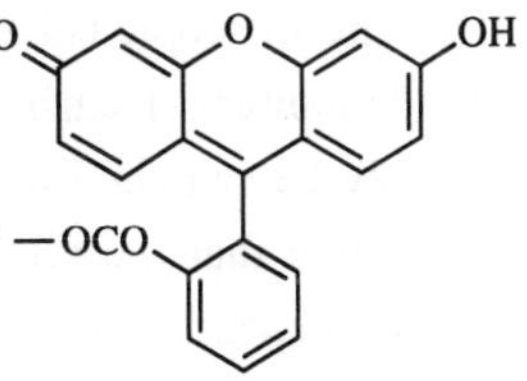

3.44c R = —OCO

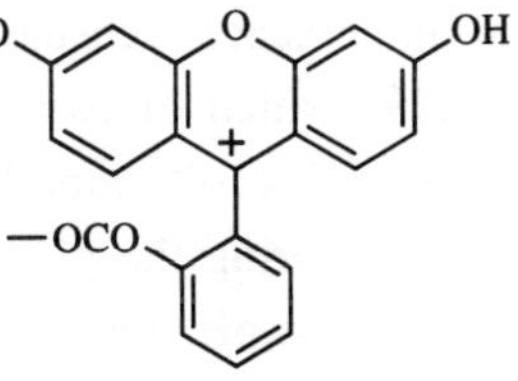

3.3. Molecular Sensors Based on Guest-Induced Dissociation of Intramolecular Complexes

Guest complexation according to each of the modes illustrated in Scheme 3.1 involves a change in the environment of the CD substituent or substituents. This normally results in alterations to the physical and spectroscopic properties of the substituted CDs, such that the complexation is readily monitored. The modified CDs then behave as molecular sensors of guest complexation. The complexation modes most often exploited in the development of molecular sensors involve guest-induced substituent displacement (Schemes 3.1b and 3.1e), in part because systems of this type are relatively straightforward to construct, and in part because the changes in environment and physical and spectroscopic properties of the substituents on dissociation of the intramolecular complexes are relatively large. A number of the substituted CDs described above have been developed as molecular sensors. For example [27], in aqueous solution the intramolecular complex of the C(2) substituted CD **3.13** dissociates through competitive binding of dodecyltrimethylammonium bromide. Since this displacement results in quenching of the phosphorescence of the bromonaphthyl group, the CD **3.13** may be considered to be a sensor of guest complexation. The CD **3.14** is unsuitable for use as a sensor in the manner described for the regioisomer **3.13** because it does not form an intramolecular complex. With the dimethylaminobenzamido-substituted CDs **3.1-3.3**, the fluorescence of the substituent decreases markedly on displacement from within a CD annulus outside to an aqueous environment, and thereby indicates complexation of a guest through the equilibria shown in Scheme 3.4 [1-3].

As discussed above, formation of intramolecular complexes may be influenced by pH and temperature. The intramolecular complexes of the dansylated CDs **3.28** and **3.29** are insensitive to other organic molecules in water at neutral pH. However, under acidic conditions when the dansyl substituent is protonated, the intramolecular complexes dissociate in response to guest binding, which results in quenching of the substituent fluorescence (Scheme 3.5) [30]. This spectroscopic behaviour of the CDs **3.28** and **3.29**, which allows detection of adamantane-1-carboxylic acid at a concentration of 5×10^{-7} mol dm^{-3} at pH 1, may be switched off or on again by increasing or lowering the pH of the solution, such that the CDs **3.28** and **3.29** may be regarded as pH sensors. The temperature-dependent behaviour of the CD

3.26 (Scheme 3.2) may be monitored using fluorescence spectroscopy [35,36]. Displacement of the substituent dye from the annulus of the CD **3.27** at lower temperatures (Scheme 3.3) changes its extent of protonation and spectral characteristics, such that its colour changes from yellow to red at pH 1.6 [34]. On the basis of these temperature-induced spectral changes, the CDs **3.26** and **3.27** may be regarded as thermosensors.

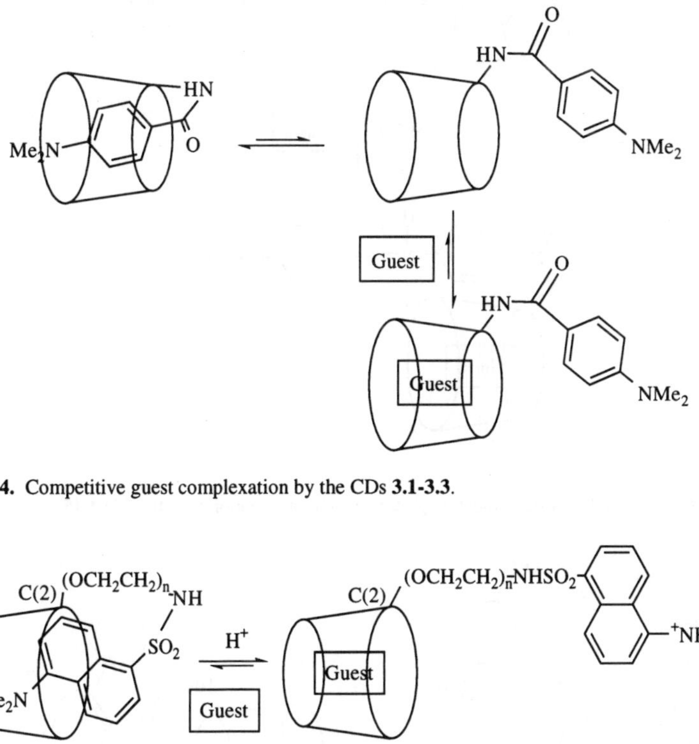

Scheme 3.4. Competitive guest complexation by the CDs **3.1-3.3**.

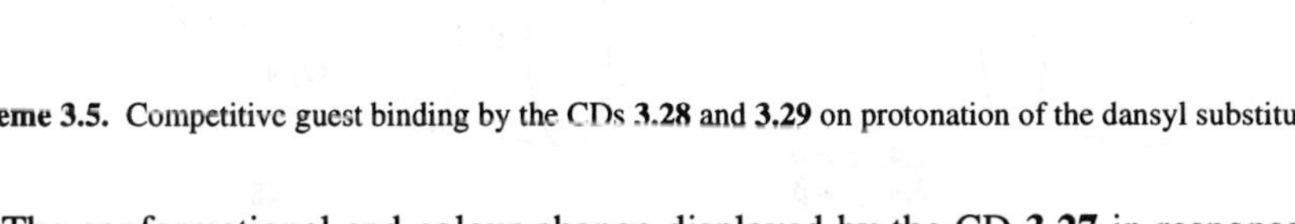

Scheme 3.5. Competitive guest binding by the CDs **3.28** and **3.29** on protonation of the dansyl substituent.

The conformational and colour change displayed by the CD **3.27** in response to the variation in temperature also occurs simply as a result of competitive guest complexation [59,60]. In a closely related system, at pH 2.5, guest-induced dissociation of the intramolecular complex of the CD **3.45** results in protonation of the substituent and a change of its colour from orange to red (Scheme 3.6) [60]. The ionisation of the substituents of the modified CDs **3.27** and **3.45**, when they are displaced from the CD annuli to an aqueous environment, reflects the greater ease of hydration of the dissociated forms and the preferred complexation of neutral species by CDs.

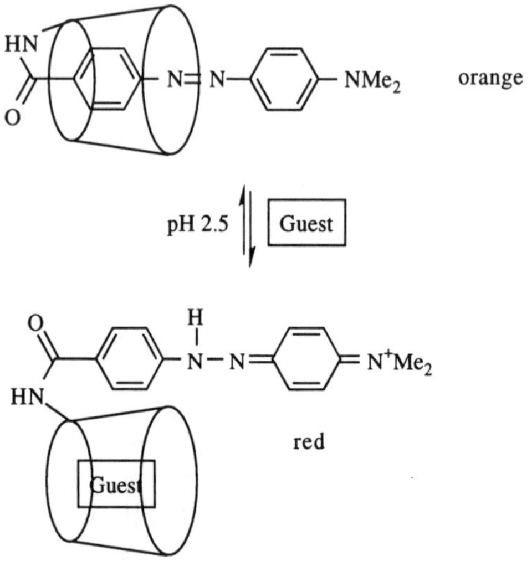

Scheme 3.6. Guest-induced substituent displacement and protonation with the CD **3.45**.

Similar behaviour has been observed with a number of other common pH indicators that have been attached covalently to CDs, such that their dissociation from intramolecular complexes on binding of other guests is readily observed [61-64].

Accordingly, the CDs **3.46-3.48** change from yellow to colourless, at pH 5.1, 5.1 and 6.5, respectively [61,64], while a phenolphthalein-modified βCD has been shown to change from colourless to pink at pH 9.7 [62].

3.4. Towards Photochemical Frequency Switches and Related Molecular Devices

The majority of the modified CDs discussed so far in this chapter possess aryl substituents which absorb photochemical irradiation. Consequently they undergo photochemically-induced reactions, where the outcome is often determined by geometrical constraints imposed by the CD annulus. An early example [65,66] of this involved irradiation of each of the $6^A,6^B$-, $6^A,6^C$-, $6^A,6^D$-, and $6^A,6^E$-bis(anthracene-9-carbonyl)-γCDs, which resulted in reversible dimerisation of the anthracene substituents, due to their proximity when attached to, and possibly complexed within, the CD annuli. More recently [2+2]-photocycloaddition reactions of a polycinnamoylCD have been used to trap *N*-methylpyrrolidin-2-one in the CD cavity [67]. The cyclobutane moieties introduced through this process are acid-labile, so it is possible to release the guest through reaction with methanolic hydrogen chloride.

Table 3.5. Products Resulting from the Irradiation at 312 nm of (*E*)-4,4'-Bis(dimethylaminomethyl)-stilbene Dihydrochloride in the Presence of CDs in Water.

CD	time/h			Product %		
		(*E*)-stilbene	(*Z*)-stilbene	*cis*-cyclobutane	*trans*-cyclobutane	phenanthrene
none	24	10	62	7	2	19
αCD	24	20	60	0	0	20
βCD	24	16	83	0	0	1
γCD	72[a]	0	0	79	19	2

[a]For shorter irradiation times (*Z*)-stilbene is found among the reaction products.

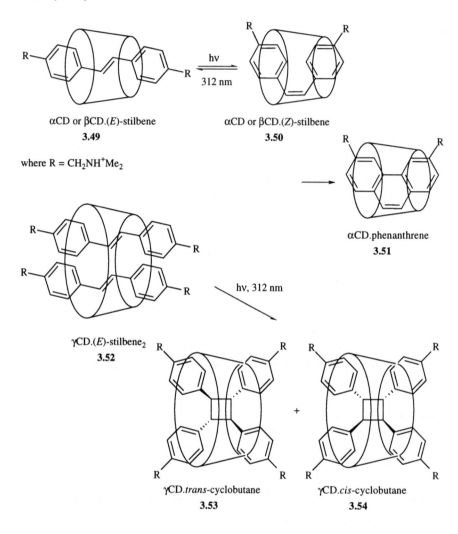

Scheme 3.7. Effect of CDs on photochemical reactions of a stilbene.

Even the natural CDs affect the course of photochemical reactions of complexed guests, by constraining reactant geometry [68-70]. In a recent example, the effect of αCD, βCD and γCD on the photochemistry of (*E*)-4,4'-bis(dimethylammonio-methyl)stilbene was examined, and found to be substantially dependent on the size of the CD annulus, as seen from Table 3.5 (Scheme 3.7) [71]. This effect may be understood when the complexes **3.49** and **3.52** are considered. The αCD and βCD 1:1 complexes of the stilbene, **3.49**, hinder the formation of the cyclobutanes while the complexation of two stilbenes in **3.52** greatly increases their formation. The βCD complex, **3.49**, participates in a photochemically driven equilibrium with **3.50** (βCD), but the analogous effect does not occur with the αCD **3.49** complex which does, however, allow the oxidation of the (Z)-stilbene to produce the phenanthrene (**3.51**). The effect of βCD is greater in that it increases the yield of the (Z)-stilbene and hinders the production of the phenanthrene.

Table 3.6. Parameters for the Formation of CD Complexes in Aqueous Solution.[a]

Complex	K_{11}	ΔH^{o}	$T\Delta S^{o}$
	dm^3 mol^{-1}	kJ mol^{-1}	kJ mol^{-1}
3.49 (αCD)	1.52×10^3	-27	-8.8
3.50 (αCD)	3.6×10^2	-7.9	+6.7
3.49 (βCD)	7.05×10^2	-11	+5.0
3.50 (βCD)	5.4×10^2	-41	-25
3.52 (γCD)	3.85×10^2	-4.6	+10
3.52 (γCD)	2.73×10^3 (K_{12})	-30	-10
3.53 (γCD)	1.8×10^4	-30	-5.4
3.54 (γCD)	5.2×10^2	+5.4	+38

[a]In ammonium acetate buffer at pH 5.7 and 298.2 K

The origin of the product distribution becomes more clear from the parameters for complex formation (Table 3.6). For both αCD and βCD, **3.49** is more stable than **3.50**, but the stability of **3.50** is greater for βCD than for αCD. The fit of the stilbenes into αCD is closer than is the case for βCD. Evidently these differences control the outcome of the photochemistry. The greater stability of ternary complex

3.52 than that of its binary precursor is consistent with γCD engendering production of the cyclobutanes through the reaction of an excited state stilbene with its complexed neighbour in **3.52**. Furthermore, the much greater stability of **3.53** by comparison with that of **3.54** is consistent with the geometry of the γCD cavity favouring the *trans*-cyclobutane product.

Another example of CD annular size preorganising a reactant to give a particular product occurs with the antiparallel (**3.55a**) and parallel (**3.55b**) forms of 2,2'-dimethyl-3,3'-(perfluorocyclopentene-1,2-diyl)bis(benzo[b]thiophenesulfonate [72] (Scheme 3.8). In aqueous solution, the ratio of **3.55a**, which photocyclises to **3.56** on irradiation at 313 nm, to non-photoreactive **3.55b** is 64:36 as shown by ^{1}H NMR. This ratio is unchanged in the presence of αCD, probably because the αCD annulus is too small to form complexes, while both βCD and γCD shift the equilibrium toward **3.55a** evidently because of a better fit of **3.55a** to their annuli than **3.55b**. As a consequence the cyclisation quantum yield for a combined **3.55a** and **3.55b** concentration of 5.0×10^{-5} mol dm^{-3} in the absence of a CD is 0.38, and in the presence of 8.0×10^{-3} mol dm^{-3} αCD, βCD and γCD is 0.37, 0.58 and 0.53, respectively.

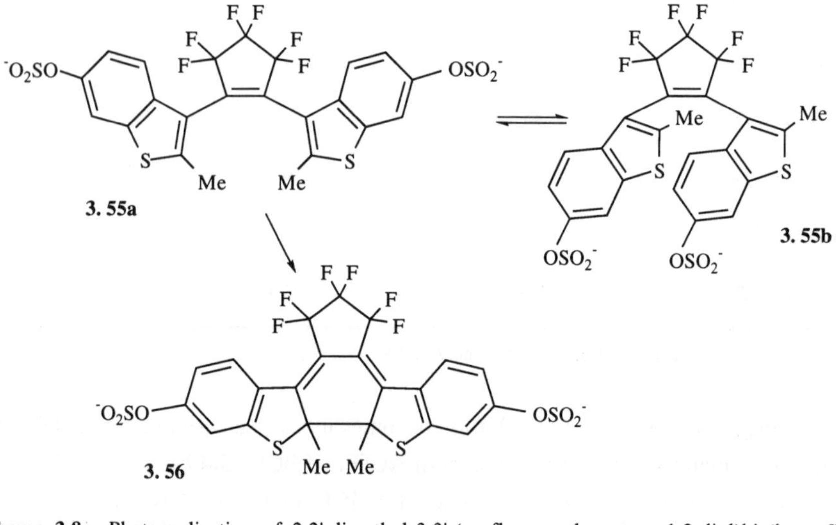

Scheme 3.8. Photocyclisation of 2,2'-dimethyl-3,3'-(perfluorocyclopentene-1,2-diyl)bis(benzo[b]-thiophenesulfonate.

Other effects of irradiation on CDs and their complexes involve photochemically-induced changes to CD substituents which affect their intramolecular complexation and, as a result, the extent of competitive intermolecular guest binding by the modified CDs [73-75]. This phenomenon is clearly demonstrated with azobenzene-modified CDs such as **3.57**, where exposure to ultraviolet and visible light results in *trans-cis* and *cis-trans* isomerisation, respectively [74] (Scheme 3.9). The *trans*-azobenzene **3.57a** forms an association dimer which limits competitive guest complexation, whereas the annulus of the *cis*-azobenzene **3.57b** is unencumbered. Thus, both dimer formation and guest complexation may be regulated by light in an on-off fashion. This selective complexation by CDs of the *trans*-isomer of azobenzenes forms the basis of a device for the electrochemical transduction of optical signals, which is discussed in Chapter 8.

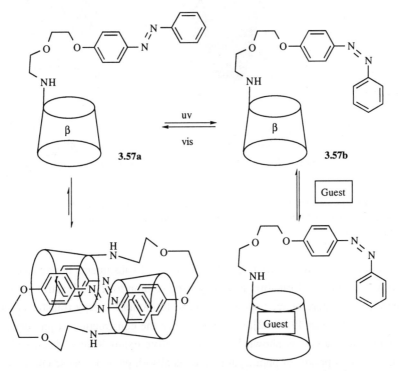

Scheme 3.9. Photochemical response of the CD **3.57** and its effect on guest complexation.

CD substituents have also been used as photosenzitizers for the production of triplet oxygen, to facilitate the oxidation of included guests [76,77]. Triplet oxygen has a half-life of approximately 2 μsec and travels less than 50 Å in aqueous media before deactivation. Therefore it is difficult to maintain a high concentration in free solution. However, when triplet oxygen is generated using, as a photosensitizer, rose bengal or one of its derivatives covalently attached to a CD, its effective concentration in the vicinity of a guest complexed in the annulus is increased and oxidation of such species is greatly enhanced.

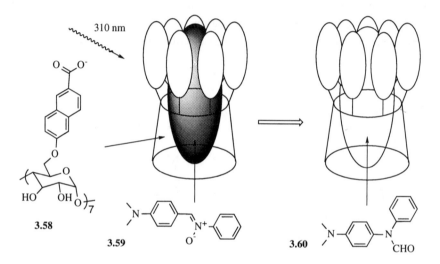

Fig. 3.1. Energy transfer from the antennae of 6^{A-G}-heptanaphthoate-βCD (**3.58**) to α-(4-dimethylaminophenyl)-*N*-phenylnitrone (**3.59**) which photolyses to *N*-(4-dimethylaminophenyl)-formanilide (**3.60**).

As described above, the change in fluorescence of substituents bound to CDs and of guests on complex formation is widely used to monitor complexation processes. Observation of fluorescence may also be used to monitor energy transfer processes within CD complexes [9,78-82], and exploited in the construction of photochemical frequency switches as is shown by the innovative example now discussed. The antennae chromophores of photosynthetic units absorb photons whose energy is then transferred to other components in the photosynthetic process. This process has been modelled using 6^{A-G}-heptanaphthoate-βCD (**3.58**) where naphthoate antennae are attached to βCD (Fig. 3.1) as one component of an energy transfer system, and which is an example of a range of similarly modified βCDs [83-85].

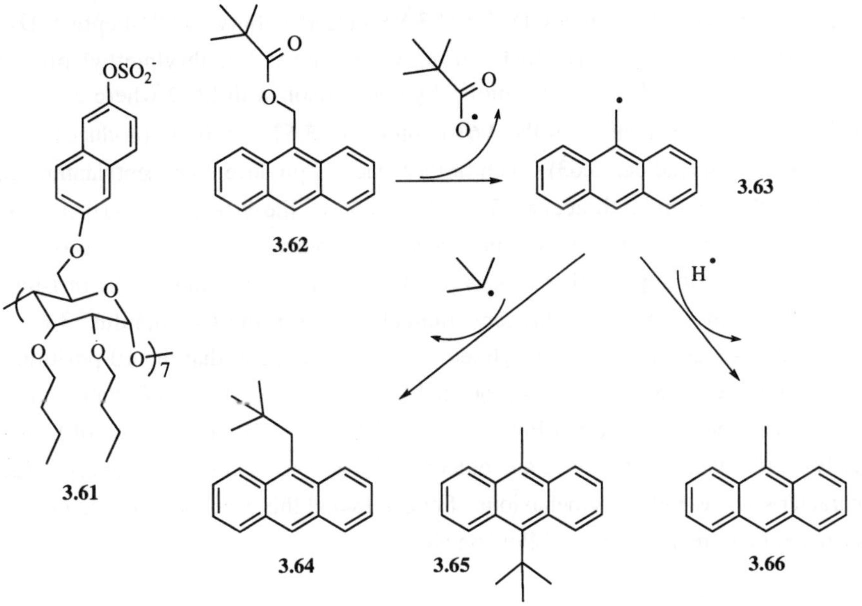

Scheme 3.10. Photochemically-induced reaction of the pivalate **3.62** in the annulus of the antenna CD **3.61**.

When **3.58** complexes the dye 4-(dicyanomethylene)-2-methyl-6-(4-(bis(hydroxyethyl)aminostyryl-4*H*-pyran and the complex is irradiated at 300 nm, the **3.58** emission band ($\lambda_{max} = 355$) overlaps the absorption band of the dye which in turn fluoresces in the range 550-750 nm. This energy transfer from **3.58** to the dye has an efficiency close to unity. The high K_{11} of 1.2 x 10^5 dm^3 mol^{-1} for the complex of **3.58** and the dye is attributed to the increased hydrophobicity of **3.58** over βCD arising from the naphthalene rings of the attached chromophores. This principle of energy transfer has been applied in the photoisomerisation of α-(4-dimethylaminophenyl)-*N*-phenylnitrone (**3.59**) to *N*-(4-dimethylaminophenyl)-formanilide (**3.60**) as shown in Fig. 3.1 [86]. Compared with this photoisomerisation in the absence of **3.58**, a 6-fold acceleration of its rate occurs when it proceeds through the **3.58** complex. This is attributed to the efficient transfer of energy gathered by the heptanaphthoate antennae of **3.58** to the complexed **3.59** so that **3.58** acts as a photochemical microreactor.

A closely related antenna CD, $2^{A-G},3^{A-G}$-tetradecabutyl-6^{A-G}-hepta(6-*O*-2-naphthalenesulfonate)-βCD (**3.61**) strongly complexes 9-anthrylmethyl pivalate (**3.62**) with $K_{11} = 7.8 \times 10^4$ dm^3 mol^{-1} by comparison with βCD where $K_{11} = 80$ dm^3 mol^{-1} [87]. Irradiation of the **3.61** complex of **3.62** at 300 nm produces the 9-methylanthracene radical (**3.63**) which further reacts to produce 9-neopentylanthracene (**3.64**) and 9-methylanthracene (**3.66**) but none of the 9-*tert*-butyl-10-methyl-anthracene (**3.65**), which is also produced by photolysis in the absence of **3.61** (Scheme 3.10). It appears that **3.64** is produced by recombination of the *tert*-butyl radical with the 9-methylanthracene radical (**3.63**) in the CD annulus, **3.64** by hydrogen abstraction by the anthrylmethyl radical **3.63**, and that the 10-position of 9-anthrylmethyl pivalate (**3.62**) is protected from reaction within the CD annulus.

The reactions illustrated in Fig. 3.1 and Scheme 3.10 involve the use of CDs as catalysts, to affect the efficiency and outcome of reactions of complexed guests. Such characteristics resemble the behaviour of enzymes and this aspect of CD chemistry is discussed in more detail in the following chapter.

3.5. References

1. K. Hamasaki, A. Ueno and F. Toda, *J. Chem. Soc., Chem. Commun.* (1993) 331.
2. K. Hamasaki, H. Ikeda, A. Nakamura, A. Ueno, F. Toda, I. Suzuki and T. Osa, *J. Am. Chem. Soc.* **115** (1993) 5035.
3. K. Hamasaki, A. Ueno, F. Toda, I. Suzuki and T. Osa, *Bull. Chem. Soc. Jpn.* **67** (1994) 516.
4. K. A. Connors, *Chem. Rev.* **97** (1997) 1325.
5. A. Ueno, S. Minato, I. Suzuki, M. Fukushima, M. Ohkubo, T. Osa, F. Hamada and K. Murai, *Chem. Lett.* (1988) 605.
6. F. Hamada, Y. Kondo, R. Ito, I. Suzuki, T. Osa and A. Ueno, *J. Inclusion Phenom. Mol. Recogn. Chem.* **15** (1993) 273.
7. A. Ueno, Y. Tomita and T. Osa, *J. Chem. Soc., Chem. Commun.* (1983) 976.

8. A. Ueno, Y. Tomita and T. Osa, *Chem. Lett.* (1983) 1635.
9. A. Ueno, F. Moriwaki, Y. Tomita and T. Osa, *Chem. Lett.* (1985) 493.
10. F. Hamada, Y. Kondo, K. Ishikawa, H. Ito, I. Suzuki, T. Osa and A. Ueno, *J. Inclusion Phenom. Mol. Recogn. Chem.* **17** (1994) 267.
11. F. Hamada, K. Ishikawa, R. Ito, H. Shibuya, S. Hamai, I. Suzuki, T. Osa and A. Ueno, *J. Inclusion Phenom. Mol. Recogn. Chem.* **20** (1995) 43.
12. A. Ueno, F. Moriwaki, T. Osa, F. Hamada and K. Murai, *J. Am. Chem. Soc.* **110** (1988) 4323.
13. A. Ueno, F. Moriwaki, A. Azuma and T. Osa, *Carbohydr. Res.* **192** (1989) 173.
14. A. Ueno, F. Moriwaki, T. Osa, F. Hamada and K. Murai, *Tetrahedron Lett.* **26** (1985) 3339.
15. A. Ueno, F. Moriwaki, T. Osa, F. Hamada and K. Murai, *Bull. Chem. Soc. Jpn.* **59** (1986) 465.
16. F. Hamada, K. Murai, A. Ueno, I. Suzuki and T. Osa, *Bull. Chem. Soc. Jpn.* **61** (1988) 3758.
17. S. Minato, T. Osa and A. Ueno, *J. Chem. Soc., Chem. Commun.* (1991) 107.
18. A. Ueno, S. Minato and T. Osa, *Anal. Chem.* **64** (1992) 1154.
19. F. Hamada, S. Minato, T. Osa and A. Ueno, *Bull. Chem. Soc. Jpn.* **70** (1997) 1339.
20. A. Ueno, I. Suzuki and T. Osa, *J. Chem. Soc., Chem. Commun.* (1988) 1373.
21. A. Ueno, I. Suzuki and T. Osa, *Chem. Lett.* (1989) 1059.
22. I. Suzuki, A. Ueno and T. Osa, *Chem. Lett.* (1989) 2013.
23. A. Ueno, I. Suzuki and T. Osa, *J. Am. Chem. Soc.* **111** (1989) 6391.
24. A. Ueno, I. Suzuki and T. Osa, *Anal. Chem.* **62** (1990) 2461.
25. Y. Wang, T. Ikeda, A. Ueno and F. Toda, *Chem. Lett.* (1992) 863.
26. Y. Wang, T. Ikeda, H. Ikeda, A. Ueno and F. Toda, *Bull. Chem. Soc. Jpn.* **67** (1994) 1598.
27. M. A. Mortellaro, W. K. Hartmann and D. G. Nocera, *Angew. Chem., Int. Ed. Engl.* **35** (1996) 1945.
28. K. Hamasaki, S. Usui, H. Ikeda, T. Ikeda and A. Ueno, *Supramolecular Chem.* **8** (1997) 125.
29. S. Hanessian, A. Benalil and M. T. P. Viet, *Tetrahedron* **51** (1995) 10131.

30. H. F. M. Nelissen, F. Venema, R. M. Uittenbogaard, M. C. Feiters and R. J. M. Nolte, *J. Chem. Soc., Perkin Trans. 2* (1997) 2045.

31. H. Parrot-Lopez, H. Galons, A. W. Coleman, F. Djedaïni, N. Keller and B. Perly, *Tetrahedron: Asymm.* **1** (1990) 367.

32. A. V. Eliseev, G. A. Iacobucci, N. A. Khanjin and F. M. Menger, *J. Chem. Soc., Chem. Commun.* (1994) 2051.

33. H. Ikeda, M. Nakamura, N. Ise, F. Toda and A. Ueno, *J. Org. Chem.* **62** (1997) 1411.

34. T. Kuwabara, A. Nakamura, A. Ueno and F. Toda, *J. Chem. Soc., Chem. Commun.* (1994) 689.

35. S. R. McAlpine and M. A. Garcia-Garibay, *J. Am. Chem. Soc.* **118** (1996) 2750.

36. S. R. McAlpine and M. A. Garcia-Garibay, *J. Org. Chem.* **61** (1996) 8307.

37. S. R. McAlpine and M. A. Garcia-Garibay, *J. Am. Chem. Soc.* **120** (1998) 4269.

38. W. Saka, Y. Yamamoto, Y. Inoue, R. Chûjô, K. Takahashi and K. Hattori, *Bull. Chem. Soc. Jpn.* **63** (1990) 3175.

39. K. Takahashi, Y. Ohtsuka and K. Hattori, *Chem. Lett.* (1990) 2227.

40. K. Takahashi, Y. Ohtsuka, S. Nakada and K. Hattori, *J. Inclusion Phenom. Mol. Recogn. Chem.* **10** (1991) 63.

41. K. Takahashi, *Bull. Chem. Soc. Jpn.* **66** (1993) 550.

42. K. Takahashi and K. Hattori, *Supramolecular Chem.* **2** (1993) 305.

43. H. Ikeda, M. Nakamura, N. Ise, N. Oguma, A. Nakamura, T. Ikeda, F. Toda and A. Ueno, *J. Am. Chem. Soc.* **118** (1996) 10980.

44. C. J. Easton and S. F. Lincoln, *Chem. Soc. Rev.* (1996) 163.

45. Y. Liu, Y.-M. Zhang, A.-D. Qi, R.-D. Chen, K. Yamamoto, T. Wada and Y. Inoue, *J. Org. Chem.* **62** (1997) 1826.

46. R. Dhillon, C. J. Easton, S. F. Lincoln and J. Papageorgiou, *Aust. J. Chem.* **48** (1995) 1117.

47. S. E. Brown, J. H. Coates, P. A. Duckworth, S. F. Lincoln, C. J. Easton and B. L. May, *J. Chem. Soc., Faraday Trans.* **89** (1993) 1035.

48. Y. Liu, B. Li, B.-H. Han, Y.-M. Li and R.-T. Chen, *J. Chem. Soc., Perkin Trans. 2* (1997) 1275.

49. T. Murakami, K. Harata and S. Morimoto, *Chem. Lett.* (1988) 553.

50. Y. Kuroda, Y. Suzuki, J. He, T. Kawabata, A. Shibukawa, H. Wada, H. Fujima, Y. Go-oh, E. Imai and T. Nakagawa, *J. Chem. Soc., Perkin Trans. 2* (1995) 1749.

51. K. B. Lipkowitz, G. Pearl, R. Coner and M. A. Peterson, *J. Am. Chem. Soc.* **119** (1997) 600.

52. K. B. Lipkowitz, R. Coner, M. A. Peterson, A. Morreale and J. Shackelford, *J. Org. Chem.* **63** (1998) 732.

53. K. Kano, T. Kitae, H. Takashima and Y. Shimofuri, *Chem. Lett.* (1997) 899.

54. T. Kitae, T. Nakayama and K. Kano, *J. Chem. Soc., Perkin Trans. 2* (1998) 207.

55. B. Hamelin, L. Jullien, F. Guillo, J.-M. Lehn, A. Jardy, L. De Robertis and H. Driguez, *J. Phys. Chem.* **99** (1995) 17877.

56. A. V. Eliseev and H.-J. Schneider, *J. Am. Chem. Soc.* **116** (1994) 6081.

57. Y. Wang, T. Ikeda, A. Ueno and F. Toda, *Tetrahedron Lett.* **34** (1993) 4971.

58. F. Hamada, K. Ishikawa, Y. Higuchi, Y. Akagami and A. Ueno, *J. Inclusion Phenom. Mol. Recogn. Chem.* **25** (1996) 283.

59. A. Ueno, T. Kuwabara, A. Nakamura and F. Toda, *Nature* **356** (1992) 136.

60. T. Kuwabara, A. Nakamura, A. Ueno and F. Toda, *J. Phys. Chem.* **98** (1994) 6297.

61. T. Kuwabara, A. Matsushita, A. Nakamura, A. Ueno and F. Toda, *Chem. Lett.* (1993) 2081.

62. T. Kuwabara, M. Takamura, A. Matsushita, A. Ueno and F. Toda, *Supramolecular Chem.* **8** (1996) 13.

63. T. Aoyagi, A. Nakamura, H. Ikeda, T. Ikeda, H. Mihara and A. Ueno, *Anal. Chem.* **69** (1997) 659.

64. A. Matsushita, T. Kuwabara, A. Nakamura, H. Ikeda and A. Ueno, *J. Chem. Soc., Perkin Trans. 2* (1997) 1705.

65. F. Moriwaki, A. Ueno, T. Osa, F. Hamada and K. Murai, *Chem. Lett.* (1986) 1865.

66. A. Ueno, F. Moriwaka, A. Azuma and T. Osa, *J. Org. Chem.* **54** (1989) 295.

67. R. Arad-Yellin and B. S. Green, *Nature* **371** (1994) 320.

68. K. Takahashi and K. Hattori, *J. Inclusion Phenom. Mol. Recogn. Chem.* **17** (1994) 1.

69. T. Osa and I. Suzuki, *Reactivity of Included Guests*, in *Comprehensive Supramolecular Chemistry*; eds. J. L. Atwood, J. E. D. Davies, D. D. MacNicol, F. Vögtle and J.-M. Lehn, Vol. 3, eds. J. Szejtli and T. Osa (Pergamon, Oxford, 1996) 367.

70. K. Takahashi, *Chem. Rev.* **98** (1998) 2013.

71. W. Herrmann, S. Wehrle and G. Wenz, *J. Chem. Soc., Chem. Commun.* (1997) 1709.

72. M. Takeshita, C. Nam Choi and M. Irie, *J. Chem. Soc., Chem. Commun.* (1997) 2265.

73. A. Ueno, Y. Tomita and T. Osa, *Tetrahedron Lett.* **24** (1983) 5245.

74. M. Fukushima, T. Osa and A. Ueno, *J. Chem. Soc., Chem. Commun.* (1991) 15.

75. M. Fukushima, T. Osa and A. Ueno, *Chem. Lett.* (1991) 709.

76. D. C. Neckers and J. Paczkowski, *Tetrahedron* **42** (1986) 4671.

77. D. C. Neckers and J. Paczkowski, *J. Am. Chem. Soc.* **108** (1986) 291.

78. I. Tabushi, K. Fujita and L. C. Yuan, *Tetrahedron Lett.* (1977) 2503.

79. D. M. Gravett and J. E. Guillet, *J. Am. Chem. Soc.* **115** (1993) 5970.

80. M. N. Berberan-Santos, J. Pouget, B. Valeur, J. Canceill, L. Jullien and J.-M. Lehn, *J. Phys. Chem.* **97** (1993) 11376.

81. B. K. Hubbard, L. A. Beilstein, C. E. Heath and C. A. Abelt, *J. Chem. Soc., Perkin Trans. 2* (1996) 1005.

82. M. N. Berberan-Santos, J. Canceill, E. Gratton, L. Jullien, J.-M. Lehn, P. So, J. Sutin and B. Valeur, *J. Phys. Chem.* **100** (1996) 15.

83. M. N. Barberon-Santos, J. Canceill, J.-C. Brochon, L. Jullien, J.-M. Lehn, J. Pouget, P. Tauc and B. Valeur, *J. Am. Chem. Soc.* **114** (1992) 6427.

84. L. Jullien, J. Canceill, B. Valeur, E. Bardez and J.-M. Lehn, *Angew Chem., Int. Ed. Engl.* **33** (1994) 2438.

85. M. Nowakowska, N. Loukine, D. M. Gravett, N. A. D. Burke and J. E. Guillet, *J. Am. Chem. Soc.* **119** (1997) 4364.

86. L. Jullien, J. Canceill, B. Valeur, E. Bardez, J.-P. Lefèvre, J.-M. Lehn, V. Marchi-Artzner and R. Panzu, *J. Am. Chem. Soc.* **118** (1996) 5432.

87. P. F. Wang, L. Jullien, B. Valeur, J.-S. Filhol, J. Canceill and J.-M. Lehn, *New J. Chem.* **20** (1996) 895.

CHAPTER 4
MODIFIED CYCLODEXTRINS

CYCLODEXTRIN CATALYSTS AND ENZYME MIMICS

4.1. Introduction

Cyclodextrins (CDs) have long attracted attention in catalysis and as enzyme mimics, due to the way in which they act as hosts to complex guest molecules and induce reactions of the complexed species (Scheme 4.1) [1-3]. The reactions exhibit kinetic characteristics, such as saturation, non-productive binding and competitive inhibition, that are typical of enzyme-catalysed processes. In addition, the discrimination displayed by CDs in binding guests and promoting their reactions is analogous to the substrate selectivity displayed by enzymes.

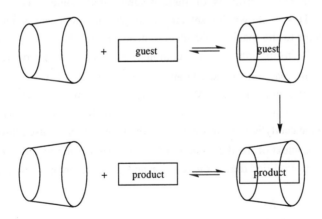

Scheme 4.1. CD-induced reaction of an included guest.

With the natural CDs, hydroxy groups are the only functionality available to promote reactions of included guests. However, the introduction of a diverse range of new functional groups, through modification of the natural CDs, results in catalysts which mimic the entire range of enzyme behaviour. The CD nucleus serves as a scaffold on which functional groups can be assembled. In some cases this has been accomplished with controlled alignment of both the functional groups and the CD annulus, to optimise the geometry for binding and reaction of a particular guest. This is an important factor in the catalytic activity of modified CDs, as it is with enzymes where the geometry at the active site is determined by the three dimensional structure of the protein. It is this catalytic activity of CDs which is the subject of this chapter.

4.2. Cyclodextrin Hydrolases

The natural CDs induce the alkaline hydrolysis of esters and other carboxylic acid derivatives, as illustrated in Scheme 4.2. Therefore they may be classed as CD hydrolases, by analogy with enzymes which catalyse reactions of this type. Bender and co-workers [4,5], in the original papers in this area, explored the CD mediation of hydrolysis of phenyl acetates. They found that in basic aqueous solution αCD and βCD greatly accelerate the rates of hydrolysis of a range of 3-substituted phenyl acetates, while the hydrolysis of their 4-substituted isomers is only slightly accelerated. This is shown for selected examples in Table 4.1 where the listed parameters refer to the mechanistic scheme of Eq. (4.1). Here, K_{11} is the stability constant for the formation of a 1:1 complex, as discussed in Chapter 1, and k_r characterises the reaction of the CD and guest in the complex. This mechanistic scheme resembles the Michaelis-Menten scheme through which enzyme catalysis proceeds, where $1/K_{11}$ is equivalent to the Michaelis constant, K_M, and k_r is equivalent to the catalytic rate constant k_{cat}. The latter terms are used throughout this chapter in the context of the discussion of CDs as enzyme mimics, although some of the CDs become modified during the reaction and are therefore not true catalysts.

$$CD + G \; \underset{\text{fast}}{\overset{K_{11}}{\rightleftharpoons}} \; CD{\cdot}G \; \overset{k_r}{\longrightarrow} \; CD + \text{products} \qquad (4.1)$$

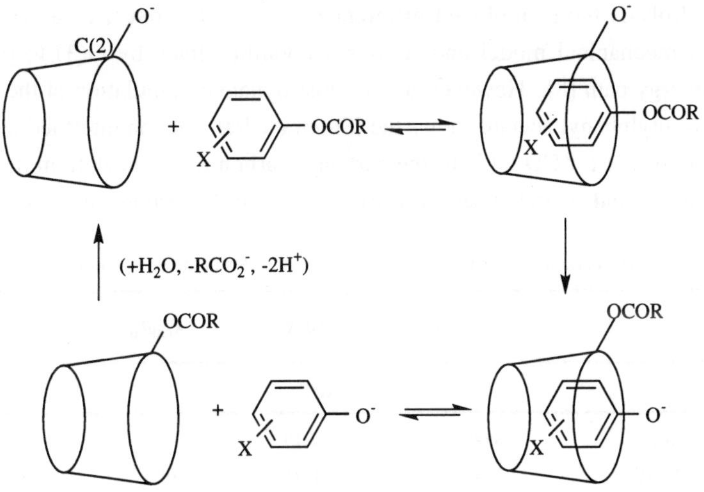

Scheme 4.2. Alkaline hydrolysis of phenyl esters by CDs.

The substrate selectivity of the mediation of the hydrolysis of selected phenyl acetates by αCD and βCD is seen in Table 4.1. In the original publications [4,5], this selectivity was attributed to a correlation with the proximity, in the host-guest complex, of the carbonyl of the phenyl acetate to the CD deprotonated secondary hydroxy group which makes the nucleophilic attack leading to the phenoxide and the acylated CD as shown in Scheme 4.2. This interpretation presupposes complexation of the phenyl moiety inside the CD annulus. The observation that the acceleration of phenyl acetate hydrolysis by αCD and βCD is slowed or inhibited in the presence of competing guests supports this interpretation [4,6]. It is notable that the acceleration of hydrolysis is independent of the stability of the complex. The αCD accelerated hydrolysis of the 3-methyl-, 4-methyl-, 3-nitro- and 4-nitro-phenyl acetates is enthalpy controlled, and it is argued that the closer proximity of the deprotonated secondary hydroxy group of the CD to the carbonyl carbon of the 3-substituted phenyl acetates than to that of their 4-substituted analogues in the host-guest complexes leads to smaller $\Delta H^{\ddagger}$ for the hydrolysis of the former [7]. (The effect of γCD on phenyl acetate hydrolysis rate is small as is its substrate selectivity by comparison with those of αCD and βCD. This probably arises from the looser fit of the guests into the γCD annulus.) Recent theoretical studies of βCD acceleration of ester hydrolysis have employed different models. The first uses a semiempirical quantum mechanical model and finds nucleophilic attack by O(3) to provide the lowest energy path [8]. However, a molecular dynamics simulation of the hydrolysis of 3-*tert*-butylphenyl acetate shows the (*S*)-tetrahedral reaction intermediate resulting from bonding of a βCD O(2) to the carbonyl carbon of the acetate moiety to be of lowest energy and characterised by a $\Delta G^{\ddagger}$ close to the experimental value [9].

Table 4.1. Parameters for Hydrolysis of Phenyl Acetates in the Presence of α- and β-CD.[a]

Acetate	$10^4 k_u/s^{-1}$ [b]	$10^2 k_{cat}/s^{-1}$	k_{cat}/k_u	$K_M^{-1}/dm^3\,mol^{-1}$
		αCD		
Phenyl	8.04	2.19	27	45
2-Tolyl	3.84	0.72	19	53
3-Tolyl	6.96	6.58	95	59
4-Tolyl	6.64	0.22	3.3	91
3-*t*-Butylphenyl	4.90	12.9	260	500
4-*t*-Butylphenyl	6.07	0.067	1.1	154
3-Nitrophenyl	46.4	42.5	300	53
4-Nitrophenyl	69.4	2.43	3.4	83
		βCD		
3-Nitrophenyl	46.4	44.4	96	125
4-Nitrophenyl	69.4	6.3	9.1	164

[a]In pH 10.60 carbonate buffer, $I = 0.2$ mol dm^{-3} at 298.2 K in 0.5% acetonitrile-water. [b]In the absence of CD.

The model illustrated in Scheme 4.2 for ester hydrolysis by CDs is likely to be too simplistic, however, to apply in all cases. In contrast to that of the 3-substituted phenyl acetates, the hydrolysis of the 4-substituted isomers is often little reduced by the presence of other CD guests which might be expected to act as inhibitors [6,10,11]. In some cases the rate of hydrolysis modestly increases in the presence of the potential inhibitor (or spectator) consistent with its occupancy of the CD annulus stabilising the orientation of the host-guest complex required for hydrolysis. This has been interpreted to indicate that for the hydrolysis of 4-substituted phenyl acetates, the phenyl group lies outside the CD annulus with the carbonyl carbon adjacent to the deprotonated secondary hydroxy group which makes the nucleophilic attack. Alternatively, a more recent study of acyl transfer reactions of 4-nitrophenyl alkanoates (acetate to decanoate) is consistent with the complexation of the 4-nitrophenyl end of the guest within the annulus by αCD and βCD for the acetate to the hexanoate, but for the guest orientation to reverse from heptanoate to decanoate so than the alkyl group occupies the annulus [12]. It also appears that the entry of a spectator guest into the CD annulus can greatly change the orientation of a reactive guest which also occupies the annulus. Thus, in a study of the βCD mediated hydrolysis of 3-nitrophenyl hexanoate the reaction kinetics are consistent with simple alcohols reversing the orientation of 3-nitrophenyl hexanoate from complexation of its 3-nitrophenyl moiety in the annulus to complexation of the hexyl group [13]. Other indications that the achievement of a reactive guest ester orientation in the CD annulus may sometimes be possible in different ways come from a study of the βCD mediated hydrolysis of phenyl trifluoroacetate and 4-methylphenyl trifluoroacetate [14]. Here the modest acceleration of hydrolysis at pH > 8 is attributed to complexation of the trifluoromethyl group in the annulus and nucleophilic attack by a deprotonated βCD secondary hydroxy group. The retardation of hydrolysis at pH < 8, where the rate of hydrolysis is pH independent, is thought to proceed through the complexes with the reversed orientation of the guests. At pH < 8 there is no deprotonation of the secondary hydroxy groups of βCD and so the only route for hydrolysis is nucleophilic attack by water from which the carbonyl group is shielded through is complexation in the βCD annulus. The same explanation may apply to the retardation of the hydrolysis of benzaldehyde dimethyl acetal by αCD, βCD and γCD and of a range of alkyl nitrites in aqueous hydrochloric acid [15,16]. Under alkaline conditions alkyl nitrite hydrolysis is accelerated through a similar mechanism to that deduced for the esters.

Regardless of the exact geometry of the reactive complex, it is important to realise that this orientation is not necessarily the most thermodynamically stable. Thus, in basic solution 6^A-amino-6^A-deoxy-α- and β-CD both accelerate the hydrolysis of 4-nitrophenyl acetate to produce their 6^A-acetamido-6^A-deoxy analogues and 4-nitrophenoxide through attack of the CD amino nucleophile on the acetate carbonyl carbon [17]. To achieve this the acetate group of 4-nitrophenyl acetate must be in the vicinity of the amino nucleophile at the primary face of the CD. This is the opposite orientation to that required for the accelerated hydrolysis by αCD and βCD. As the K_M for the 6^A-amino-6^A-deoxy-α- and β-CD mediated reactions of 4-nitrophenyl acetate are similar to those for the αCD and βCD mediated reactions, it appears likely that the isomeric complexes with opposed orientations of 4-nitrophenyl acetate are formed by αCD and βCD, but only those with the acetate group at the secondary face are reactive.

4.1a   **4.1b**

4.2a R^1 = Me, R^2 = H
4.2b R^1 = H, R^2 = Me

In principle, either the complexation of chiral guests by a CD (K_M), or any subsequent reaction between the CD and the guests (k_{cat}), or both, may display chiral discrimination in the Michaelis-Menten reaction scheme of Eq. (4.1). This facet of the hydrolytic activity of CDs continues to attract attention and varying degrees of chiral discrimination are observed [18-25]. These are exemplified by the reactions of the βCD complexed ferrocenyl esters **4.1a** and **4.1b** where the 62-fold difference between their k_{cat} values appears to be the highest enantioselectivity reported for ester hydrolyses catalysed by βCD [20]. For **4.1a** and **4.1b**, K_M = 1.75 × 10^{-2} and 2.12 × 10^{-2} mol dm^{-3}, and k_{cat} = 9.2 × 10^{-2} and 1.49 × 10^{-3} s^{-1}, respectively, at pH 10.0 and 303 K, in dimethyl sulfoxide/water (60/40 v/v). The acceleration of the reactions by βCD is 5.8 × 10^{-6} and 9.4 × 10^{-4}, respectively, as a comparison with the uncatalysed reaction rate constant, k_u = 1.58 × 10^{-8} s^{-1}, shows. The reactions proceed through the Michaelis-Menten mechanism and k_{cat} characterises the nucleophilic attack of a deprotonated secondary hydroxy group of βCD on the carbonyl carbon of **4.1a** and **4.1b**. It is seen that the greatest chiral discrimination occurs in the reaction of the complexed species (k_{cat}) rather than in the initial complexation step (K_M), and this appears to be often the case in CD catalysed reactions. Molecular modelling of the βCD complexes of **4.1a** and **4.1b**, the tetrahedral intermediate for the reaction, and the resulting acylated βCD provide substantial insight into the reaction mechanism [26,27]. The catalysis of the reactions of the enantiomers of the phenylpropionate **4.2a** and **4.2b** by βCD provide another example of the second step in the Michaelis-Menten scheme being the more chirally discriminating as shown by the enantioselectivity of complexation, $K_{M(R)}/K_{M(S)} = 0.8$ and that of reaction, $k_{cat(R)}/k_{cat(S)} = 15.5$ [22].

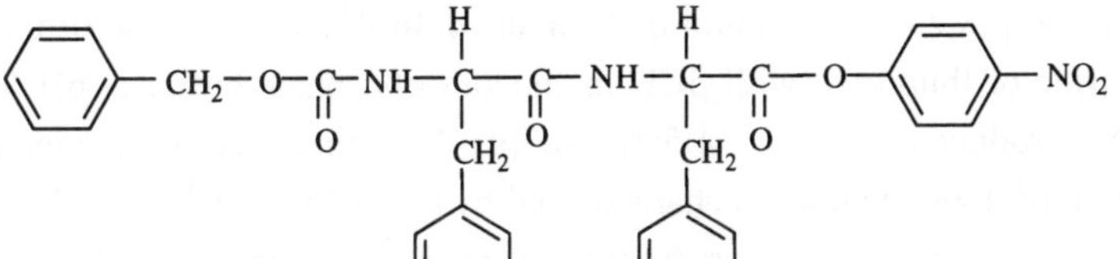

4.3

The extent of chiral discrimination may vary considerably with the CD as is exemplified by the catalysed hydrolysis of the stereoisomers of the dipeptide ester **4.3** [24]. While the variation of K_M with CD and dipeptide guest is quite small, k_{cat} for the DL diastereomer is much greater than that for the LL diastereomer (Table 4.2), and k_{cat} for βCD and γCD is much greater than that for αCD.

Table 4.2. Parameters for the Catalysis of the Hydrolysis of the Dipeptide Ester **4.3** by CDs.[a]

CD	Parameter	LL	DL	DL / LL
αCD	k_{cat}/s^{-1}	0.00104	0.0230	22
	$K_M^{-1}/dm^3\ mol^{-1}$	28.7	22.2	0.8
βCD	k_{cat}/s^{-1}	0.0180	0.758	42
	$K_M^{-1}/dm^3\ mol^{-1}$	23.0	26.4	1.1
γCD	k_{cat}/s^{-1}	0.0170	0.778	46
	$K_M^{-1}/dm^3\ mol^{-1}$	32.7	79.9	2.4
none	k_u/s^{-1}	0.000595	0.00547	9.2

[a] In 3% acetonitrile 97% water by volume at pH 9.5 and 298.2 K.

The immediate products of the reactions of the esters discussed above are the released phenoxides and the $O(2)$-acylated CDs. Depending on the acyl group, the rate of hydrolysis of the modified CD to regenerate the natural CD and complete the catalytic cycle varies considerably. However, it should be noted that chiral discrimination may be expected to apply in both the nucleophilic attack by the CD and the subsequent hydrolysis of the resulting modified CD. Few detailed studies of both steps have been reported. However, the βCD diastereomeric esters of the nonsteroidal antiinflammatory drug Ibuprofen **4.4a** and **4.4b** are of particular interest as they are both prodrugs and show complementarity in their formation and deacylation to give a *ca.* 50-fold chiral disrimination in favour of *(R)*-Ibuprofen over *(S)*-Ibuprofen [28,29]. Formation of **4.4a** and **4.4b** through reaction of the racemic acid chloride of Ibuprofen with βCD in aqueous phosphate buffer at pH 6.0 gave a **4.4a:4.4b** product ratio of *ca* 4.5:1. At pH 11.5 in aqueous carbonate buffer, deacylation of **4.4a** and **4.4b** is characterised by $k = 3.16 \times 10^{-4}$ s^{-1} and 2.97×10^{-5} s^{-1}, respectively, at 310 K. Apart from their intrinsic interest, the formation of βCD prodrugs exemplified by those of Ibuprofen **4.4a** and **4.4b** provides opportunities for directed oral drug delivery. While neither CDs nor their prodrugs are either appreciably absorbed through the intestinal wall or therapeutically active, release from the prodrug of the drug at a selected site restores its activity. This has been exploited in the design of potential large intestine targeting prodrugs as exemplified by 6^A-O-[(4-biphenyl)acetyl]-α-, β - and γ-CD and 6^A-O-{[(4-biphenyl)-acetyl]amino}-6^A-deoxy-α-, β- and γ-CD [30-32]. These prodrugs hydrolyse very slowly in the small intestine but the CD component is metabolised by the microflora of the large intestine to release the drug and thereby effectively target the large intestine for drug delivery.

4.4a R^1 = Me, R^2 = H
4.4b R^1 = H, R^2 = Me

Despite the interest in this area, it must be acknowledged that the catalytic efficiency of the natural CDs in hydrolysing esters is limited, because alkaline reaction conditions are required to deprotonate the CD secondary hydroxy groups, for rapid reaction, and the intermediate acylated CDs are generally slow to hydrolyse under these circumstances. An early improvement on the natural CDs, as catalysts for ester hydrolysis, was the hydroxamic acid derivative **4.5** [33]. Through deprotonation at near neutral pH, the hydroxamic acid moiety provides a nucleophile which is more reactive than the CD hydroxy groups, and the adjacent tertiary amine acts intramolecularly as a base, to promote hydrolysis of the intermediate acylated CD **4.6** and regeneration of the catalyst **4.5**. As a result the modified CD **4.5** is more than three orders of magnitude more effective than αCD in promoting reaction of *p*-nitrophenyl acetate at pH 7.80, and the acylated CD intermediate **4.6** (R = Me) hydrolyses 10^5 times faster than the corresponding $O(2)$-acylated αCD **4.7**.

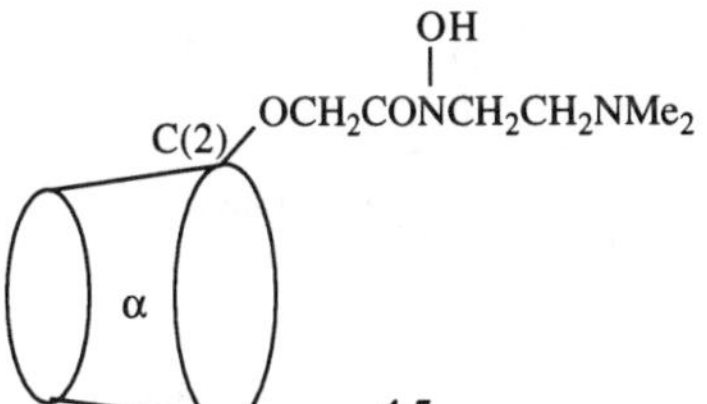

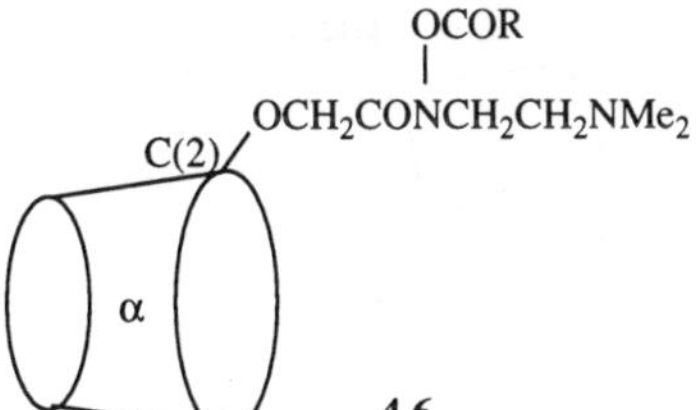

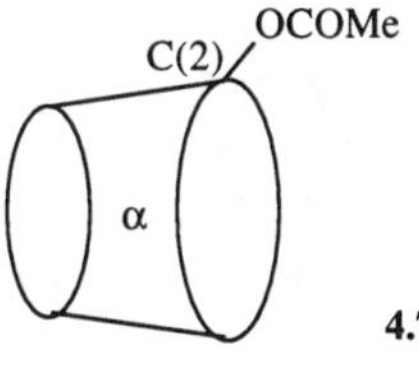

A variety of imidazole-substituted CDs have also been synthesised, as catalysts of ester hydrolysis at near neutral pH [34-40]. For example, the CD **4.8** was shown to accelerate the hydrolysis of *p*-nitrophenyl acetate at pH 6.80 and 298 K, with a catalytic rate constant (k_{cat}) of 0.82 x 10^{-3} s^{-1} and a binding constant (K_M) of 4.4 x 10^{-3} mol dm^{-3} [36]. The efficiency of this process approaches that for hydrolysis of the same ester by chymotrypsin, for which the k_{cat} and K_M values are 6.5 x 10^{-3} s^{-1} and 7.7 x 10^{-3} mol dm^{-3}, respectively. Such a comparison is of limited validity however, as *p*-nitrophenyl acetate bears little resemblance to the natural substrates of the enzyme, and the enzyme and the CD **4.8** function through different mechanisms. With the natural CDs, the hydroxamic acid derivative **4.5**, and chymotrypsin, an hydroxy group of the catalyst acts first as a nucleophile, in a transacylation reaction of the substrate, while the CD **4.8** functions as a general base catalyst, with the imidazole delivering water to the bound species [41,42]. In this regard, the mechanism of the reaction catalysed by the CD **4.8** is more closely related to that of the hydrolase enzyme ribonuclease A, which is discussed in more detail below.

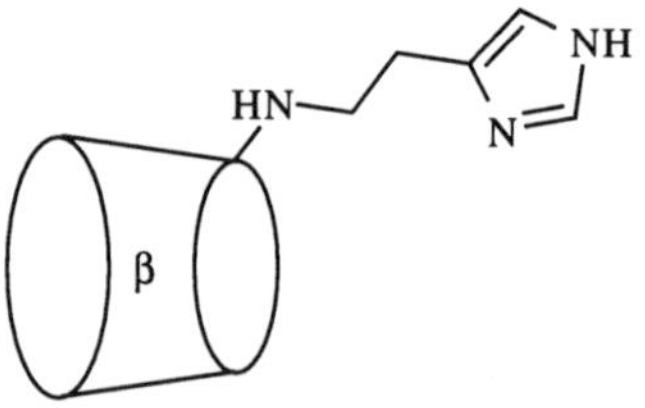

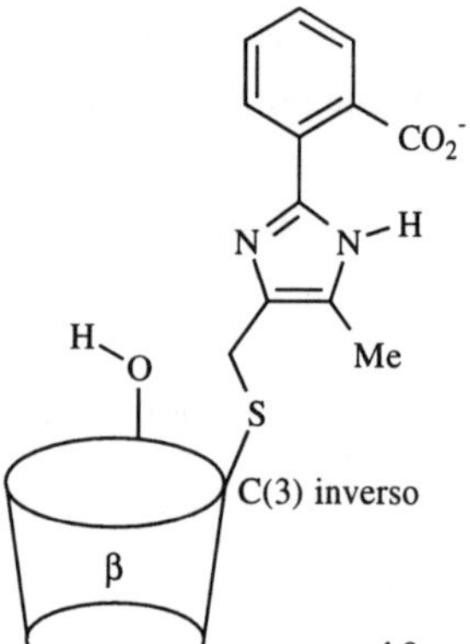

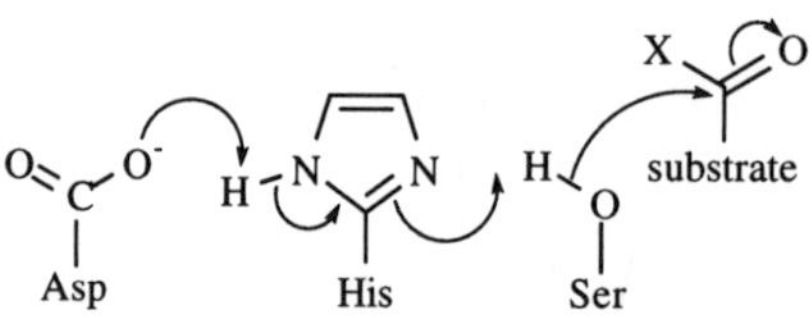

Fig. 4.1. Schematic representation of a reaction involving the catalytic triad of functional groups at the active site of chymotrypsin.

Chymotrypsin and other enzymes possess a catalytic triad of an aspartate carboxylate, a histidine imidazole and a serine hydroxy group at the active site, to promote hydrolysis of amides and esters under neutral conditions (Fig. 4.1). The modified CD **4.9** was synthesised as an artificial enzyme possessing this combination of functional groups, and the early reports indicated that it was an efficient catalyst of ester hydrolysis [43,44]. However, subsequent studies established that it was less effective than native βCD, and that the hydrolysis probably involves a deprotonated secondary hydroxy group but not the imidazole-carboxylate moiety [45,46]. This highlights the fact that the mere presence of multiple functional groups is not sufficient for them to act in concert in catalysis. The relative geometry and alignment of those groups must also be controlled.

A significant degree of such geometric control has been achieved by Breslow *et al.* [41,42,47-55], in their construction of a series of bis-imidazole-substituted CDs. These were made and studied as mimics of the enzyme ribonuclease A, which catalyses the cleavage of RNA through the cooperative function of an imidazole base and an imidazolium ion. The disubstituted CDs prepared in the course of this work include the $6^A,6^B$-, $6^A,6^C$-, and $6^A,6^D$-disubstituted βCD isomers **4.10, 4.13** and **4.14** [48], and the $6^A,6^B$-disubstituted αCD and γCD derivatives **4.11** and **4.12** [55].

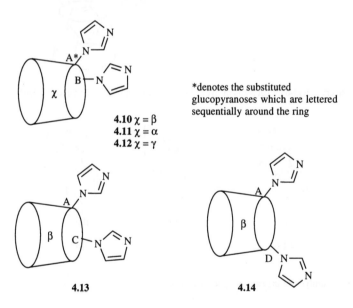

4.10 χ = β
4.11 χ = α
4.12 χ = γ

*denotes the substituted glucopyranoses which are lettered sequentially around the ring

4.13   4.14

The βCDs **4.10, 4.13** and **4.14** are catalysts in the hydrolysis of 4-*tert*-butylcatechol cyclic phosphate **4.15** (Scheme 4.3), at near neutral pH, with the $6^A,6^B$-isomer **4.10** being the most effective [48]. The enhanced activity of this isomer

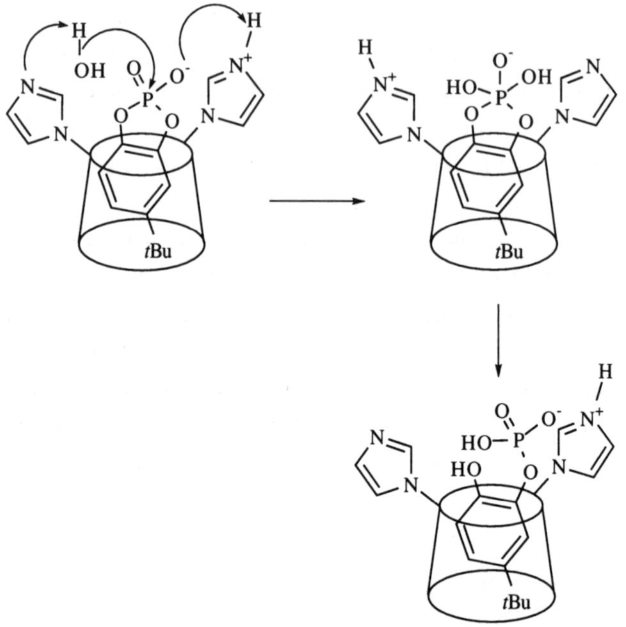

Fig. 4.2. Concerted acid-base catalysis of hydrolysis of the phosphate **4.15** by the CD **4.10**.

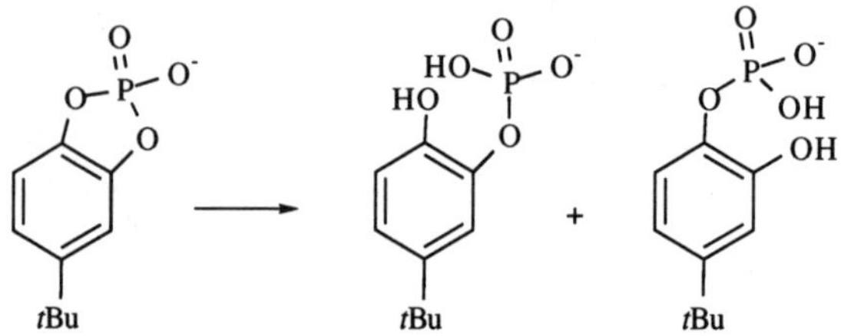

Scheme 4.3. Hydrolysis of the catechol phosphate **4.15**.

indicates a preferred reaction geometry. Together with results of deuterium isotope studies [41], it provides strong evidence that the reaction mechanism involves simultaneous bifunctional acid-base catalysis, where one imidazole ring acts as a base and the other, which is protonated and therefore present as the imidazolium ion, acts as an acid, to promote formation of the intermediate phosphorane (Fig. 4.2). The CD

4.10 also affects the regioselectivity of oxygen-phosphorus bond cleavage in the phosphorane. Whereas hydrolysis of the catechol **4.15** in imidazole or other buffers affords the products **4.16** and **4.17** in nearly equal yields, the product ratio from the reaction catalysed by the CD **4.10** is > 99:1.

At 298 K in 0.05 mol dm^{-3} phosphate buffer at pH 6.2, the k_{cat} and K_M values for the interaction of the catechol **4.15** with the CD **4.10** are 1.2 x 10^{-3} s^{-1} and 0.41 x 10^{-3} mol dm^{-3}, respectively. By comparison, the values obtained for the corresponding γCD derivative **4.12** are 0.09 x 10^{-3} s^{-1} and 4.8 x 10^{-3} mol dm^{-3}, respectively, while the αCD analogue **4.11** complexes the catechol **4.15** but does not induce reaction of the bound species [55]. For the interactions of the methyl-substituted catechol **4.18** with the CDs **4.10-4.12**, the k_{cat} and K_M values are 2.8 x 10^{-5} s^{-1} and 1.7 x 10^{-3} mol dm^{-3}, 3.8 x 10^{-5} s^{-1} and 5.5 x 10^{-3} mol dm^{-3}, and 0.83 x 10^{-5} s^{-1} and 7.6 x 10^{-3} mol dm^{-3}, respectively. Therefore, taking into account both complexation and reaction of the bound species (k_{cat}/K_M), the βCD **4.10** has the best geometry for catalysis with these substrates. It also exhibits the highest regioselectivity of phosphorus-oxygen bond cleavage in reaction of each of the intermediate phosphoranes [55]. Even so, the orientation of the catalytic imidazole groups is still flexible and more efficient catalysts might be expected in systems where their geometry is restricted and optimised.

Me   **4.18**

4.3. Cyclodextrin Isomerases

Concerted acid-base catalysis is not only a feature of hydrolase enzymes; some isomerase enzymes also function in this manner. It follows that bis-imidazole substituted CDs which provide acid-base catalysis as hydrolases, as described above, can also function as isomerase mimics [51-54,56]. As a CD isomerase, at pD 6.2, the 6^A,6^D-disubstituted CD **4.14** catalyses enolisation of the ketone **4.19** (Scheme 4.4). This was investigated by measuring deuterium incorporation into the substrate when the reaction was carried out in D_2O. The $6^A,6^B$- and $6^A,6^C$-disubstituted βCDs **4.10** and **4.13** are much less effective as catalysts in this case, showing activity similar to that displayed by a corresponding mono-substituted CD. On this basis it seems likely that the CDs **4.10** and **4.13** act as general base catalysts, while the $6^A,6^D$-isomer **4.14** provides bifunctional catalysis (Fig. 4.3). Stereoelectronic arguments have been advanced to account for this particular activity of the $6^A,6^D$-isomer **4.14**.

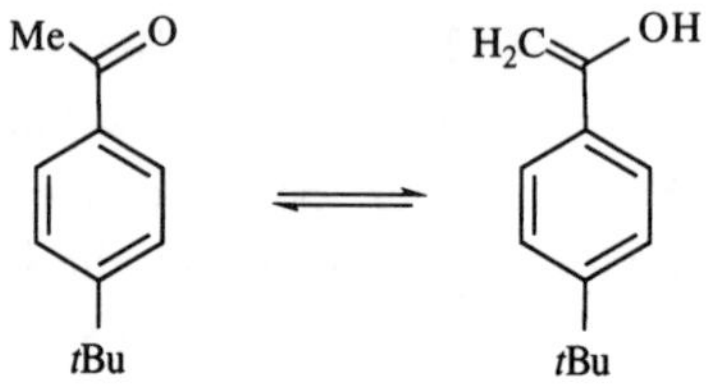

Scheme 4.4. Enolisation of the ketone **4.19**.

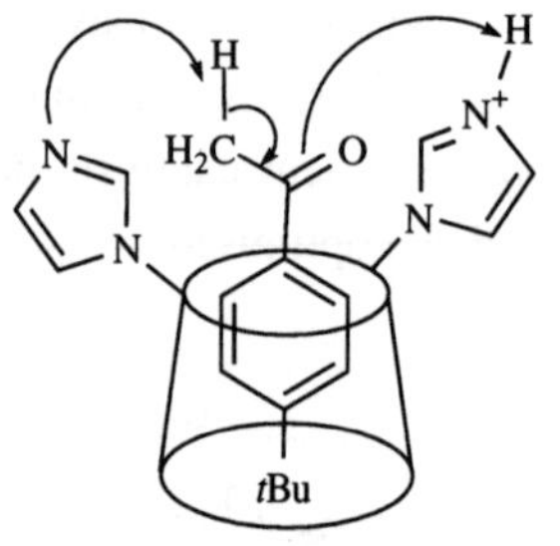

Fig. 4.3. Simultaneous acid-base catalysis of the enolisation of the ketone **4.19** by the CD **4.14**.

A number of other modified CDs which may be classified as isomerases act through general base catalysis alone. At 310 K and pH 9.0, the three isomeric monophosphates of βCD, substituted at either C(2), C(3) or C(6), catalyse enolisation of the ketoalcohol **4.20** (Scheme 4.5), with k_{cat} and K_M values of 1.98 x 10^{-2} s^{-1} and 3 x 10^{-3} mol dm^{-3}, 3.18 x 10^{-2} s^{-1} and 2 x 10^{-3} mol dm^{-3}, and 3.0 x 10^{-2} s^{-1} and 3 x 10^{-3} mol dm^{-3}, respectively [57]. By comparison, the corresponding values for native βCD are 0.024 x 10^{-2} s^{-1} and 1.2 x 10^{-3} mol dm^{-3}, so the phosphate group of the modified CDs has little effect on guest complexation but increases the rate of reaction of the bound species, by approximately two orders of magnitude, in each case. More recently, enolisation of acetaldehyde and a variety of ketones and α-ketoacids catalysed by heptakis-6^A-amino-6^A-deoxy-βCD has also been reported [58].

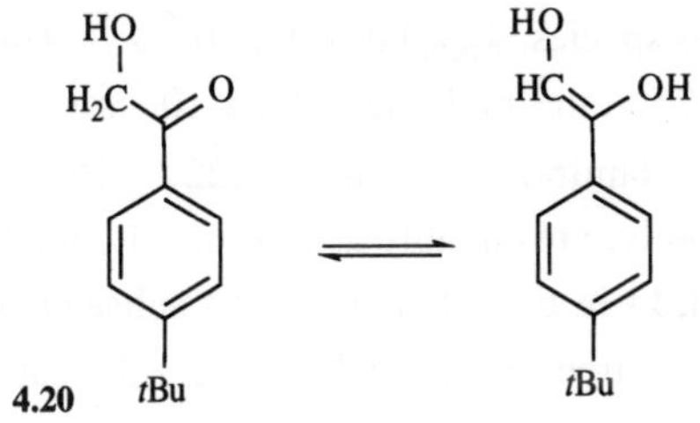

Scheme 4.5. Enolisation of the ketoalcohol **4.20**.

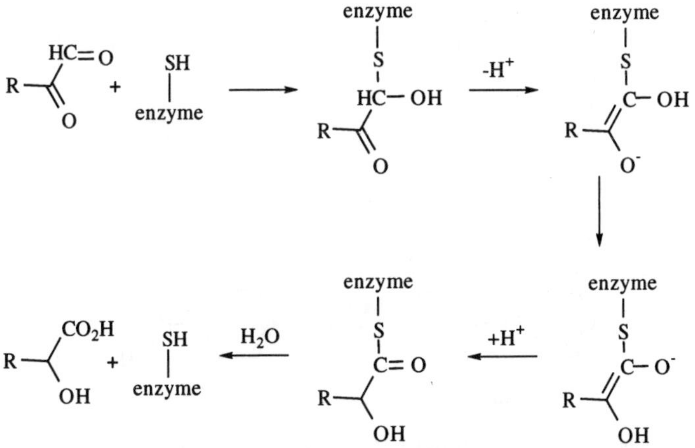

Scheme 4.6. Mechanism for the conversion of α-ketoaldehydes to the corresponding α-hydroxycarboxylic acids catalysed by glyoxalase enzymes.

Glyoxalase enzymes catalyse the conversion of α-ketoaldehydes to the corresponding α-hydroxycarboxylic acids. The isomerisation involves formation of an enzyme-bound α-keto hemithioacetal, followed by deprotonation to form an hydroxyenolate, rearrangement of the hydroxyenolate, and then protonation of the rearranged species, to give a thioester. Hydrolysis of the thioester affords the α-hydroxycarboxylic acid, which is released from the enzyme (Scheme 4.6). The modified CD **4.21** has been investigated as a glyoxalase mimic [59]. As a catalyst of the reaction of the naphthylglyoxal **4.23** to give the naphthyl-α-hydroxyacetic acid **4.24** (Scheme 4.7), at pH 9.75 and 298 K the equilibrium constant for the formation of the α-keto hemithioacetal (K_{hemi}) is 1100 dm^3 mol^{-1}, while the rate constant for the rearrangement of this species (k_{rear}) is 1.2 x 10^{-2} s^{-1}. This compares with values for K_{hemi} and k_{rear} of 250 dm^3 mol^{-1} and 2.0 x 10^{-2} s^{-1} for the analogous reaction catalysed by 2-(dimethylamino)ethanethiol **4.22**. The CD **4.21** forms the hemithioacetal more readily, presumably as a result of complexation of the naphthyl moiety of the glyoxal **4.23** in the CD cavity, but subsequent rearrangement of the bound species is less favoured and partially offsets the catalytic advantage of CD complexation.

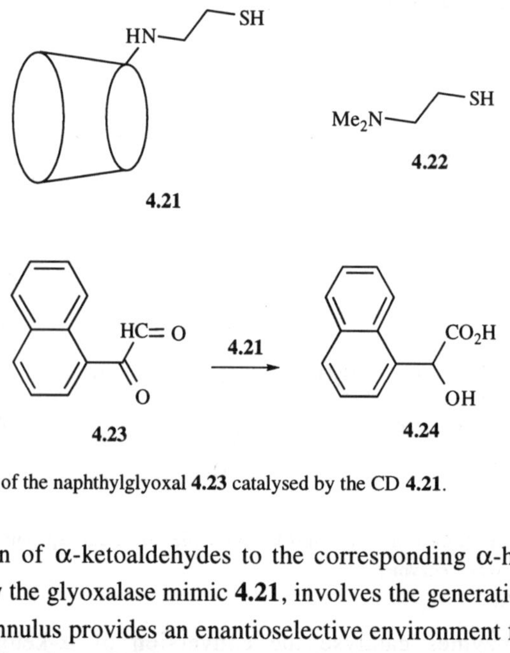

Scheme 4.7. Reaction of the naphthylglyoxal **4.23** catalysed by the CD **4.21**.

The conversion of α-ketoaldehydes to the corresponding α-hydroxycarboxylic acids, catalysed by the glyoxalase mimic **4.21**, involves the generation of a new chiral centre. The CD annulus provides an enantioselective environment for the reaction of phenylglyoxal **4.25** and, at pH 8.0 and 283 K, affords (*S*)-mandelic acid **4.26** in 46% enantiomeric excess (Scheme 4.8) [60]. This stereoselectivity is much greater than that obtained using combinations of 2-(dimethylamino)ethanethiol **4.22** with the native CDs, or their hydroxypropyl or methyl derivatives. Thus, the geometric constraints resulting from covalent attachment of the thiol functional group to the CD **4.21** appear to contribute to the enantioselectivity, although the system remains quite flexible and it is reasonable to expect that greater enantioselectivity is possible through optimisation of the molecular recognition in this system.

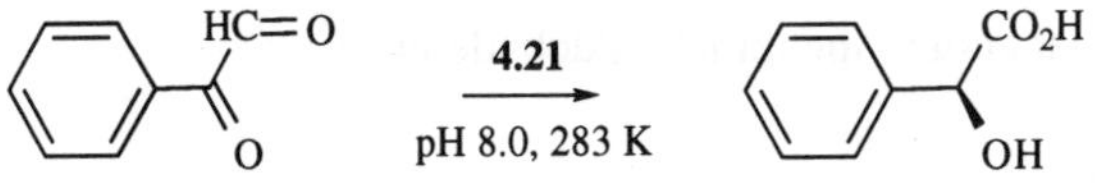

Scheme 4.8. Reaction of phenylglyoxal **4.25** catalysed by the CD **4.21**.

4.4. Cyclodextrin Lyases

The modified CDs and in particular the $6^A,6^D$-disubstituted CD derivative **4.14**, which catalyse enolisation, have also been examined as catalysts for aldol condensations [51,53,54,61]. From a mechanistic viewpoint, enolisation and aldol condensation reactions of carbonyl compounds are similar, but in the former context the CDs behave as isomerases, while in the latter context they may be regarded as CD lyases. The CD **4.14** catalyses the intramolecular aldol condensation of the keto aldehyde **4.27** (Scheme 4.9) [61]. The first step of the reaction involves enolisation of the ketone moiety. Cyclisation of the enol **4.28** then occurs, to give the *trans*-isomer of the β-hydroxyketone **4.29**, with minimal enantioselectivity. Catalysis of the aldol reaction by the CD **4.14** may arise from the effect of the CD imidazole groups on either the formation of the enol **4.28** or the cyclisation of that species or both. In any event, the CD **4.14** increases the rate of the overall process by approximately 4,000-fold [53]. It also catalyses the slow dehydration of the hydroxyketone **4.29** to the corresponding enone **4.30**.

While the CD **4.14** does not alter the regioselectivity of the aldol condensation of the keto aldehyde **4.27**, it and the corresponding $6^A,6^B$-disubstituted CD isomer **4.10** do affect the regioselectivity of cyclisation of the dialdehyde **4.31** (Scheme 4.10), in a striking example of the control of this aspect of a chemical reaction by an enzyme mimic [62]. The reaction of the dialdehyde **4.31** induced by imidazole buffer, at 298 K, afforded approximately equal quantities of the three β-hydroxy aldehydes **4.32**-**4.34**. By contrast, the analogous reaction carried out in the presence of the CD **4.14** gives a 1:6:17 mixture of the products **4.32-4.34**, while the corresponding ratio is 1:9:24 for the reaction catalysed by the CD **4.10**. This indicates that the CDs **4.10** and **4.14** catalyse the enolisation of the benzylic aldehyde group of the substrate **4.31** and the addition of that enol to the remote aldehyde group.

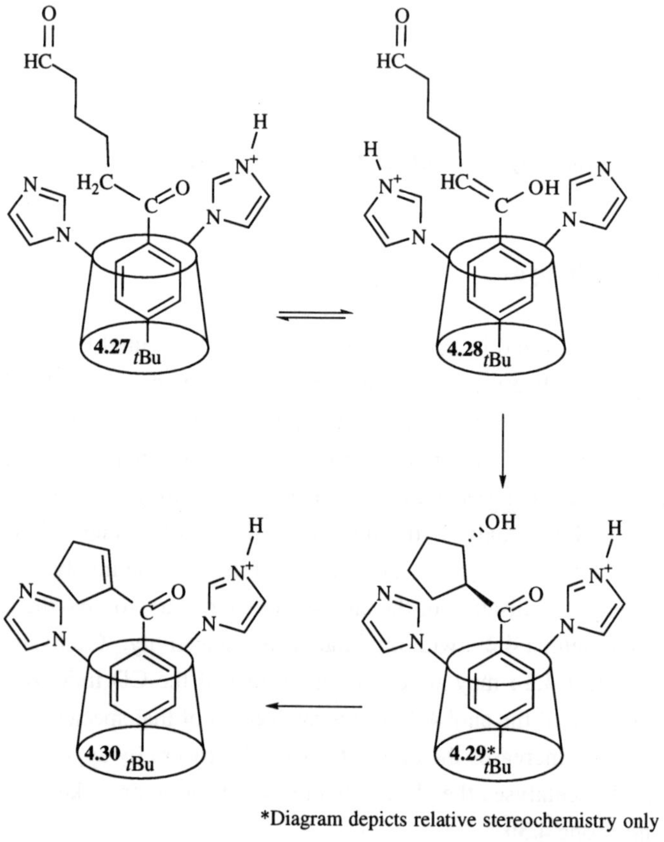

4.27
tBu
4.28
tBu
4.29*
tBu
4.30
tBu
*Diagram depicts relative stereochemistry only

Scheme 4.9. Aldol reaction of the keto aldehyde **4.27** catalysed by the CD **4.14**.

Other enzyme cofactors have also been attached to CDs, to produce CD lyases. CDs substituted with thiazolium salts have been investigated as catalysts of the benzoin condensation of benzaldehyde **4.39** (Scheme 4.11) [54,63,64]. At pH 8 and 323 K, the second order rate constants for the reaction of benzaldehyde **4.39** are found to be 0.22 and 3.7 mol dm^{-3} s^{-1}, when the condensation is carried out in the presence of 2 x 10^{-2} mol dm^{-3} of the CDs **4.35** and **4.36**, respectively. The corresponding rate constants measured for the reactions carried out in the presence of the thiazolium salts **4.37** and **4.38**, lacking a CD annulus, are almost an order of magnitude lower, at 0.025 and 0.52 mol dm^{-3} s^{-1}, respectively. By comparison with the γCDs **4.35** and **4.36**, the analogous βCD derivatives have much less effect on the rate of reaction, presumably because the larger CD annulus is required to simultaneously complex two molecules of benzaldehyde **4.39**, for effective catalysis.

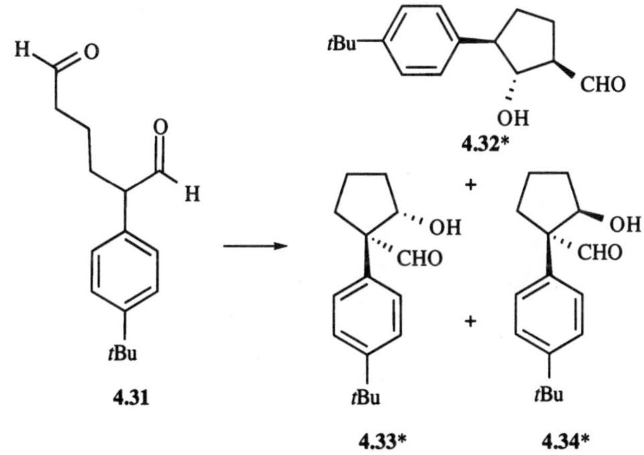

*Diagram depicts relative stereochemistry only

Scheme 4.10. Aldol condensation of the dialdehyde **4.31**.

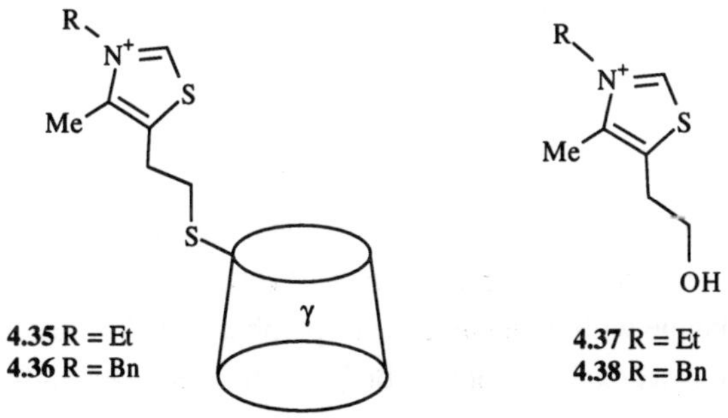

4.35 R = Et
4.36 R = Bn

4.37 R = Et
4.38 R = Bn

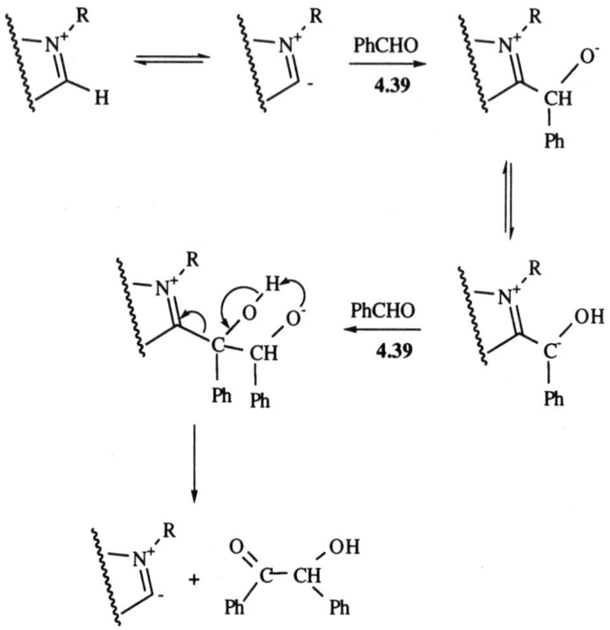

Scheme 4.11. Benzoin condensation of benzaldehyde **4.39** catalysed by the CDs **4.35** and **4.36**.

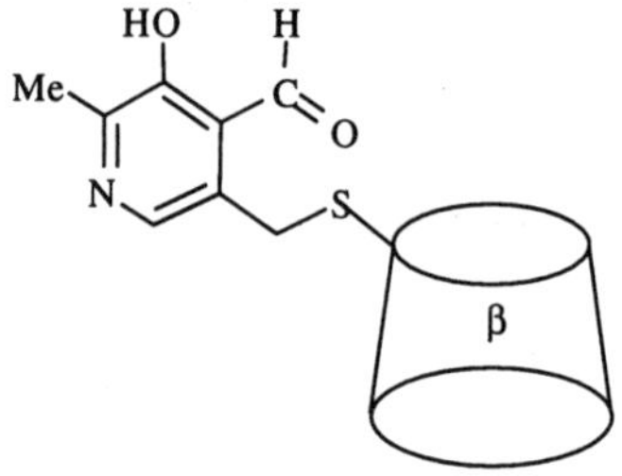

4.40

Pyridoxamine phosphate and pyridoxal phosphate are cofactors of numerous enzymes involved in amino acid metabolism. Alone, pyridoxal induces the reaction of α,β-dehydroalanine with indole to give tryptophan, but 3-5 times more tryptophan is produced using pyridoxal attached to βCD, in the conjugate **4.40**, under otherwise identical conditions [65] (Fig. 4.4). In this regard, the pyridoxal derivative **4.40** is another example of a CD lyase.

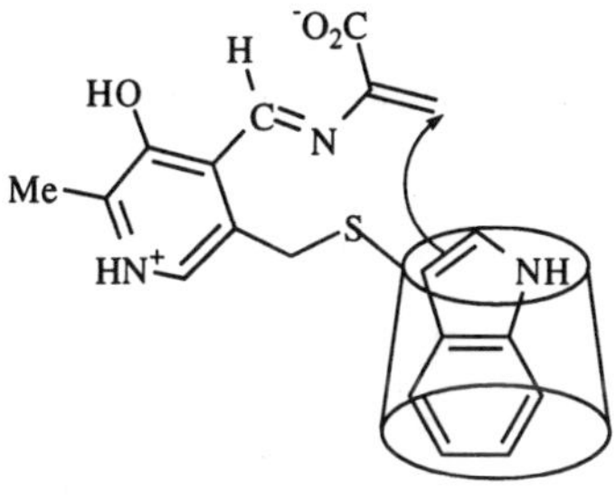

Fig. 4.4. Reaction of indole with α,β-dehydroalanine bound to the CD **4.40**.

4.5. Cyclodextrin Transferases

In amino acid metabolism, pyridoxamine phosphate and pyridoxal phosphate are also involved as cofactors of enzymes which catalyse the transfer of the amino group between amino acids and keto acids (Scheme 4.12). Pyridoxamine derivatives of CDs have been investigated as models of these transferase enzymes [54,66-72]. Pyruvic acid **4.41**, phenylpyruvic acid **4.42** and indolepyruvic acid **4.43** react equally well with pyridoxamine, to give the corresponding amino acids, alanine **4.44**, phenylalanine **4.45** and tryptophan **4.46**. The C(6)-pyridoxamine-substituted βCD **4.47** has little effect beyond that of pyridoxamine itself on the reaction of pyruvic acid, but accelerates the reactions of phenylpyruvic acid **4.42** and indolepyruvic acid **4.43**, by factors of 15 and 12, respectively [66,70,71]. The C(3)-modified CD **4.48** displays a similar selectivity for the aromatic substrates, and causes a 20-fold increase in the rates of reaction of both phenylpyruvic acid **4.42** and indolepyruvic acid **4.43** [67,70,71]. The reactions with the CD **4.47** afford mainly the (*S*)-enantiomers of phenylalanine **4.45** and tryptophan **4.46**, in 66% and 33% enantiomeric excess, respectively. By comparison, the reactions with the CD **4.48** occur without enantioselectivity in the case of phenylpyruvic acid **4.42**, and give the (*R*)-isomer of tryptophan **4.46** in 29% enantiomeric excess.

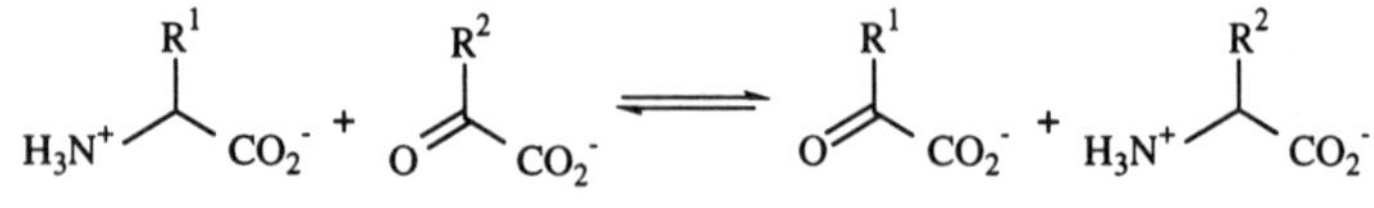

Scheme 4.12. Transamination of α-amino acids and α-keto acids.

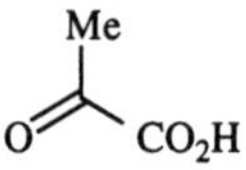

4.41

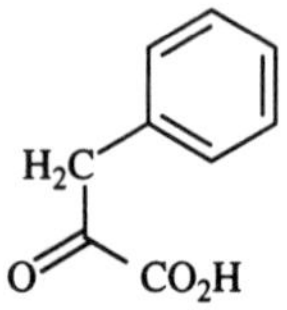

4.42

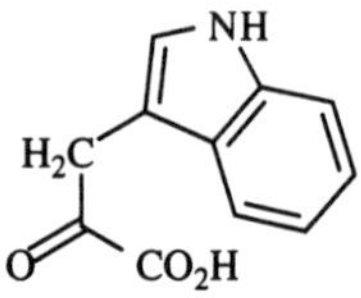

4.43

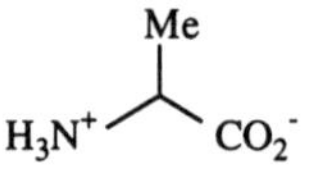

4.44

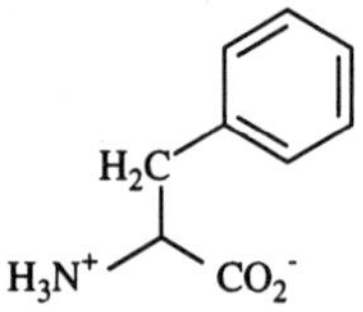

4.45

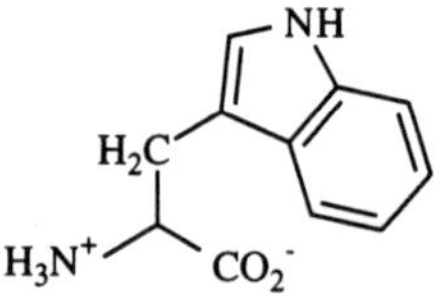

4.46

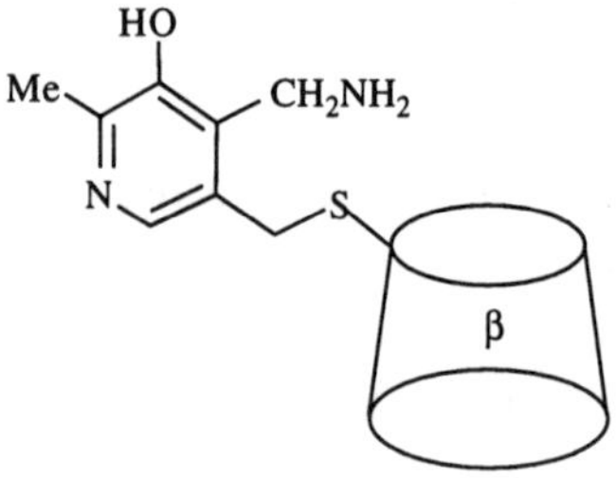

4.47

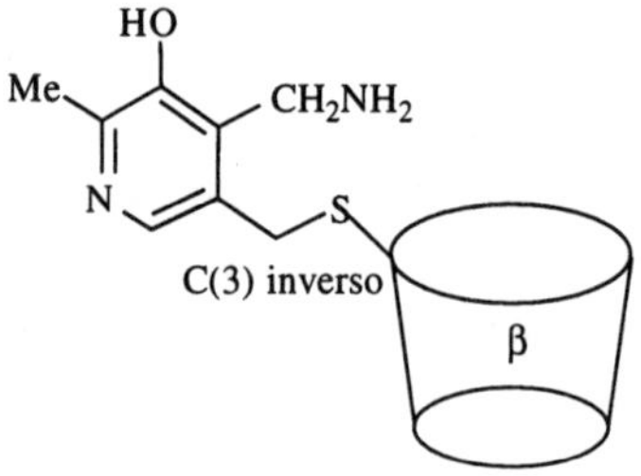

4.48

The substrate selectivity of the modified CD **4.47** is further demonstrated through the use of the *tert*-butyl-substituted phenylpyruvic acids **4.49** and **4.50** [72]. In competition experiments with pyruvic acid **4.41**, the effect of complexation in the annulus of the CD **4.47** results in a 28,000-fold selectivity for reaction of 4'-*tert*-butylphenylpyruvic acid **4.49**, but neglible selectivity for reaction of the 5'-*tert*-butylphenylpyruvic acid derivative **4.50**. Using the CD **4.51**, where the orientation of the pyridoxamine moiety with respect to the CD annulus is more rigidly controlled, leads to a decrease in the selectivity of transamination of 4'-*tert*-butylphenylpyruvic acid **4.49**, but an increase in the selectivity for reaction of the 5'-*tert*-butylphenylpyruvic acid derivative **4.50**. Again this demonstrates the importance of complementary host and guest geometry for optimum catalysis.

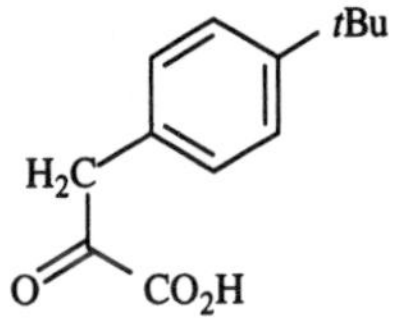

4.49

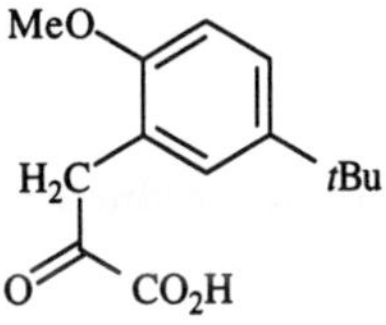

4.50

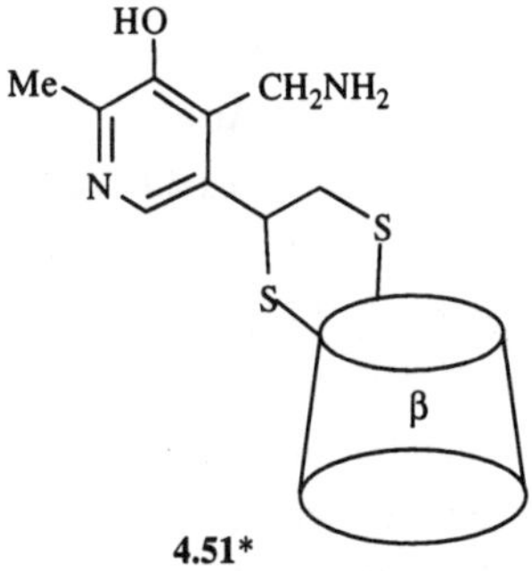

4.51*

*Mixture of $6^A,6^B$- and $6^B,6^A$-regioisomers

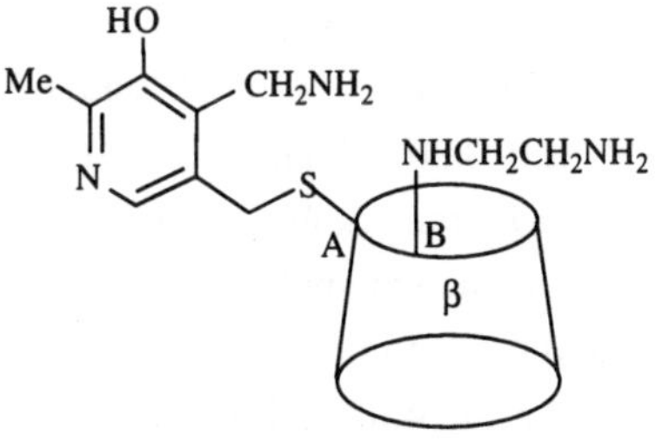

4.52

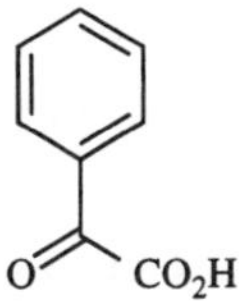

4.53

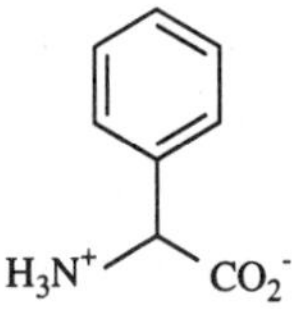

4.54

Tabushi *et al.* [69], constructed the disubstituted CD **4.52**, possessing a pyridoxamine substituent adjacent to an aminoethylamino group. The production of phenylalanine **4.45**, tryptophan **4.46** and phenylglycine **4.54**, from the corresponding keto acids **4.42**, **4.43** and **4.53**, occurs approximately 2,000 times faster with the CD **4.52** in place of pyridoxamine. The reactions are also enantioselective, affording mainly the (S)-isomers of the amino acids **4.45, 4.46** and **4.54**, each in greater than 90% enantiomeric excess. The enantioselectivity is attributed to the stereoselective participation of the diamine moiety in proton transfer to the complexed guests. The rate acceleration is probably due to a combination of the catalytic effect of this group and guest complexation in the annulus of the CD **4.52**.

4.6.   Cyclodextrin Oxidoreductases

Another principal class of enzymes are oxidoreductases, and CD analogues of these have also been prepared and investigated. The dihydronicotinamide-substituted CD **4.55** reduces complexed ninhydrin [73]. At pH 7 and 298 K, the measured k_{cat} and K_M values are 2.0 x 10^{-2} s^{-1} and 2.1 x 10^{-5} mol dm^{-3}, respectively. The flavocyclodextrins **4.56-4.59** have also been assembled [74-77]. At pH 7.4 and 298 K, the αCD **4.56** reduces the N-alkyldihydronicotinamide derivatives **4.60-4.62**, with k_{cat} and K_M values of 0.5 s^{-1} and 4.0 x 10^{-4} mol dm^{-3}, 0.36 s^{-1} and 3.8 x 10^{-3} mol dm^{-3}, and 0.06 s^{-1} and 9.5 x 10^{-4} mol dm^{-3}, respectively [74]. The values are particularly remarkable in that they indicate a facile electron transfer from the complexed nicotinamides **4.60-4.62** to the flavin substituent of the CD **4.56**. The CDs **4.57** and **4.58** do not show saturation kinetics for reactions with the nicotinamides **4.60, 4.62** and **4.63**, neither do the reactions of the CD **4.59** with the substrates **4.60** and **4.62** [76]. It appears that the complexes formed between these species do not have the orientation required for reaction. The reaction of the CD **4.59** with the napthylamine derivative **4.63** is characterised by k_{cat} and K_M values of 2.8 x 10^{-2} s^{-1} and 3.7 x 10^{-3} mol dm^{-3}, respectively [76]. The CDs **4.57-4.59** have also been investigated as catalysts for the oxidation of thiols [77]. The k_{cat} and K_M values for reaction of the CD **4.59** with phenylmethanethiol **4.64**, and the o-, m- and p-chlorophenylmethanethiols **4.65-4.67**, are 1.1 x 10^{-3} s^{-1} and 1.9 x 10^{-3} mol dm^{-3}, 0.18 x 10^{-3} s^{-1} and 1.5 x 10^{-3} mol dm^{-3}, 3.5 x 10^{-3} s^{-1} and 8.9 x 10^{-3} mol dm^{-3}, and 1.2 x 10^{-3} s^{-1} and 2.9 x 10^{-3} mol dm^{-3}, respectively. As is the case with riboflavin, with each of the CDs **4.57-4.59**, the reduced form of the flavin moiety reoxidises readily in air, indicating the true catalytic nature of these species [76].

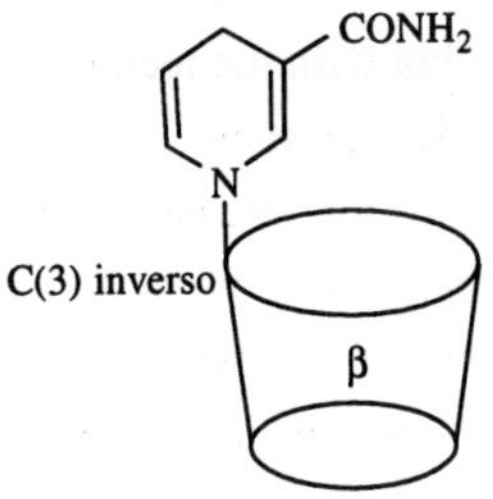

4.55

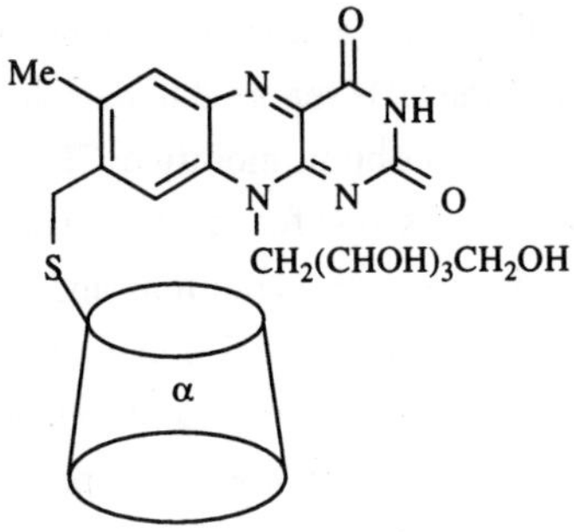

4.56

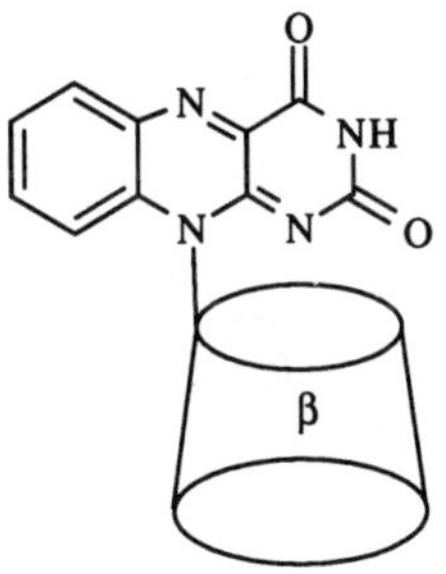

4.57

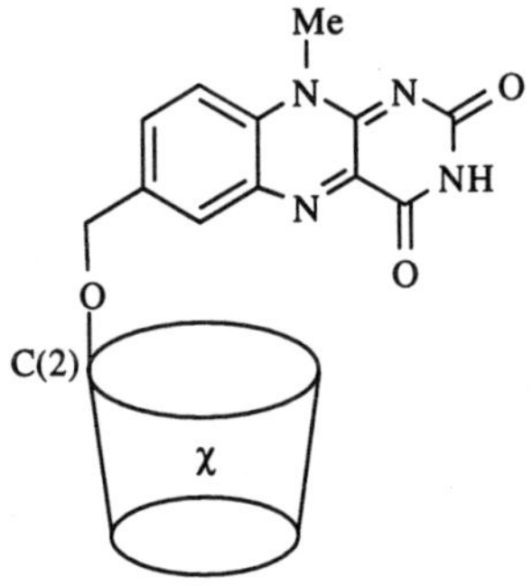

4.58 χ = α
4.59 χ = β

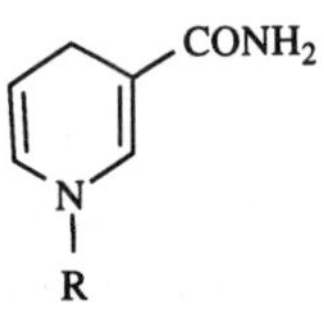

4.60 R = C$_6$H$_{13}$
4.61 R = CHMe$_2$
4.62 R = Bn
4.63 R = 1-naphthylmethyl

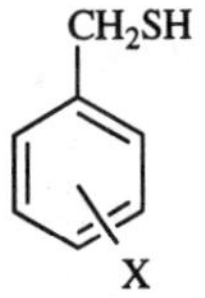

4.64 R = H
4.65 R = 2-Cl
4.66 R = 3-Cl
4.67 R = 4-Cl

4.7. Future Prospects

The reactions described above demonstrate ways in which CDs and particularly their modified forms may be regarded as enzyme mimics, which catalyse reactions and affect the regio- and stereo-chemical outcomes of those processes. The catalytic efficiency of the CDs reported to date is often modest, but with the techniques that are now available to obtain modified CDs of predetermined geometry (see Chapter 2), there is plenty of scope for the design and synthesis of even better catalysts. It is also possible to incorporate metal ions and multiple binding sites in the modified CDs, as is discussed in Chapters 5 and 6, respectively. This provides opportunities to introduce new functionality and enhance molecular recognition. Together these aspects of CD chemistry hold great potential for the development of new catalysts for use in chemical processing.

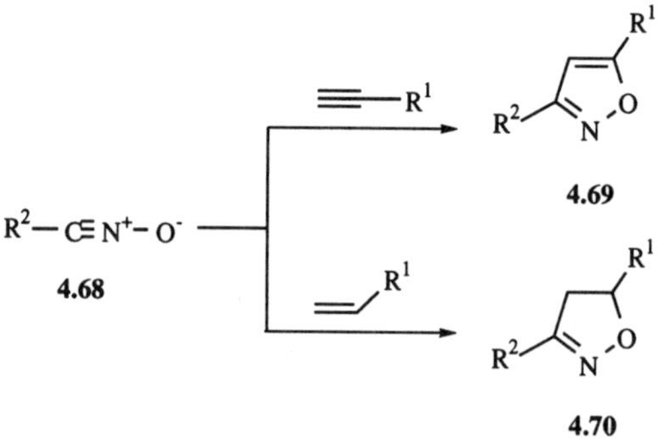

Scheme 4.13. Usual regioselectivity of cycloaddition reactions of nitrile oxides with monosubstituted alkenes and alkynes.

Significant advantages can be expected with CD-based catalysts when compared to their enzyme counterparts. They should be more robust and less susceptible to chemical and biological degradation. Perhaps even more importantly, they can be designed to induce tranformations for which there is no enzyme catalyst, and to bring about reactions which are entirely different to those which normally occur in free solution. The deployment of βCD as a purpose designed molecular scaffold for the reversal of the regioselectivity of nitrile oxide cycloadditions [78,79] illustrates the potential use of CDs in this area, although at this stage it has yet to be developed into a catalytic process. Nitrile oxides (1,3-dipoles) (**4.68**) undergo efficient [3+2] cycloadditions with alkynes and alkenes (dipolarophiles) to produce isoxazoles and 4,5-dihydroisoxazoles, respectively [80]. When the dipolarophile is unsymmetrical the possibility of producing regioisomeric mixtures arises, but it is usually found that steric effects determine the regioselectivity so that the more crowded end of the dipolarophile becomes attached to the oxygen of the nitrile oxide. This is exemplified by terminal alkynes and alkenes affording predominantly 5-substituted isoxazoles **4.69** and dihydroisoxazoles **4.70,** respectively (Scheme 4.13). However, by attaching the dipolarophiles to βCD, the modified CDs then predetermine the regioselectivity of the cycloaddition reactions by controlling the relative orientations of the dipolarophiles and the nitrile oxides in the 1:1 complexes. This is illustrated by the reactions of the CD **4.71** with 4-*tert*-butylbenzonitrile oxide **4.72a** and 4-phenyl-benzonitrile oxide **4.72b** (Scheme 4.14), which are summarised in Table 4.3. In water, where complexation of the nitrile oxides **4.72a** and **4.72b** by the CD **4.71** is favoured, the corresponding 4-substituted isoxazoles **4.73a** and **4.73b** are formed as the dominant products. In *N,N*-dimethylformamide, where the CD complexes are less stable than those formed in water, the regioselectivity is reversed such that the reactions favour formation of the 5-substituted isoxazoles **4.74a** and **4.74b**, respectively. By comparison, the reactions of methyl propynoate and the nitrile oxides **4.72a** and **4.72b** in both water and *N,N*-dimethylformamide produce only the corresponding 5-substituted cycloaddition regioisomers. Thus, attachment of the dipolarophile to a CD reverses the regioselectivity of cycloaddition reactions with nitrile oxides in a predetermined manner.

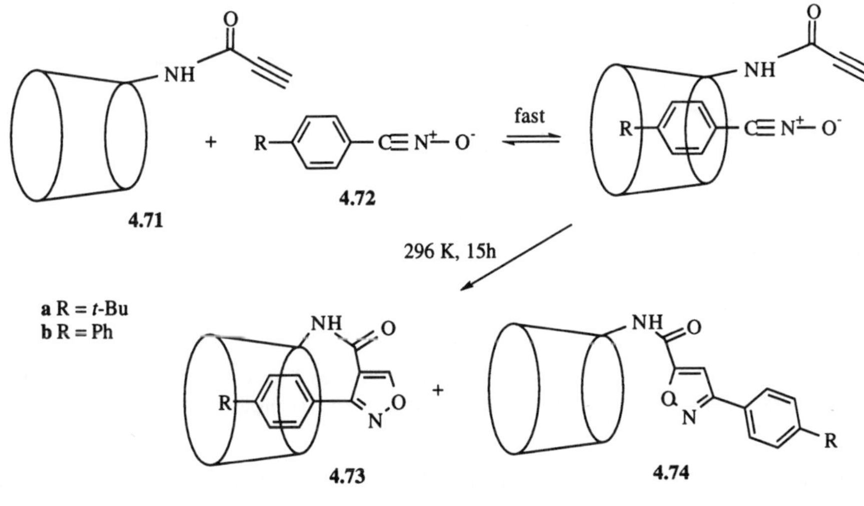

Scheme 4.14. Use of CDs to reverse the regioselectivity of nitrile oxide cycloaddition reactions.

Table 4.3. Products of Reactions of the Modified βCD **4.71** with the Nitrile Oxides **4.72a** and **4.72b**.

Nitrile Oxide	Solvent	Ratio of regioisomers	Yield %
4.72a	H_2O	**4.73a**:**4.74a**, 15:1	71
4.72a	DMF	**4.73a**:**4.74a**, 1:1.5	86
4.72b	H_2O	**4.73b**:**4.74b**, 5:1	100
4.72b	DMF	**4.73b**:**4.74b**, 1:5	93

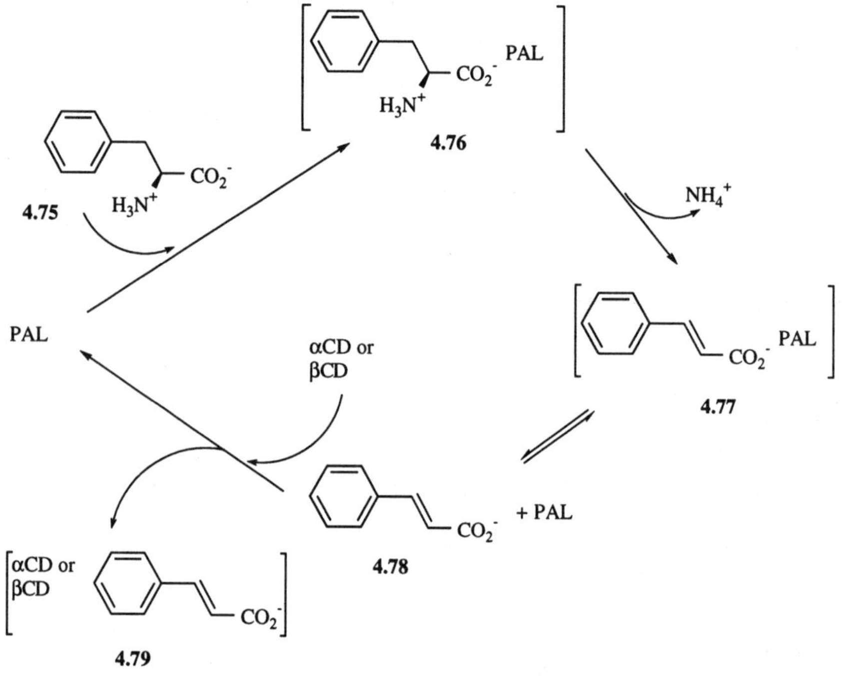

Fig. 4.5. The catalysis of reaction of (*S*)-phenylalanine **4.75** by PAL to ammonium ion and *trans*-cinnamate **4.78**, and the sequestration of the latter as an αCD or a βCD complex **4.79**.

Alternatively, CDs may be used in conjunction with enzymes to improve the efficiency of reactions brought about by these catalysts. Substrate and product inhibition are important methods of biological control of enzymes but they limit the utility of enzymes *in vitro*. However, CDs may be used to selectively complex either the substrate or the product of an enzyme catalysed reaction, thus reducing both the concentration of that species in solution and the associated enzyme inhibition. This has been clearly demonstrated with the enzyme (*S*)-phenylalanine ammonia lyase (PAL), which catalyses the elimination of ammonium ion from (*S*)-phenylalanine **4.75**, proceeding through the (*S*)-phenylalanine complex **4.76** and the *trans*-cinnamate complex **4.77**, to free *trans*-cinnamate **4.78** and PAL, as shown in Fig. 4.5. As the *trans*-cinnamate **4.78** builds up the equilibrium between **4.78** and PAL and **4.77** moves to the right, thereby inhibiting the formation of the catalytic complex **4.76** and decreasing the catalytic effect of PAL. However, as αCD and βCD complex *trans*-cinnamate **4.78** much more strongly than (*S*)-phenylalanine **4.75**, the addition of either CD decreases this inhibiting formation of **4.77** by sequestering *trans*-cinnamate **4.77** as the αCD or βCD complex **4.79** [81].

4.8. References

1. M. Komiyama and H. Shigekawa, *Cyclodextrins as Enzyme Models*, in *Comprehensive Supramolecular Chemistry*, eds. J. L. Atwood, J. E. D. Davies, D. D. MacNicol, F. Vögtle and J.-M. Lehn, Vol. 3, eds., J. Szetjli and T. Osa (Pergamon, Oxford, 1996) p. 401.
2. R. Breslow and S. D. Dong, *Chem. Rev.* **98** (1998) 1997.
3. R. Breslow, *J. Chem. Educ.* **75** (1998) 705.
4. R. L. VanEtten, J. F. Sebastian, G. A. Clowes and M. L. Bender, *J. Am. Chem. Soc.* **89** (1967) 3242.
5. R. L. VanEtten, G. A. Clowes, J. F. Sebastian and M. L. Bender, *J. Am. Chem. Soc.* **89** (1967) 3253.
6. O. S. Tee and J. J. Hoeven, *J. Am. Chem. Soc.* **111** (1989) 8318.
7. M. Komiyama and M. L. Bender, *J. Am. Chem. Soc.* **100** (1978) 4576.
8. V. B. Luzhkov and C. A. Venanzi, *J. Phys. Chem.* **99** (1995) 2312.
9. V. Luzhkov and J. Åqvist, *J. Am. Chem. Soc.* **120** (1998) 6131.
10. O. S. Tee, C. Mazza and X.-X. Du, *J. Org. Chem.* **55** (1990) 3603.
11. O. S. Tee, M. Bozzi, J. J. Hoeven and T. A. Gadosy, *J. Am. Chem. Soc.* **115** (1993) 8990.

12. T. A. Gadosy and O. S. Tee, *J. Chem. Soc., Perkin Trans. 2* (1995) 71.

13. O. S. Tee and J. B. Giorgi, *J. Chem. Soc., Perkin Trans. 2* (1997) 1013.

14. M. A. Fernandez and R. H. de Rossi, *J. Org. Chem.* **62** (1997) 7554.

15. O. S. Tee, A. A. Fedortchenko and P. L. Soo, *J. Chem. Soc, Perkin Trans. 2* (1998) 123.

16. E. Iglesias and A. Fernández, *J. Chem. Soc, Perkin Trans. 2* (1998) 1691.

17. C. J. Easton, S. Kassara, S. F. Lincoln and B. L. May, *Aust. J. Chem.* **48** (1995) 269.

18. K. Flohr, R. M. Paton and E. T. Kaiser, *J. Am. Chem. Soc.* **97** (1975) 1209.

19. G. L. Trainor and R. Breslow, *J. Am. Chem. Soc.* **103** (1981) 154.

20. R. Breslow, G. Trainor and A. Ueno, *J. Am. Chem. Soc.* **105** (1983) 2739.

21. Y. Ihara, E. Nakanishi, M. Nango and J. Koga, *Bull. Chem. Soc. Jpn.* **59** (1986) 1901.

22. R. Fornasier, F. Reniero, P. Scrimin and U. Tonellato, *J. Chem. Soc., Perkin Trans. 2* (1987) 193.

23. U. Tonellato, *Bull. Chim. Soc. Fr.* (1988) 277.

24. R. Ueoka, Y. Matsumoto, K. Harada, H. Akahoshi, Y. Ihara and Y. Kato, *J. Am. Chem. Soc.* **114** (1992) 8339.

25. C. J. Easton and S. F. Lincoln, *Chem. Soc. Rev.* (1996) 163.

26. F. M. Menger and M. J. Sherrod, *J. Am. Chem. Soc.* **110** (1988) 8606.

27. H.-J. Thiem, M. Brandl and R. Breslow, *J. Am. Chem. Soc.* **110** (1988) 8612.

28. J. H. Coates, C. J. Easton, S. J. van Eyk, B. L. May, P. Singh and S. F. Lincoln, *J. Chem. Soc., Chem. Commun.* (1991) 759.

29. J. H. Coates, C. J. Easton, N. L. Fryer and S. F. Lincoln, *Chem. Lett.* (1994) 1153.

30. F. Hirayama, K. Minami and K. Uekama, *J. Pharm. Pharmacol.* **48** (1996) 27.

31. K. Uekama, K. Minami and F. Hirayama, *J. Med. Chem.* **40** (1997) 2755.

32. K. Minami, F. Hirayama and K. Uekama, *J. Pharm. Sci.* **87** (1998) 715.

33. Y. Kitaura and M. L. Bender, *Bioorg. Chem.* **4** (1975) 237.

34. F. Cramer and G. Mackensen, *Angew. Chem., Int. Ed. Engl.* **5** (1966) 601.

35. Y. Iwakura, K. Uno, F. Toda, S. Onozuka, K. Hattori and M. L. Bender, *J. Am. Chem. Soc.* **97** (1975) 4432.

36. T. Ikeda, R. Kojin, C. Yoon, H. Ikeda, M. Iijima, K. Hattori and F. Toda, *J. Inclusion Phenom.* **2** (1984) 669.

37. T. Ikeda, R. Kojin, C. Yoon, H. Ikeda, M. Iijima and F. Toda, *J. Inclusion Phenom.* **5** (1987) 93.

38. H. Ikeda, R. Kojin, C. Yoon, T. Ikeda and F. Toda, *J. Inclusion Phenom. Mol. Recogn. Chem.* **7** (1989) 117.

39. K. Rama Rao, T. N. Srinivasan, N. Bhanumathi and P. B. Sattur, *J. Chem. Soc., Chem. Commun.* (1990) 10.

40. K. Hamasaki and A. Ueno, *Chem. Lett.* (1995) 859.

41. E. Anslyn and R. Breslow, *J. Am. Chem. Soc.* **111** (1989) 8931.

42. R. Breslow, J. B. Doherty, G. Guillot and C. Lipsey, *J. Am. Chem. Soc.* **100** (1978) 3227.

43. M. L. Bender, *J. Inclusion Phenom.* **2** (1984) 433.

44. V. T. D'Souza and M. L. Bender, *Acc. Chem. Res.* **20** (1987) 146.

45. R. Breslow and S. Chung, *Tetrahedron Lett.* **30** (1989) 4353.

46. S. C. Zimmerman, *Tetrahedron Lett.* **30** (1989) 4357.

47. R. Breslow, P. Bovy and C. L. Hersh, *J. Am. Chem. Soc.* **102** (1980) 2115.

48. E. Anslyn and R. Breslow, *J. Am. Chem. Soc.* **111** (1989) 5972.

49. R. Breslow, *Acc. Chem. Res.* **24** (1991) 317.

50. R. Breslow, *Isr. J. Chem.* **32** (1992) 23.

51. R. Breslow, *J. Mol. Cat.* **91** (1994) 161.

52. R. Breslow, *Pure Appl. Chem.* **66** (1994) 1573.

53. R. Breslow, *Recl. Trav. Chim. Pays-Bas* **113** (1994) 493.

54. R. Breslow, *Acc. Chem. Res.* **28** (1995) 146.

55. R. Breslow and C. Schmuck, *J. Am. Chem. Soc.* **118** (1996) 6601.

56. R. Breslow and A. Graff, *J. Am. Chem. Soc.* **115** (1993) 10988.

57. B. Siegel, A. Pinter and R. Breslow, *J. Am. Chem. Soc.* **99** (1977) 2309.

58. W. H. Binder and F. M. Menger, *Tetrahedron Lett.* **37** (1996) 8963.

59. S. Tamagaki, A. Katayama, M. Maeda, N. Yamamoto and W. Tagaki, *J. Chem. Soc., Perkin Trans. 2* (1994) 507.

60. S. Tamagaki, J. Narikawa and A. Katayama, *Bull. Chem. Soc. Jpn.* **69** (1996) 2265.

61. J. M. Desper and R. Breslow, *J. Am. Chem. Soc.* **116** (1994) 12081.

62. R. Breslow, J. Desper and Y. Huang, *Tetrahedron Lett.* **37** (1996) 2541.

63. D. Hilvert and R. Breslow, *Bioorg. Chem.* **12** (1984) 206.

64. R. Breslow and E. Kool, *Tetrahedron Lett.* **29** (1988) 1635.

65. W. Weiner, J. Winkler, S. C. Zimmerman, A. W. Czarnik and R. Breslow, *J. Am. Chem. Soc.* **107** (1985) 4093.

66. R. Breslow, M. Hammond and M. Lauer, *J. Am. Chem. Soc.* **102** (1980) 421.

67. R. Breslow and A. W. Czarnik, *J. Am. Chem. Soc.* **105** (1983) 1390.

68. A. W. Czarnik and R. Breslow, *Carbohydr. Res.* **128** (1984) 133.

69. I. Tabushi, Y. Kuroda, M. Yamada and H. Higashimura, *J. Am. Chem. Soc.* **107** (1985) 5545.

70. R. Breslow, A. W. Czarnik, M. Lauer, R. Leppkes, J. Winkler and S. Zimmerman, *J. Am. Chem. Soc.* **108** (1986) 1969.

71. R. Breslow, J. Chmielewski, D. Foley, B. Johnson, N. Kumabe, M. Varney and R. Mehra, *Tetrahedron* **44** (1988) 5515.

72. R. Breslow, J. W. Canary, M. Varney, S. T. Waddell and D. Yang, *J. Am. Chem. Soc.* **112** (1990) 5212.

73. M. Kojima, F. Toda and K. Hattori, *Tetrahedron Lett.* **21** (1980) 2721.

74. I. Tabushi and M. Kodera, *J. Am. Chem. Soc.* **109** (1987) 4734.

75. D. Rong, H. Ye, T. R. Boehlow and V. T. D'Souza, *J. Org. Chem.* **57** (1992) 163.

76. H. Ye, D. Rong, W. Tong and V. T. D'Souza, *J. Chem. Soc., Perkin Trans. 2*, (1992) 2071.

77. H. Ye, W. Tong and V. T. D'Souza, *J. Chem. Soc., Perkin Trans. 2* (1994) 2431.

78. A. G. Meyer, C. J. Easton, S. F. Lincoln and G. W. Simpson, *J. Chem. Soc., Chem. Commun.* (1997) 1517.

79. A. G. Meyer, C. J. Easton, S. F. Lincoln and G. W. Simpson, *J. Org. Chem.* **63** (1998) 9069.

80. C. J. Easton, C. M. M. Hughes, G. P. Savage and G. W. Simpson, *Adv. Heterocyclic Chem.* **60** (1994) 261.

81. C. J. Easton, J. B. Harper and S. F. Lincoln, *J. Chem. Soc., Perkin Trans. 1* (1995) 2525 .

CHAPTER 5

MODIFIED CYCLODEXTRINS

METALLOCYCLODEXTRINS

5.1. Cyclodextrins, Metal Complexes and Metallocyclodextrins

Metallocyclodextrins (metalloCDs) may be viewed as coordination compounds or metal complexes in which a modified CD acts as a ligand. In some cases this modification may simply amount to one or more of the secondary hydroxy groups of a native CD losing a proton to produce an alkoxide which coordinates a metal ion to form the simplest type of metalloCD. However, the majority of metalloCDs are formed from functionalised or modified CDs which incorporate one or more metal ion coordinating groups of varying degrees of complexity. Before considering the multitude of such metalloCDs that have appeared in the literature, it should be noted that native CDs can form complexes where only secondary bonding exists between the CD host and the metal complex guest and some of these are now discussed.

Square-planar cyclobutane-1,1-dicarboxylatodiamineplatinum(II) and αCD form a 1:1 complex in water where K_{11} = 60 kg mol^{-1}, ΔH^0 = -25.3 kJ mol^{-1}, and ΔS^0 = -42 J K^{-1} mol^{-1} as determined by microcalorimetry [1]. This complexation is also detected by ^{1}H NMR spectroscopy. In the solid state, X-ray crystallography shows cyclobutane-1,1-dicarboxylatodiamineplatinum(II) oriented with the cyclobutane ring protruding into the αCD annulus with its plane parallel to the αCD pseudo C_6 axis [2]. The two amine ligands are singly hydrogen bonded to secondary O(3)H groups of adjacent glucopyranose moieties so that Pt^{2+} is approximately in the plane of the six O(3) Hs. It is probable that the hydrophobic nature of the cyclobutane moiety makes a major contribution towards the stabilisation of this αCD complex in water. The cycloocta-1,5-diene (cod) ligands of square-planar cycloocta-1,5-dienediaminerhodium(I), [Rh(NH$_3$)$_2$cod], and cycloocta-1,5-diene-ethane-1,2-diaminerhodium(I), [Rh(en)cod], probably similarly stabilise the 1:1 complexes which they form with αCD in water. The latter complex is characterised by $K_{11} = 520$ kg mol^{-1}. Even so, the X-ray structure of αCD.[Rh(NH$_3$)$_2$cod] shows that the cod penetration of the αCD annulus is shallow in the crystalline state [3]. Thus, cod is positioned centrally in the αCD annulus with two carbons lying 0.88 Å and 1.08 Å below the mean plane of the twelve αCD secondary hydroxy groups and the bulk of [Rh(NH$_3$)$_2$cod] rests outside the αCD annulus.

Sometimes more than one metal complex may associate with a CD. Thus, electron spin resonance studies are consistent with bis(2-pyridylcarbinolato)copper(II) forming a 1:1 complex with αCD, a 2:1 complex with γCD, and complexes of both stoichiometries with βCD in frozen aqueous solution [4]. Circular dichroic studies at room temperature are also consistent with the complexation of bis(2-pyridyl-carbinolato)copper(II) by αCD and γCD in aqueous solution at room temperature.

The metalloCD Δ and Λ diastereomers of (αCD or βCD)bis(ethane-1,2-diamino)-cobalt(III), Δ and Λ[Co(αCD or βCD)(en)$_2$]$^+$, are formed when Co^{3+} coordinates to O$^-$(2) and O$^-$(3) of a single glucopyranose unit of doubly deprotonated αCD and βCD under basic conditions [5]. Thus, the coordination of three bidentate ligands by octahedral Co^{3+} results in [Co(αCD)(en)$_2$]$^+$, where αCD acts as a bidentate ligand. This confers either Δ or Λ chirality on the metalloCD which, combined with the homochirality of the CDs, produces diastereomers as shown by the resulting d-d band circular dichroic spectra. A similar situation holds for Δ and Λ[Co(βCD)(en)$_2$]$^+$, and both diastereomers of each complex may be separated by fractional crystallisation. A similar mode of coordination applies in (αCD or βCD)(1,4,7,10-tetraazacyclodo-decane)cobalt(III), [Co(αCD or βCD)(cyclen)]$^+$ whose circular dichroic spectra arise from the singular chiralities conferred by αCD or βCD.

The formal oxidation state of the cobalt centre can greatly influence the formation of metal complexes. Thus, the Co^{2+}-centred cobaltacene and carboxycobaltacene are complexed by βCD with $K_{11} = 2.0 \times 10^3$ and 1.8×10^3 dm^3 mol^{-1} at 298.2 K in aqueous solution, but their oxidised Co^{3+}-centred cobalticinium forms are not, probably because their increased positive charge decreases their hydrophobicity [6]. A similar explanation may apply to the anionic Fe^{2+}-centred ferrocenecarboxylic acid βCD complex where $K_{11} = 2.2 \times 10^3$ dm^3 mol^{-1} at 293.2 K in aqueous solution at pH 9.2 ($\Delta H^o = -12$ kJ mol^{-1}, $\Delta S^o = 21$ J K^{-1} mol^{-1}, and dissociation constant $k =$

2.1×10^4 s^{-1}) while the neutral oxidised Fe^{3+} ferrociniumcarboxylic acid βCD complex is not detected [7].

The complexation of deprotonated CDs by Cu^{2+} and Pb^{2+} in alkaline solution produces metalloCDs where more than one metal ion is bound per CD [8-12]. In the solid state, X-ray crystallography shows several such metalloCDs to consist of two CDs linked together through O(2) and O(3) atoms coordinated to metal ions. This linkage occurs in blue $Li_3[Li_3(H_2O)_3Cu_3(\alpha CDH_{-6})_2]$ (**5.1**) where alternating Cu^{2+} and Li^+ form a coordinating circle binding the secondary faces of two αCDH_{-6} together as shown schematically in Fig. 5.1 [12]. (The formalism αCDH_{-6} indicates the loss of six hydroxy protons.) Hydrogen bonding between adjacent secondary hydroxy and alkoxy groups forms two hydrogen bonding circles. A similar structure is found for $Na_3[Na_3Cu_3(\alpha CDH_{-6})_2]$. However, in $K_4[Cu_2(\alpha CDH_{-4})_2]$ and its Rb^+ analogue two Cu^{2+} coordinated to the A and D glucopyranose moieties of both αCDs

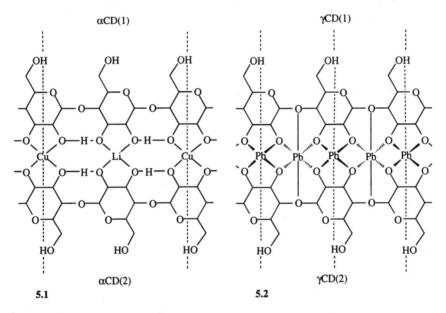

Fig. 5.1. One repeating unit of $[Li_3(H_2O)_3Cu_3(\alpha CDH_{-6})_2]Li_3$ is shown between the vertical broken lines of **5.1**. Each Li^+ has a H_2O bound to it and pointing out from the annulus (not shown) so that the coordination stereochemistry of Li^+ is midway between square pyramidal and trigonal bipyramidal. The coordination stereochemistry about Cu^{2+} is approximately tetrahedral. The three unbound Li^+ are not shown. Two repeating units of $[Pb_{16}(\gamma CDH_{-16})_2]$ incorporating equal numbers of Pb^{2+} in alternating sites are shown between the vertical broken lines of **5.2**.

and extensive hydrogen bonding holds the αCDH-4 secondary faces together. The K+ and Rb+ show little if any interaction with αCD in these structures. In Li11[Cu4(βCDH-11.5)2] four Cu2+ and seven Li+ coordinate O(2) and O(3) atoms to link two βCDH-11.5 together through their secondary faces [9]. Two adjacent Li+ interpose between pairs of Cu2+ except in one case where a single Li+ interposes which is a consequence of βCD being composed of seven glucopyranose moieties.

The highest ratio of metal ions to CD is found in [Pb16(γCDH-16)2] (**5.2**) whose structure is shown schematically in Fig. 5.1 [11]. The repeating units of [Pb16(γCDH-16)2] incorporate equal numbers of alternating Pb2+ sites. In one site Pb2+ is above a square array of two O(2) and two O(3) alkoxide groups and points outwards from the annulus, while in the other site Pb2+ is below a square array of two O(2) and two O(3) alkoxy groups and points inwards to the annulus with two O(1) atoms also in bonding distance. This results in a rather beautiful eight-pointed planar star-like array of Pb2+ which binds the two γCDH-16 together through their secondary faces.

5.2. Metallocyclodextrins of Modified Cyclodextrins

Most metalloCD studies concern the coordination of a metal ion by a modified CD to produce a binary metalloCD. Subsequently, a guest may be complexed in the CD annulus and may also be coordinated by the metal centre to give a ternary metalloCD as shown in Fig. 5.2. Under these circumstances an opportunity arises to study the effects of the metal centre and CD interactions on metalloCD stability and guest complexation. This is exemplified by the binary metallo-6A-(3-aminopropylamino)-6A-deoxy-βCD, [M(βCDpn)]2+, and metallo-6A-(2-(bis(2-aminoethyl)amino)ethylamino)-6A-deoxy-βCD, [M(βCDtren)]2+, and their complexation of tryptophan anion, Trp-, to form the ternary metalloCDs [M(βCDpn)Trp]+ and [M(βCDtren)Trp]+ [13-15]. The substitution of βCD at C(6) by -NH(CH2)3NH2 and -NH(CH2)2N((CH2)2NH2)2 results in strong M2+ coordination in the binary metalloCDs (Table 5.1) which, nevertheless, is not as strong as that in [M(pn)]2+ and [M(tren)]2+ where pn is 1,3-diaminopropane and tren is tris(2-aminoethyl)amine [16]. This is probably a consequence of differences in the electron donating powers of the secondary amine groups in βCDpn and βCDtren and the primary amine groups in pn and tren and the greater steric hindrance to M^{2+} coordination caused by βCDpn and βCDtren. The greater stabilities of $[M(\beta CDtren)]^{2+}$, by comparison with those of $[M(\beta CDpn)]^{2+}$, result from the tetra-dentate nature of βCDtren. This increase in metalloCD stability as the number of donor atoms of similar type increases in the chelating group(s) substituted on CDs is consistent with expectations arising from coordination chemistry. This is well illustrated for Cu^{2+} where $\log(K/dm^3 \ mol^{-1}) = 7.35, 13.79$ and 17.29 for $[Cu(\beta CDpn)]^{2+}$, $[Cu(\beta CDdien)]^{2+}$ and $[Cu(\beta CDtren)]^{2+}$, respectively, where tridentate βCDdien is 6^A-(5-amino-3-azapentylamino)-6^A-deoxy-βCD [14,15,17].) The variations of stability with the nature of M^{2+} for both binary metallocyclo-dextrins arise through a combination of M^{2+} size and ligand-field variations.

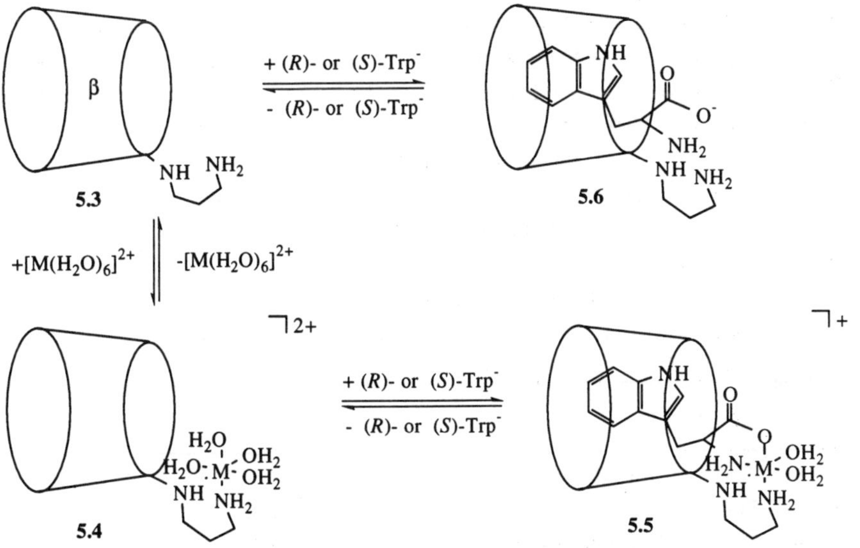

Fig. 5.2. The coordination of βCDpn (**5.3**) by M^{2+} to form a binary metalloCD $[M(\beta CDpn)]^{2+}$ (**5.4**) and subsequently a ternary metalloCD (**5.5**), $[M(\beta CDpn)(S)\text{-}Trp]^+$ and $[M(\beta CDpn)(R)\text{-}Trp]^+$, through the complexing of either (S)- or (R)-tryptophan anion. These anions may also be complexed by βCDpn to form either βCDpn.(S)-Trp⁻ or βCDpn.(R)-Trp⁻ (**5.6**).

Table 5.1. Stability Constants (K) for MetalloCDs of βCDpn and βCDtren[a] and Related Species.

Equilibrium	M^{2+} and $\log(K/dm^3\ mol^{-1})$[b]			
	Co^{2+}	Ni^{2+}	Cu^{2+}	Zn^{2+}
$M^{2+} + pn \rightleftharpoons [M(pn)]^{2+}$		6.31	9.75	
$M^{2+} + tren \rightleftharpoons [M(tren)]^{2+}$	12.7	14.6	18.5	14.5
$M^{2+} + βCDpn \rightleftharpoons [M(βCDpn)]^{2+}$	4.22	5.2	7.35	4.96
$M^{2+} + βCDtren \rightleftharpoons [M(βCDtren)]^{2+}$		11.65	17.29	12.25
$M^{2+} + βCDpnH^+ \rightleftharpoons [M(βCDpnH)]^{3+}$	2.5	3.1	3.09	3.0
$M^{2+} + βCDtrenH^+ \rightleftharpoons [M(βCDtrenH)]^{3+}$		8.46	11.56	7.92
$M^{2+} + Trp^- \rightleftharpoons [M(Trp)]^+$	4.41	5.42	8.11	4.90
$[M(βCDpn)]^{2+} + (R)\text{-}Trp^- \rightleftharpoons [M(βCDpn)(R)\text{-}Trp]^+$	4.04	4.1	7.85	5.3
$[M(βCDpn)]^{2+} + (S)\text{-}Trp^- \rightleftharpoons [M(βCDpn)(S)\text{-}Trp]^+$	4.32	5.1	8.09	5.3
$[M(βCDtren)]^{2+} + (R)\text{-}Trp^- \rightleftharpoons [M(βCDtren)(R)\text{-}Trp]^+$		8.2	9.5	8.1
$[M(βCDtren)]^{2+} + (S)\text{-}Trp^- \rightleftharpoons [M(βCDtren)(S)\text{-}Trp]^+$		8.1	9.4	8.3
$[M(βCDtren)]^{2+} + (R)\text{-}TrpH \rightleftharpoons$ $[M(βCDtren)(R)\text{-}TrpH]^{2+}$		4.6	4.3	
$[M(βCDtren)]^{2+} + (S)\text{-}TrpH \rightleftharpoons$ $[M(βCDtren)(S)\text{-}TrpH]^{2+}$		4.3	4.2	
$[M(βCDtrenH)]^{3+} + (R)\text{-}TrpH \rightleftharpoons$ $[M(βCDtrenH)(R)\text{-}TrpH]^{3+}$		3.56	4.4	4.82
$[M(βCDtrenH)]^{3+} + (S)\text{-}TrpH \rightleftharpoons$ $[M(βCDtrenH)(S)\text{-}TrpH]^{3+}$		3.6	4.4	4.96

Equilibrium not involving M^{2+}	$\log(K/dm^3\ mol^{-1})$
$βCD + (R)\text{-}Trp^- \rightleftharpoons βCD.(R)\text{-}Trp^-$	2.33
$βCD + (S)\text{-}Trp^- \rightleftharpoons βCD.(S)\text{-}Trp^-$	2.33
$βCDpn + (R)\text{-}Trp^- \rightleftharpoons βCDpn.(R)\text{-}Trp^-$	3.41
$βCDpn + (S)\text{-}Trp^- \rightleftharpoons βCDpn.(S)\text{-}Trp^-$	3.40
$βCDtren + (R)\text{-}Trp^- \rightleftharpoons βCDtren.(R)\text{-}Trp^-$	6.36
$βCDtren + (S)\text{-}Trp^- \rightleftharpoons βCDtren.(S)\text{-}Trp^-$	6.5

[a]CDpn is 6^A-(3-aminopropylamino)-6^A-deoxy-βCD and βCDtren is 6^A-(2-(*N*,*N*-bis(2-aminoethyl)-amino)ethylamino)-6^A-deoxy-βCD. [b]In aqueous solution at 298.2 K and $I = 0.10$ (NaClO$_4$).

The complexation of (R)-Trp$^-$ and (S)-Trp$^-$ by [Ni(βCDpn)]$^{2+}$ shows a 10-fold chiral discrimination in favour of [Ni(βCDpn)(S)-Trp]$^+$ over [Ni(βCDpn)(R)-Trp]$^+$ while the Co^{2+} and Cu^{2+} analogues show lesser discrimination, and the Zn^{2+} analogue shows none [13,14]. The influence of M^{2+} on chiral discrimination coincides with the variation in the ionic radii of six-coordinate Co^{2+}, Ni^{2+}, Cu^{2+} and Zn^{2+}, which are 0.745, 0.69, 0.73 and 0.74 Å, respectively, and the stereochemical constraints arising from ligand field effects in Co^{2+}, Ni^{2+} and Cu^{2+}. It is noteworthy that [Zn(βCDpn)(R)-Trp]$^+$ and [Zn(βCDpn)(S)-Trp]$^+$ are of the same stability, while the analogous diastereomeric complexes of the other three metal ions differ in stability. This may indicate that the absence of ligand field generated stereochemical constraints on d^{10} Zn^{2+} allows more flexibility in the structures adopted by [Zn(βCDpn)(R)-Trp]$^+$ and [Zn(βCDpn)(S)-Trp]$^+$ and as a result enantioselectivity is negligible. In contrast, the d^9 electronic configuration for similar-sized Cu^{2+} imposes a tetragonally distorted octahedral stereochemistry. This probably places greater constraints on the interaction of the chiral centres of (R)-Trp$^-$ and (S)-Trp$^-$ with the βCDpn moiety and decreases the stability of [Cu(βCDpn)(R)-Trp]$^+$ by comparison with that of [Cu(βCDpn)(S)-Trp]$^+$. Similar arguments apply in the cases of d^7 Co^{2+} and d^8 Ni^{2+} whose six-coordinate stereochemistries more closely approach regular octahedra. The greater enantioselectivity observed for Ni^{2+} may indicate that the size of the metal centre is important, and that a difference of 0.04 Å can produce a substantial change in the degree of enantioselectivity. The major influence of M^{2+} in chiral discrimination in these systems is demonstrated by the lack of chiral discrimination in the βCDpn.(S)-Trp$^-$ and βCDpn.(R)-Trp$^-$ complexes. Similar variations in chiral discrimination are seen in the analogous phenylalanine anion metalloCDs [18].

The major factors affecting the stability of the ternary metalloCDs appear to be: (i) the hydrophobic interactions between the interior of the βCD annulus and the guest, (ii) the coordination of the guest to the metal centre, and (iii) the interaction of the guest's chiral centre with the chirality of βCD. Significant thermodynamic chiral discrimination is only likely to occur when (iii) makes a significant and different contribution to the stabilities of the diastereomeric ternary metalloCDs for enantiomeric guests. Thus, the absence of chiral discrimination in [M(βCDtren)(R)-Trp]$^+$ and [M(βCDtren)(S)-Trp]$^+$ is probably a consequence of factors (i) and (ii) dominating despite a considerable increase in stability over that of $[M(\beta CDpn)(R)\text{-}Trp]^+$ and $[M(\beta CDpn)(S)\text{-}Trp]^+$.

The effect of protonation of the guest is shown by $[M(\beta CDtren)(R)\text{-}TrpH]^{2+}$ where the monodentate tryptophan (TrpH) coordinates less strongly than bidentate Trp^- in the more stable $[M(\beta CDtren)(R)\text{-}Trp]^+$. The $[M(\beta CDpn)(R)\text{-}TrpH]^{2+}$ species is not detected. This probably reflects the lower stability of $[M(\beta CDpn)]^{2+}$ by comparison with that of $[M(\beta CDtren)]^{2+}$ where the tetradentate tren substituent coordinates M^{2+} much more strongly than does the bidentate pn substituent.

The stabilities of $\beta CDtren.(R)\text{-}Trp^-$ and $\beta CDtren.(S)\text{-}Trp^-$ are *ca.* 1000-fold greater than those of $\beta CDpn.(R)\text{-}Trp^-$ and $\beta CDpn.(S)\text{-}Trp^-$ which are *ca.* 10-fold greater than those for $\beta CD.(R)\text{-}Trp^-$ and $\beta CD.(S)\text{-}Trp^-$ as is seen from Table 5.1. These variations are attributable to the interactions of the Trp^- amino and carboxylate groups with the primary face of the βCD annulus such that Trp^- egress is hindered more than ingress with the substitution of a polyamine at $C(6)$. The stabilities of $[M(\beta CDtren)(R)\text{-}Trp]^+$ and $[M(\beta CDtren)(S)\text{-}Trp]^+$ are greater than those of the analogous $MTrp^+$ and $\beta CDtren.Trp^-$ consistent with the coordination of Trp^- by M^{2+} and the interaction of Trp^- with the βCD annulus being mutually reinforcing and stabilising $[M(\beta CDtren)\text{-}(R)\text{-}Trp]^+$ and $[M(\beta CDtren)(S)\text{-}Trp]^+$. In contrast, while the stabilities of $[M(\beta CDpn)(R)\text{-}Trp]^+$ and $[M(\beta CDpn)(S)\text{-}Trp]^+$ are greater than those of $\beta CDpn.Trp^-$, indicating the stabilising effect of coordination of Trp^- by M^{2+}, they more closely approach those of $MTrp^+$ which is consistent with significant competition between the Trp^- complexing effects of the βCD annulus and M^{2+} in these ternary metalloCDs [14].

The less labile Pt^{2+} is also coordinated by $\beta CDpn$ to form *cis*-dichloro(6^A-(3-aminopropylamino)-6^A-deoxy-βCD)platinum(II) where square-planar Pt^{2+} is coordinated by two amine nitrogens and two chlorides [19]. The closely related *cis*-dichloro($6^A,6^B$-diamino-$6^A,6^B$-dideoxy-βCD)platinum(II) [20], *cis*-dichloro(6^A-(2-aminoethylamino)-6^A-deoxy-βCD)platinum(II), aminodichloro(6^A-amino-6^A-deoxy-βCD)platinum(II) and the γCD analogues of the latter two have also been reported [19,21]. These platinum(II) metalloCDs are analogues of the cisplatin anti-cancer drug, *cis*-$[Pt(NH_3)Cl_2]$, but they have no significant anti-cancer activity probably because they do not penetrate the cell membrane.

MetalloCDs exhibit the usual forms of isomerism shown by other metal complexes. This is exemplified by $[ReBr(CO)_3(TM\beta CDbpy)]$, the Re^+ metalloCDs of TMβCDbpy, where all of the βCD hydroxy groups are replaced by methoxy groups except one at C(6) which is linked through O(6) to a bipyridyl moiety which chelates octahedral Re⁺ as shown in **5.7-5.10** [22]. Thus, **5.7** and **5.8**, **5.9** and **5.10**, **5.7** and **5.10**, and **5.8** and **5.9** are diastereomers, while **5.7** and **5.9** are rotamers, as are **5.8** and **5.10**, which may interconvert by rotation about the linker between the bipyridyl moiety and the TMβCDbpy C(6). In practice, ^{1}H NMR spectroscopy shows resonances for only two diastereomers which probably indicates that **5.7** and **5.9** and **5.8** and **5.10** are in fast exchange in CDCl$_3$ at 303.2 K. The link between O(6) and the main body of TMβCDbpy may either be extended [23] or two CD moieties may be attached to the bipyridyl coordinating group [24] as shown in **5.11** and **5.12**, respectively.

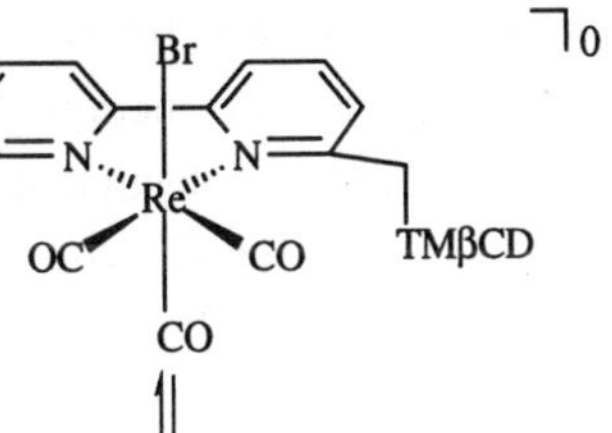

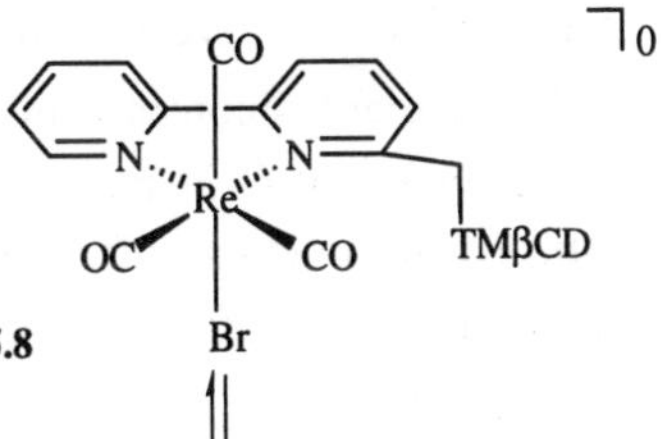

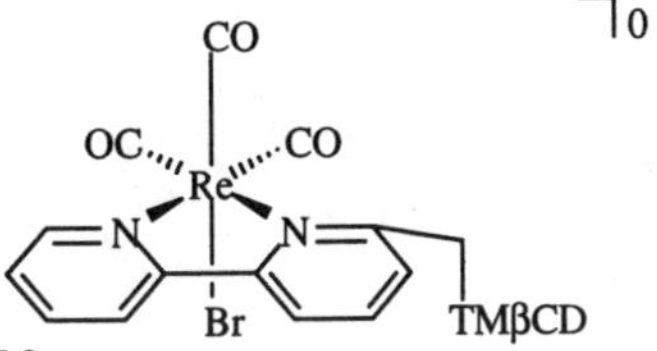

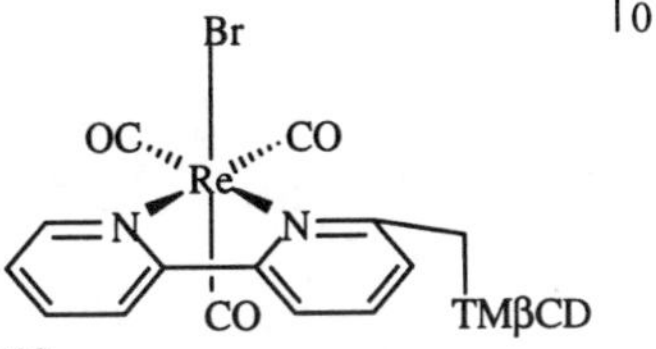

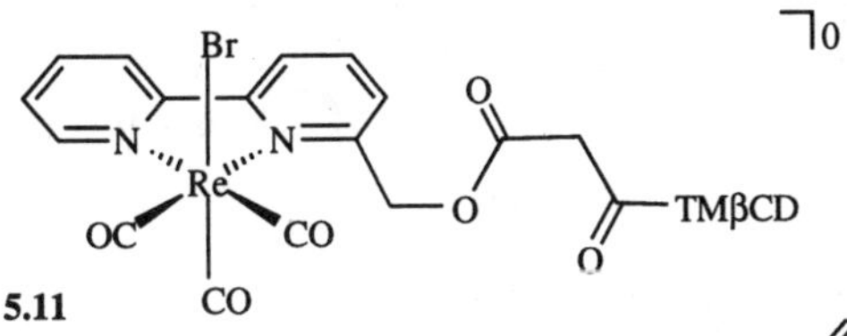

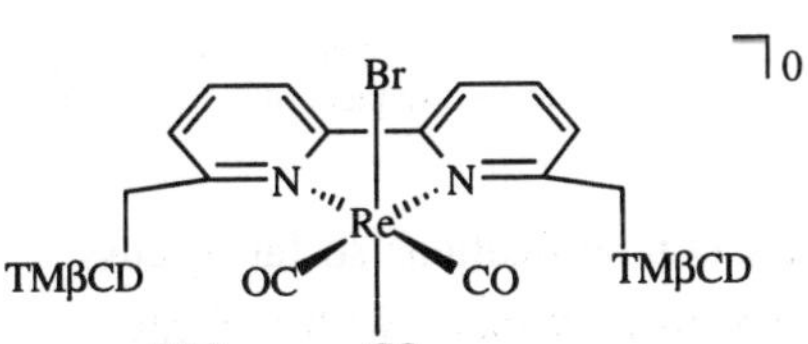

The multiple substitution of CDs with metal ion coordinating groups is illustrated by $6^A,6^C,6^E$-trideoxy-$6^A,6^C,6^E$-tris(2,3-dihydroxybenzamido)pentadeca-O-methyl-αCD **5.13** (where the three $6^B,6^D$ and 6^F methyl groups at O(6) and the twelve at O(2) and O(3) are not shown). It has been synthesised to incorporate some characteristics of natural siderophores such as enterobactin, parabactin and agrobactin by coordinating Fe^{3+} and Al^{3+} while binding a guest in the αCD annulus [25]. All six phenol groups are thought to coordinate these metal ions consistent with the very large $K_{11} = 10^{39}$ dm^3 mol^{-1} characterising the coordination of Fe^{3+} by **5.13** in aqueous solution. This compares with an even greater value of $K_{11} = 10^{52}$ dm^3 mol^{-1} for enterobactin. ^{1}H NMR studies indicate that when Al^{3+} is coordinated by **5.13**, 4-nitrophenolate binds in the αCD annulus. A related modification is seen in $6^A,6^B,6^D,6^E$-tetradeoxy-$6^A,6^B,6^D,6^E$-tetra(O-nicotinyl)tetradeca-O-methyl-αCD which in dimethyl sulfoxide coordinates two *cis*-PtCl$_2$ moieties through pairs of immediately adjacent nicontinyl nitrogens (**5.14** where the two methyl groups at O(6) and the twelve at O(2) and O(3) are not shown) [26]. Molecular modelling shows both PtCl$_2$ moieties of **5.14** to lie above and outside the CD rim as is also the case for CoCl$_2$ and CuCl$_2$ in its Co^{2+} and Cu^{2+} analogues [27].

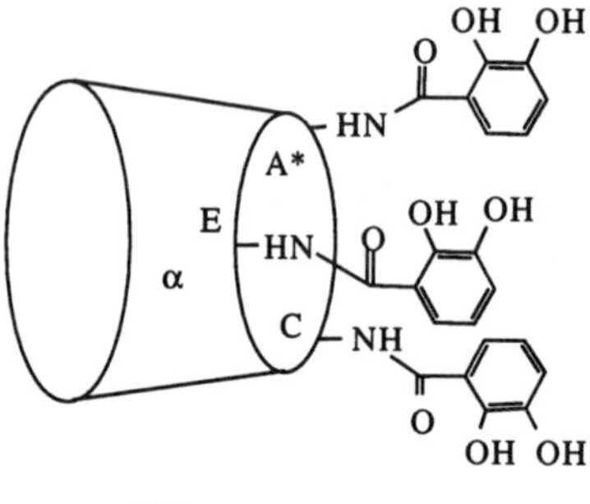

5.13 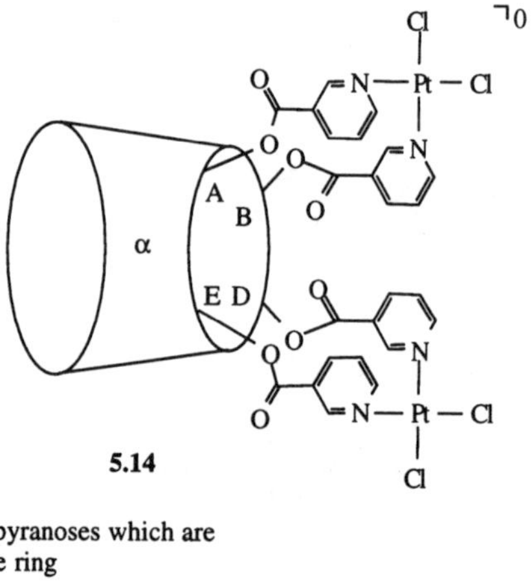   **5.14**

*denotes the substituted glucopyranoses which are lettered sequentially around the ring

The CD modifications so far discussed and the many more which are discussed below are usually made with the intention of manipulating the interactions of the coordinated metal ion and a guest complexed in the CD annulus. These interactions result in major themes that pervade metalloCD chemistry. One arises from the homochiral nature of CDs and is the tendency for chiral discrimination to manifest itself in a variety of ways. Another arises from the close proximity of the metal centre and the hydrophobic guest-complexing cavity, a characteristic found in metalloenzymes, which has lead to extensive studies of the catalytic and biomimetic aspects of metalloCDs. These same characteristics have also led to the study of electron transfer and energy transfer between the metal centre and guest species. These aspects are now explored.

5.3. Metallocyclodextrins and Chiral Discrimination

Chiral discrimination in ternary metalloCDs is very dependent on the nature of the metal ion and the coordinating group. It also shows a considerable dependence on the nature of the chiral guest. Some of these aspects are illustrated by the complexation of amino acids by 6^A-[2-(imidazol-4-yl)ethylamino]-6^A-deoxy-βCDcopper(II), [Cu(βCDhm)]$^{2+}$ [28], and its use as a chiral discriminating agent added to the mobile phase in LEC HPLC studies (LEC = ligand exchange chromatography) [29,30]. Thus, the (R)-enantiomers of tyrosine, phenylalanine and tryptophan elute ahead of the (S)-enantiomers where the ratios of their elution rates, α = 1.10, 1.12, and 1.23, respectively. This is attributed to the guest (R)-amino acid anions forming more stable ternary metalloCDs because their aromatic moieties enter the βCD annulus of the ternary metalloCDs whereas those of the (S)-enantiomers do not, as is exemplified by [Cu(βCDhm)((R)-Trp)]$^+$ (**5.15**) and [Cu(βCDhm)((S)-Trp)]$^+$ (**5.16**), respectively. This deduction is supported by circular dichroic spectroscopic solution studies which show a much stronger Cotton effect for the ternary metalloCDs formed with the aromatic (R)-amino acid anions than for their (S)-diastereomers. It is also consistent with the structure of **5.16**, determined by X-ray crystallography, where the tryptophan anion aromatic moiety is outside the βCD annulus [31]. The five-coordinate Cu^{2+} assumes a distorted square-pyramidal stereochemistry and is sited 0.17 Å above the basal plane of three nitrogen and an oxygen donor atoms with a water molecule (Cu-O distance = 2.35 Å) occupying the fifth coordination site. It may be that Cu^{2+} in **5.15** also becomes five-coordinate through coordination of a water in an apical site.

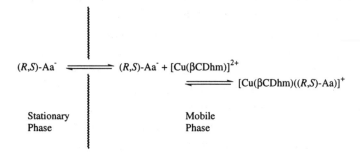

In LEC HPLC studies, the amino acid anions participate in a partitioning equilibrium between the mobile aqueous phase and the non-aqueous stationary phase, while the binary and ternary metalloCDs are insoluble in the latter phase as shown in Fig. 5.3. The enantiomers which form the most stable ternary metalloCDs spend less time in contact with the HPLC column and elute first. However, the enantiomers of the aliphatic amino acids alanine, proline, and leucine are not separated by this method which indicates the importance of the presence of an aromatic moiety in the guest to engender enantioselectivity through complexation in these systems.

$$(R,S)\text{-Aa}^- \rightleftharpoons (R,S)\text{-Aa}^- + [\text{Cu}(\beta\text{CDhm})]^{2+}$$
$$\rightleftharpoons [\text{Cu}(\beta\text{CDhm})((R,S)\text{-Aa})]^+$$

Stationary   Mobile
Phase   Phase

Fig. 5.3. The principle of ligand exchange chromatography (LEC) HPLC where $(R,S)\text{-Aa}^-$ is an amino acid anion.

The potentiometrically determined $\log(\beta/\text{dm}^6\,\text{mol}^{-2})$ values shown in parentheses for the (S)- and (R)-amino acid anions, respectively, are: alanine (15.53 and 15.51), leucine (14.89 and 14.96), norvaline (14.80 and 14.87), phenylalanine (15.68 and 15.85), tyrosine (14.82 and 15.22), tryptophan (16.12 and 16.47), and histidine

(16.78 and 16.70), where $\beta = [Cu(\beta CDhm)(guest)^+]([Cu^{2+}][\beta CDhm][guest])^{-1}$ [30]. These data independently show that a more substantial selectivity for the (R)-enantiomer over the (S)-enantiomer occurs for the aromatic amino acid anions than for the aliphatic amino acid anions.

In contrast to $[Cu(\beta CDhm)]^{2+}$, LEC HPLC studies show that 6^A-[4-(2-aminoethyl)imidazol-1-yl]-6^A-deoxy-βCDcopper(II), $[Cu(\beta CDmh)]^{2+}$, causes (S)-Trp⁻ to elute before (R)-Trp⁻ with an $\alpha = 2.4$ [32]. This reversal of chiral discrimination is attributed to the higher stability of the (S)-Trp⁻ ternary metalloCD which is thought to complex the aromatic moiety of the guest inside the βCD annulus of $[Cu(\beta CDmh)((S)\text{-Trp})]^+$ (**5.17**), whereas that of its less stable (R)-Trp⁻ analogue, $[Cu(\beta CDmh)((R)\text{-Trp})]^+$ (**5.18**) does not. The greater enantioselectivity of $[Cu(\beta CDmh)]^{2+}$ is attributed to **5.17** and **5.18** being more rigid than **5.15** and **5.16** combined with the preference for the amino groups of the modified βCD and of the amino acid anion to coordinate Cu^{2+} in *cis* positions.

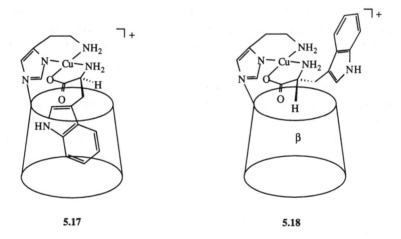

5.17   **5.18**

The subtle nature of chiral discrimination between guests is illustrated by two further metalloCD systems. The first example is 6^A-(2-aminoethylamino)-6^A-deoxy-βCDcopper(II), $[Cu(\beta CDen)]^{2+}$, which shows no thermodynamic enantioselectivity for alanine, phenylalanine, and trytophan anions in forming ternary metalloCDs [33]. However, it does cause a partial LEC HPLC separation of tryptophan anions with the (S)-enantiomer eluting first, but no separation was observed for the anions of the other two amino acids.   This is consistent with amplification of a small enantioselectivity by chromatography for tryptophan anion.

The second example is the three isomeric Cu^{2+} metalloCDs formed with $6^A,6^X$-diamino-$6^A,6^X$-dideoxy-βCD, $[Cu(βCD-A,X-(NH_2)_2)]^{2+}$, where X is either B, C or D [34].   The $[Cu(βCD-A,X-(NH_2)_2)]^{2+}$ binary metalloCDs and their ternary metalloCDs with (R)- and (S)-alanine, phenylalanine, tyrosine and tryptophan anions have been characterised in solution by ESR and circular dichroic spectroscopy.   It appears that the aromatic moieties of the guest amino acid anions can enter the βCD annulus of the ternary metalloCDs formed by $[Cu(βCD-A,B-(NH_2)_2)]^{2+}$, but not the βCD annuli of those formed by the other two isomers because Cu^{2+} coordination of both amino groups of $βCD-A,C-(NH_2)_2$ and $βCD-A,D-(NH_2)_2$ blocks their primary faces.   (Such bidentate coordination is observed in the X-ray crystal structure of *cis*-dichloro(6^A-(2-aminoethylamino)-6^A-deoxy-βCD)platinum(II) where Pt^{2+} is at the centre of a square plane delineated by the 6^A- and 6^B-amine groups and two chloride ligands [35]).   In accord with this interpretation, the (R)-enantiomers of the tyrosine, tryptophan and phenylalanine anions elute ahead of the (S)-enantiomers where the ratios of their elution rates, α = 1.50, 1.06 and 1.18, respectively, when $[Cu(βCD-A,B-(NH_2)_2)]^{2+}$ is in the mobile phase in LEC HPLC studies.   Interestingly, no enantiomeric separation is observed for *o*- and *m*-hydroxyphenylalanine which indicates the critical role of the position of the OH groups of the guest in the chiral discrimination pattern.   The aliphatic alanine and leucine anions are similarly not separated.   When either $[Cu(βCD-A,C-(NH_2)_2)]^{2+}$ or $[Cu(βCD-A,D-(NH_2)_2)]^{2+}$ is added to the mobile phase in LEC HPLC studies, no enantiomeric separation is observed.

Bearing some similarity to $[Cu(βCD-A,C-(NH_2)_2)]^{2+}$, in its ability to coordinate Cu^{2+} through donor groups sited on two C(6) centres, is 6^A6^C-*cyclo*-(L-histidyl-L-histidyl)-6^A6^C-dideoxy-βCD where the *cyclo*-L-histidyl-L-histidyl moiety links two C(6) sites across the primary face of βCD and coordinates Cu^{2+} through the imidazole nitrogens [36].   This, in turn, is related to 6^A-*cyclo*-(L-histidyl-L-histidyl)-6^A-deoxy-βCD where the *cyclo*-L-histidyl-L-histidyl moiety is bound at C(6) and also coordinates Cu^{2+} [37], and 6^A-*cyclo*-(L-histidyl-L-leucyl)-6^A-deoxy-βCD where the *cyclo*-L-histidyl-L-leucyl moiety is likewise bound at C(6) [37,38].

The site of substitution of the coordinating group of a modified CD is important in the interaction of its metalloCDs with chiral guests.   This is illustrated by the reaction of 6^A-(2-aminoethylamino)-6^A-deoxy-βCD with diethylenetriaminepenta-acetic dianhydride to produce a multidentate oxygen and nitrogen donor group which coordinates Dy^{3+} to give the binary metalloCD **5.19**, and by an analogous reaction of 2^A-(2-aminoethylamino)-2^A-deoxy-βCD which leads to **5.20** [39]. Dysprosium(III) possesses paramagnetic properties which induce large ^1H NMR chemical shift changes without significant broadening of the ^1H resonances of coordinated ligands. Its complexes are consequently used as NMR shift reagents, as are those of Eu^{3+}. Because of the homochirality of the C(6)-substituted βCD, **5.19** is a chiral shift reagent which significantly increases the ^1H NMR chemical shift differences for the enantiomers of aspartame, tryptophan, propranolol, and 1-anilino-8-naphthalenesulfonate, compared with those observed in the presence of βCD alone. The corresponding C(2)-substituted metallo-βCD **5.20** induces a greater chemical shift difference than does **5.19**. It appears that in both cases the enantiomeric guests complex with a major portion of their aromatic moieties inside the βCD annulus, and that either K_{11} for the ternary metalloCD formation is greater for **5.20** than for **5.19** so that a greater proportion of the guest is complexed, or that the stereochemistry of the ternary metalloCD formed by **5.20** produces a greater Dy^{3+} induced change in chemical shift, or both.

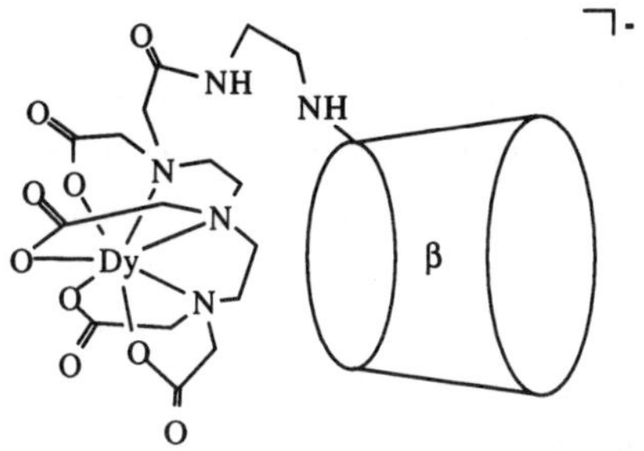

5.19 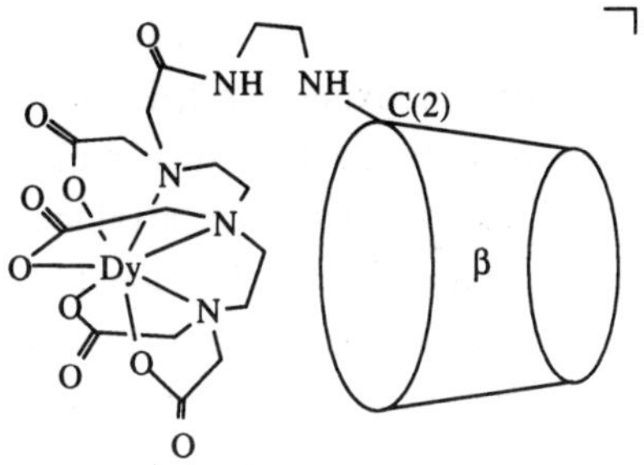   **5.20**

5.4. Catalytic Metallocyclodextrins: Mono-Cyclodextrin Systems

As in many metalloenzymes, binary metalloCDs incorporate a metal centre in close proximity to a hydrophobic cavity capable of complexing a guest to form a ternary metalloCD which resembles a Michaelis metalloenzyme complex or holoenzyme-substrate complex. Accordingly, it is expected that metalloCDs will act as artificial enzymes or as metalloenzyme mimics to some extent [40-49]. It should be noted, however, that metalloenzymes have optimised their active site-substrate stereochemistry over millions of years, and it is to be expected that substantial misalignments may occur between the CD, the metal centre and the guest in ternary metalloCDs chosen as potential enzyme mimics, and that their catalytic activities and selectivities will be relatively low as a consequence. Nevertheless, increasingly sophisticated and selective metalloCD enzyme mimics have been studied over the past thirty years or so, particularly by Breslow and co-workers [41-43,49]. This sophistication has been achieved through the attachment of a wide range of metal coordinating groups to CDs, and increasingly through the linking of two or more CDs to metal coordinating groups. Accordingly, these metalloCDs conveniently fall into two classes: those incorporating a single CD and those incorporating two or more CDs, and it is with the first group that this discussion of catalytic metalloCDs begins.

The first reported catalysis by a metalloCD appears to be that of the hydrolysis of 4-nitrophenyl acetate by the αCD-based Ni^{2+} metalloCD **5.21** [50]. Hydrolysis is accelerated > 1000-fold over the uncatalysed rate, and proceeds through acylation of the pyridinecarboxaldoxime ligand followed by deacylation of the resulting acetate. However, the catalysis by **5.21** is only 4-fold more effective than that caused by the pyridinecarboxaldoximenickel(II) complex. It appears that while the αCD annulus of **5.21** assists in the catalysis by complexing 4-nitrophenyl acetate in close proximity to the attacking pyridinecarboxaldoxime oxygen, either significant freedom of movement exists for 4-nitrophenyl acetate in the αCD annulus or the complex stereochemistry is not optimal for catalysis and that its catalytic effect is quite small.

The importance of the orientation of the metal centre and the guest in the ternary metalloCD is considerable as is shown by the ≥ 1000-fold rate acceleration of the hydrolysis of 4-nitrophenyl acetate over the uncatalysed rate ($k_u = 1.3 \times 10^{-6} \ s^{-1}$) caused by 6^A-deoxy-6^A-(1,4,7,10-tetraazadodec-1-yl)-βCDcobalt(III) (**5.22**), and the *ca.* one half less effective catalysis caused by 3^A-deoxy-3^A-(1,4,7,10-tetraazadodec-1-yl)-βCDcobalt(III) (**5.23**) at pH 7 [51,52]. The catalytic effectiveness of **5.22** is greatest at pH 7 coincident with deprotonation of a water ligand coordinated to Co^{3+} to produce the nucleophilic hydroxo ligand which attacks the carbonyl carbon of 4-nitrophenyl acetate. It appears that the lesser catalytic effectiveness of **5.23** may result from a less favourable orientation of the Co^{3+} complex moiety with respect to the 4-nitrophenyl acetate guest. (The catalytic activity of **5.23** is reported to be much decreased after column chromatography [52].)

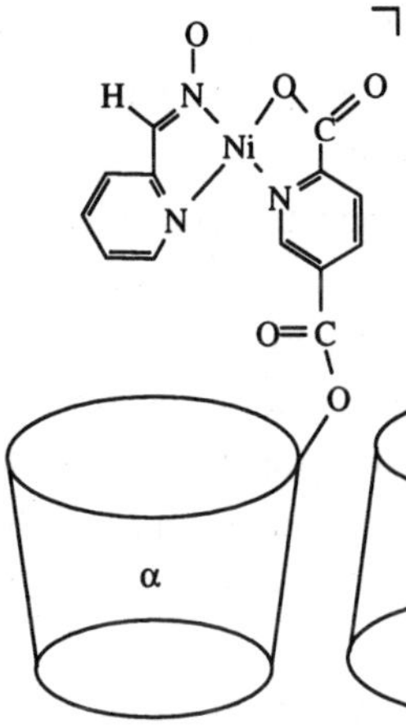

| 5.21 | 5.22 | 5.23 | 5.24 |

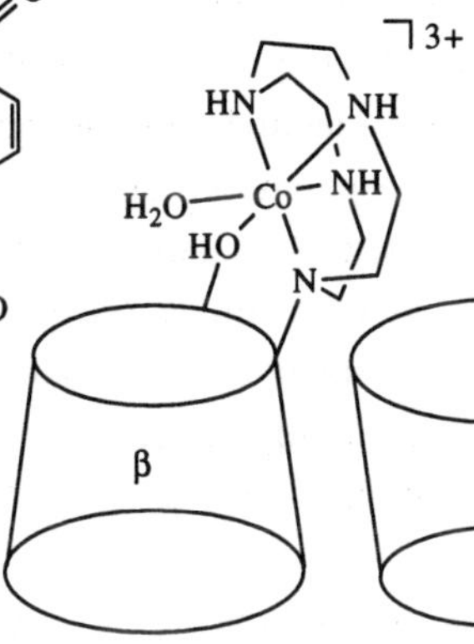

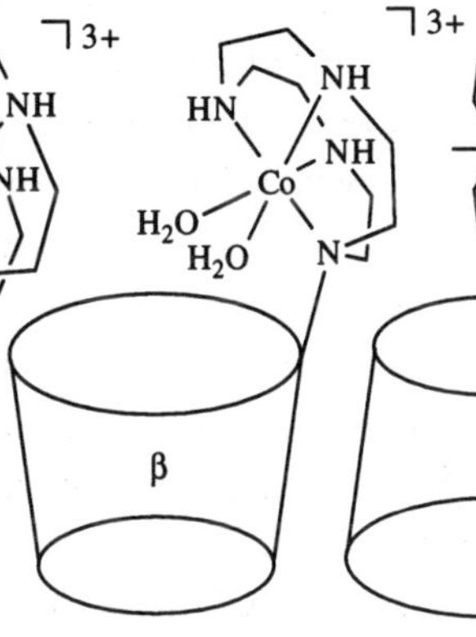

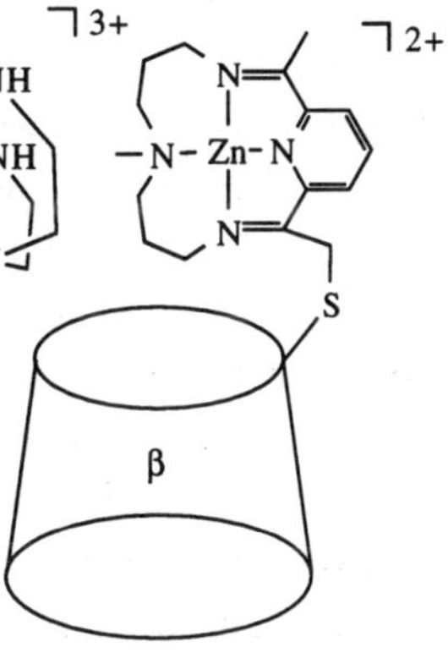

The $[Co(cyclen)(OH)(H_2O)]^{2+}$ complex alone (where cyclen is 1,4,7,10-tetraaza-cyclododecane) has no catalytic effect, but 6^A-deoxy-6^A-(1,4,7,10-tetraazadodec-1-yl)-βCD causes an 8.6-fold hydrolysis rate acceleration under similar conditions. However, in the latter case it is a nitrogen of the 1,4,7,10-tetraazadodec-1-yl moiety which acts as the nucleophile and becomes acylated [52]. In contrast to **5.22**, its Ni^{2+}, Cu^{2+} and Zn^{2+} analogues cause only 16-, 14-, and 12-fold accelerations of hydrolysis of 4-nitrophenyl acetate at pH 7 which indicate the lesser effectiveness of these metal centres in this catalysis [53]. At higher pHs these divalent analogues of **5.22** become less effective and it may be that precipitation of the hydroxides of the divalent metal ions leave predominantly 6^A-deoxy-6^A-(1,4,7,10-tetraazadodec-1-yl)-βCD in solution. Such precipitation of the hydroxides of labile transition metal ions is always a possibility in studies of their metalloCDs at high pH.

The hydrolysis of 4-nitrophenyl carbonate and 4-nitrophenyl phosphate is accelerated by factors of 2.9×10^3 and 3.7×10^3, respectively, in the presence of either 10^{-3} mol dm^{-3} **5.22** or **5.23** compared with $k_u = 1.9 \times 10^{-8}$ s^{-1} at pH = 7.0 and 298.2 K [51]. However, $[Co(cyclen)(OH)(H_2O)]^{2+}$ accelerates the hydrolysis of

4-nitrophenyl phosphate about 20 times more than either of these metalloCDs which is attributed to 6^A-deoxy-6^A-(1,4,7,10-tetraazadodec-1-yl)-βCD hindering 4-nitrophenyl phosphate coordination by Co^{3+} in the metalloCD. The hydrolysis of 4-nitrophenyldiphenyl phosphate in the presence of the Zn^{2+} metallo-βCD **5.24** shows Michaelis-Menten kinetics where $k_{cat} = 3.63 \times 10^{-4}$ s^{-1} and $K_M = 1.67 \times 10^{-3}$ mol dm^{-3} at pH 8 in 20% acetonitrile aqueous phosphate buffer at 298.2 K. It is accelerated 7-fold by comparison with the catalysis caused by the complex where the modified βCD substituent is replaced by a methyl group in the tetraaza macrocycle [54]. In the ternary metalloCD where a phenyl group is complexed in the βCD annulus, Zn^{2+} appears to act as a bifunctional catalytic centre by simultaneously providing a nucleophilic hydroxo ligand to attack the phosphorus centre and coordinating a phosphate oxygen. While the hydroxo ligand is not shown in **5.24**, the coordination number of Zn^{2+} commonly ranges from four to six, so the likelihood of Zn^{2+} assuming a coordination number of six in the ternary metalloCD catalytic intermediate is reasonable.

In a related study, the hydrolysis rates of the 2',3'-cyclic monophosphates of adenosine, guanosine, cytosine and uridine are accelerated 23-, 28-, 3.5- and 9.6-fold at 293.2 K and pH 9.5 in the presence of 10^{-2} mol dm^{-3} of the Zn^{2+} metalloCD **5.25** formed by βCD substituted by diethylenetriamine at C(6), βCDdien [55]. This variation is consistent with the purine residues of the first two 2',3'-cyclic monophosphates complexing more strongly in the βCD annulus and aiding the formation of stable ternary metalloCDs more than the pyrimidine residues of the second two. At pH 9.5, the most probable structure of the binary metalloCD **5.25** is shown with four-coordinate Zn^{2+} coordinating a hydroxo ligand which acts as the nucleophile. It is probable that Zn^{2+} expands its coordination number to five to coordinate the ribonucleoside 2',3'-cyclic phosphate guests thereby increasing the stability of the catalytic ternary metalloCD. Smaller rate accelerations occur for the hydrolysis of ribonucleotide dimers in the presence of **5.25**. The high stabilities of a range of ground state ternary metalloCDs formed by **5.25** have been attributed to Zn^{2+} coordination of the guest [56]. Thus, for the complexing of adamantan-2-one-1-carboxylate by βCD and **5.25** $K_{11} = 8.3 \times 10^2$ and 2.8×10^5 dm^3 mol^{-1}, respectively, in aqueous solution at 298.2 K. The analogous values for complexing adamantan-1-carboxylate are 2.3×10^2 and 5.3×10^3 dm^3 mol^{-1}, respectively, and for complexing 4-nitrophenoxide are 4.8×10^2 and 1.2×10^3 dm^3 mol^{-1}, which illustrate the variability of ternary metalloCD stability with the nature of the guest.

The ternary metalloCD **5.26** is thought to form when Zn^{2+} is coordinated by 6^A,6^C-bis[2-(4-imidazolyl)ethylamino]-6^A,6^C-dideoxyβCD (ACβCDdihm) in imidazole buffer at pH 7. It resembles the active site of carbonic anhydrase where Zn^{2+} is coordinated by three imidazoles at the base of a cavity formed by the protein [57,58]. For CO$_2$ hydration, **5.26** is ≥ 3 more effective as a catalyst than is Zn^{2+} alone, but is much less effective than carbonic anhydrase. Although there is evidence for the turnover of **5.26**, the formation of a carbamate by bis(histamino)-βCD steadily decreases the concentration of **5.26** as the reaction proceeds. The dehydration of HCO$_3^-$ is not catalysed by **5.26** probably because HCO$_3^-$ coordinates to Zn^{2+} too strongly.

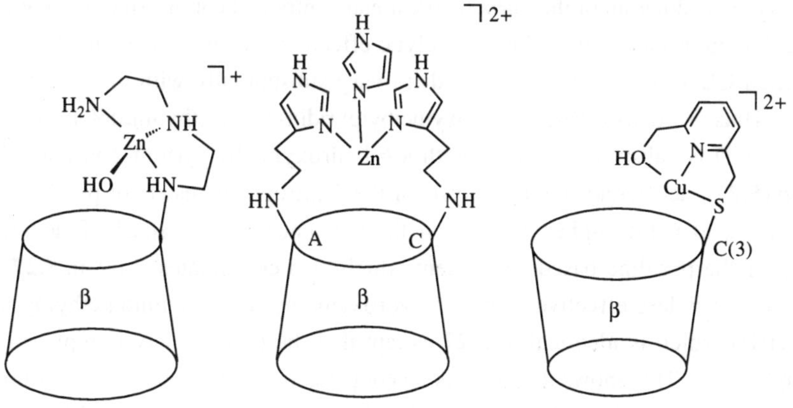

5.25   **5.26**   **5.27**

More recently, the 6^A,6^B-bis[2-(4-imidazolyl)ethylamino]-6^A,6^B-dideoxyβCD regioisomer (ABβCDdihm) has been shown to coordinate Cu^{2+} in aqueous 0.1 mol dm^{-3} KNO$_3$ at 298.2 K with $K_{11} = 1.3 \times 10^{10}$ dm^3 mol^{-1}, a high value attributable to the coordination of Cu^{2+} by four nitrogens [59,60]. The corresponding values for the mono- and diprotonated analogues are 4.2×10^7 dm^3 mol^{-1} and 1.5×10^5 dm^3 mol^{-1}, respectively, which reflect the inability of the protonated nitrogen donor atoms to coordinate. 6^A,6^B-Bis[2-(4-imidazolyl)ethylamino]-6^A,6^B-dideoxyβCDcopper(II) ([Cu(ABβCDdihm)]$^{2+}$) shows substantial superoxide dismutase activity probably because the four donor nitrogens provide a coordination sphere midway between the square-planar stereochemistry favoured by Cu^{2+} and the tetrahedral stereochemistry preferred by Cu^+ in the reduced intermediate form of the metalloCD catalysing the superoxide dismutation:

$$2O_2^- + 2H^+ \rightleftharpoons O_2 + H_2O_2$$

The order of catalytic effectiveness is: $[Cu(AB\beta CDdihm)]^{2+} > [Cu(AC\beta CDdihm)]^{2+} > [Cu(AD\beta CDdihm)]^{2+}$ which corresponds to a decreasing tendency towards tetra–hedral stereochemistry and the ability to stabilise Cu^+ in the same order [60,61].

The formation of a ternary metalloCD does not necessarily ensure that reaction of the guest will be catalysed to a greater extent than the catalysis caused by an independent component of the precursor binary metalloCD. A combination of the effects of coordination of the guest by the metal centre and complexing of the guest in the CD annulus causes the relative catalytic effectiveness of the metalloCD (and the modified CD from which it is formed) to vary substantially with the nature of the guest. Thus, 3^A-deoxy-3^A-((6-hydroxymethylpyridin-2-yl)methylthio)βCDcopper(II), where βCD is substituted at C(3) with a 6-hydroxymethylpyridin-2-yl)methylthiyl group (**5.27**), accelerates the hydrolysis of the 4-nitrophenyl esters of picolinic acid, quinaldic acid and its 6-phenyl derivative through a nucleophilic attack of the hydroxy group of the pyridine based substituent which also coordinates Cu^{2+} in **5.27** [62]. However, it is less effective than is 2-hydroxymethyl-6-methylthiomethylpyridine-copper(II) which is identical to **5.27** except that the βCD moiety is replaced by a methyl group. This shows that there is no cooperative catalytic effect of coordination of the guest by Cu^{2+} and its complexing in the βCD annulus in **5.27**, possibly because of a misalignment of the guest in the ternary metalloCD.

Sometimes a binary metalloCD and its dimer are formed where the M^{m+}:CD ratios are 1:1 and 1:2 when M^{m+} coordinates one or two modified CDs, respectively. Thus, **5.28** (βCDen) forms both $[Cu(\beta CDen)]^{2+}$ and $[Cu(\beta CDen)_2]^{2+}$ at pH 10.5, and the latter accelerates the oxidation of furoin to furil by 20-fold ($k_{cat} = 1.8 \times 10^{-2}$ s^{-1} and $K_{11}^{-1} = K_M = 2.6 \times 10^{-3}$ mol dm^{-3}) over the uncatalysed rate ($k_u = 9.5 \times 10^{-4}$ s^{-1}), whereas βCDen does not [63]. Michaelis-Menten kinetics are observed and this is attributed to the complexing of furoin simultaneously in both βCD annuli of $[Cu(\beta CDen)_2]^{2+}$ stabilising the furoin derived enolate anion which may coordinate to the Cu^{2+} centre. It appears that Cu^{2+} may be able to act as an oxidant in addition to O_2. Michaelis-Menten kinetics are also observed for the hydrolysis of 4-nitrophenyl benzoate and 4-nitrophenyl acetate catalysed by βCDen, $[Cu(βCDen)]^{2+}$ and $[Cu(βCDen)_2]^{2+}$ as shown by the data in Table 5.2 [64]. However, a change in the nature of the guest substantially changes the order of catalytic efficiency within the ternary metalloCD as indicated by k_{cat}/k_u which increases in the sequence: βCDen < $[Cu(βCDen)_2]^{2+}$ < $[Cu(βCDen)]^{2+}$ for 4-nitrophenyl benzoate but in the sequence $[Cu(βCDen)]^{2+}$ < $[Cu(βCDen)_2]^{2+}$ < βCDen for 4-nitrophenyl acetate. This probably results from changes in the proximity of the guest carbonyl carbon to the nucleophile which is an amine group for βCDen and a hydroxo ligand for $[Cu(βCDen)]^{2+}$ and $[Cu(βCDen)_2]^{2+}$. While the variation of K_M for βCDen reflects changes in interactions between this catalyst and the guest alone, the overall variation of K_M reflects variations in the competing interactions of the βCD annulus and Cu^{2+} with the two guests.

Table 5.2. Parameters for Ester Hydrolysis in Aqueous Solution.[a]

Catalyst	Guest	k_{cat} s^{-1}	k_{cat}/k_u	K_M mol dm^{-3}
none	4-nitrophenyl benzoate	$0.002 = k_u$	(deduced from [64])	
βCDen	4-nitrophenyl benzoate	0.038	19	0.00261
$[Cu(βCDen)]^{2+}$	4-nitrophenyl benzoate	0.0755	38	0.00662
$[Cu(βCDen)_2]^{2+}$	4-nitrophenyl benzoate	0.059	30	0.00155
none	4-nitrophenyl acetate	$0.0057 = k_u$	(deduced from [64])	
βCDen	4-nitrophenyl acetate	0.453	80	0.0145
$[Cu(βCDen)]^{2+}$	4-nitrophenyl acetate	0.129	23	0.0092
$[Cu(βCDen)_2]^{2+}$	4-nitrophenyl acetate	0.219	38	0.00425

[a]At pH 10.7 and 298.2 K. Data from Lineweaver-Burk plots.

In another catalysed redox system, thioanisole is oxidised to phenylmethyl sulfoxide by hydrogen peroxide in the presence of the binary metalloCD thought to be formed by oxodiperoxomolybdate, $MoO(O_2)^{2-}$ and **5.28-5.31** to give a substantial excess of (R)-phenylmethyl sulfoxide as seen from Table 5.3 [65]. In the presence of hydrogen peroxide, $MoO(O_2)^{2-}$ and βCD, racemic phenylmethyl sulfoxide is produced which indicates that the coordination of $MoO(O_2)_2^{2-}$ by **5.28-5.31** is important in forming the ternary metalloCD with thioanisole and the enantioselective oxidation which results. The effect of isomerism is seen in differing enantioselectivities of 6^A-(2-aminoethyl)amino-6^A-deoxy-βCD **5.28** and its 3^A isomer **5.29** where the latter produces a lesser enantioselectivity. This is probably because the wider secondary face of the $MoO(O_2)_2^{2-}$ ternary metalloCD results in a looser fit of the thioanisole guest. The lower enantioselectivies engendered by **5.30** and **5.31**, by comparison with that of **5.28**, may result from their bulkier substituents hindering access to the βCD annulus.

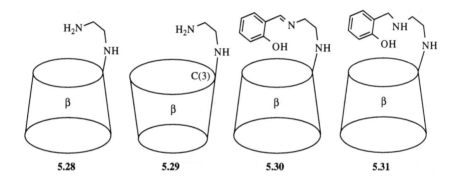

Table 5.3. Enantioselective Oxidation of Thioanisole by Hydrogen Peroxide in the Presence of $MoO(O_2)_2^-$ and Modified CDs.[a]

CD	% Yield of PhS(O)CH$_3$	% Excess of (R)-PhS(O)CH$_3$
βCD	69	0
5.28	91	59
5.29	90	17
5.30	76	24
5.31	89	20

[a]Reaction of thioanisole suspensions in HEPES buffer at pH 7 and 293.5 K.

The most highly charged metal ion studied in metalloCD catalysis is Ce^{4+} which, in the presence of γCD, acts as an effective peptidase for di- and tri-peptides in neutral aqueous solution [66]. Apart from solubilising Ce^{4+}, the nature of the interaction between Ce^{4+} and γCD is unclear, but presumably some degree of complexation of the catalytic Ce^{4+}-peptide complex occurs. It is possible that the highly charged Ce^{4+} may sufficiently polarise a secondary hydroxy group of γCD to produce an alkoxide and coordination similar to that discussed in section 5.1.

5.5. Catalytic Organometallocyclodextrins: Mono-Cyclodextrin Systems

Organometallic compounds generally incorporate metals in their lower oxidation states so that the metal centre acts as a soft acid, is coordinated by soft base donor atoms, and tends to catalyse reactions at soft base centres in organic molecules. A fascinating example of this is provided by the selective complexation, phase transfer and catalytic hydrogenation and hydroformylation of alkenes in the presence of the βCD-based Rh^+ metalloCD catalysts **5.32-5.36** (cod = 1,5-cyclooctadiene) [67]. Hydrogenation of an equimolar mixture of the alkenes **5.38** and **5.39** to 10% conversion in *N,N*-dimethylformamide in the presence of the control catalyst $[PhN(CH_2PPh_2)_2Rh(cod)]^+$ **5.37** produces an equimolar mixture of the corresponding product alkanes **5.40** and **5.41**. The same reaction in the presence of the catalysts **5.32-5.36** gives predominantly **5.40**, as shown in Table 5.4. This is attributed to preferential complexing of the phenyl group of **5.38** in the βCD annulus. This predominance of the **5.40** product increases through the interplay of preferential complexing, Rh^+ catalysis and phase transfer catalysis in a two-phase system where the organic phase is *N,N*-dimethylformamide and the aqueous phase is 30% *N,N*-dimethylformamide and 70% water, and appears to operate as shown in Fig. 5.4 Thus, **5.33** has a water soluble βCD component and an organic phase soluble Rh^+ organometallic component which allows it to operate as a phase transfer agent at the phase interface. The organic phase soluble alkene enters the hydrophobic interior of the βCD annulus where Rh^+ catalysed hydrogenation occurs to produce the corresponding alkane.

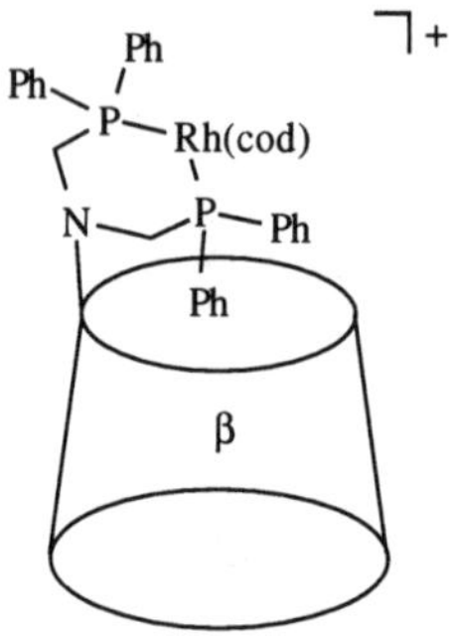

5.32

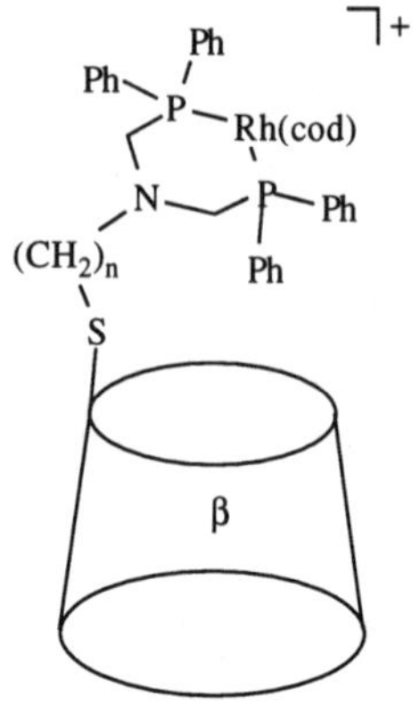

5.33 n = 2
5.34 n = 3
5.35 n = 4

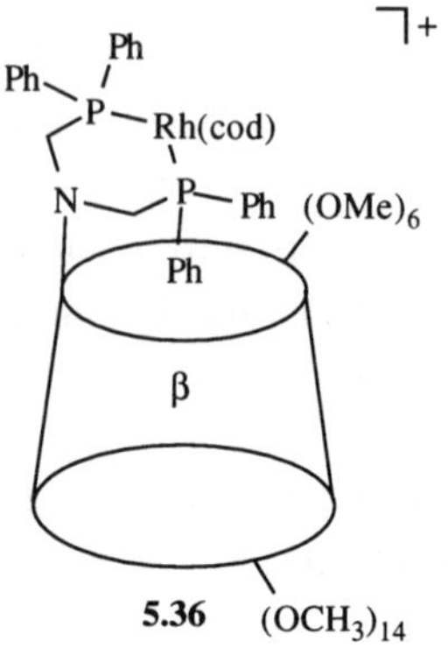

5.36 (OCH$_3$)$_{14}$

5.37

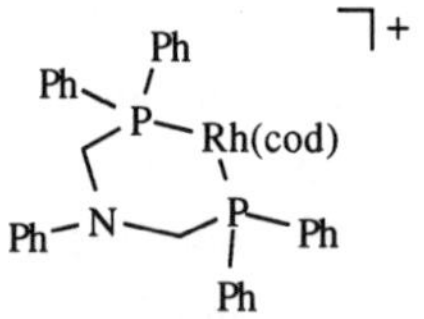

5.38

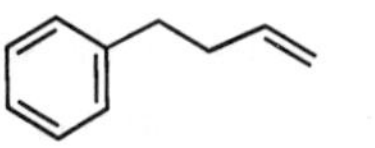

5.39 n-C$_6$H$_{13}$

5.40

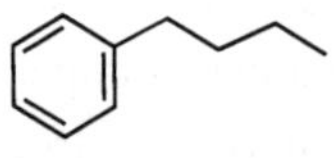

5.41 n-C$_6$H$_{13}$

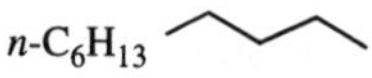

Table 5.4. Selectivity for Hydrogenation of the Alkenes **5.38** and **5.39**.[a]

Catalyst[b]	Product ratio **5.40/5.41**	Catalyst[b]	Product ratio **5.40/5.41**
5.37	50/50	**5.35**	66/34
5.32	68/32	**5.32**[c]	82/18
5.33	74/26	**5.33**[c]	81/19
5.34	71/29	**5.36**[c]	87/13

[a]At 1 Atm H$_2$ in *N,N*-dimethylformamide at 295.2 K. [b]0.5 mol% catalyst. [c]Where the organic phase is *N,N*-dimethylformamide and the aqueous phase is 30% *N,N*-dimethylformamide and 70% water.

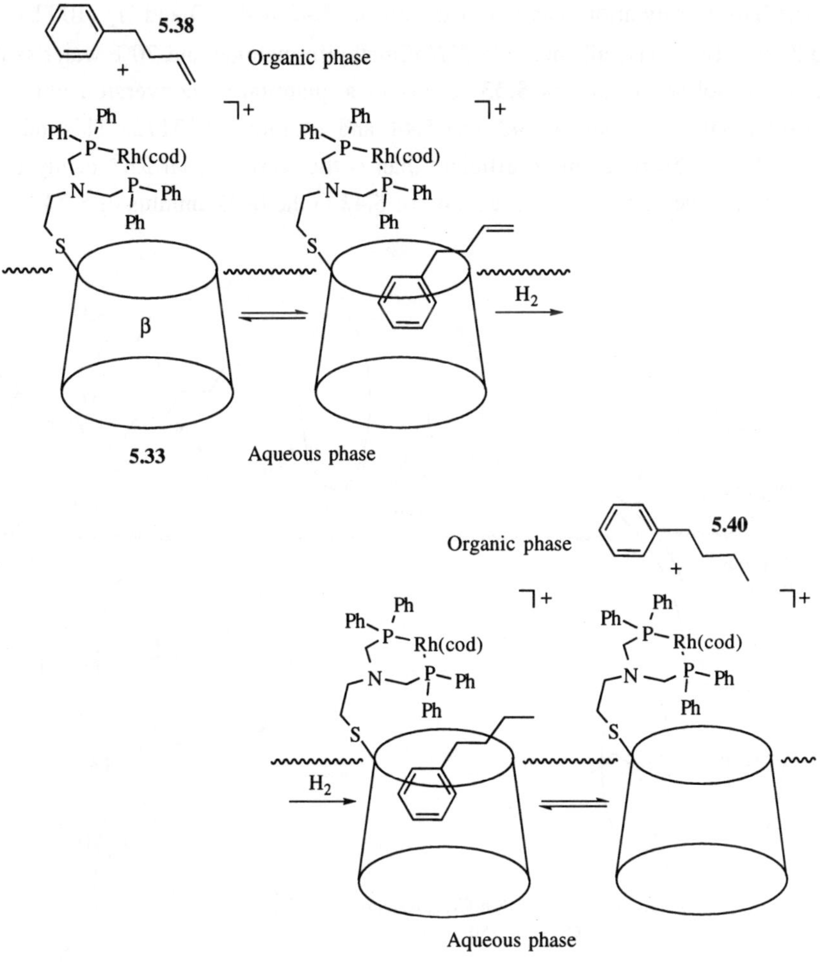

Fig. 5.4. Phase transfer and selective catalysis of alkene hydrogenation. The organic phase is *N*,*N*-dimethylformamide and the aqueous phase is 30% *N*,*N*-dimethylformamide and 70% water.

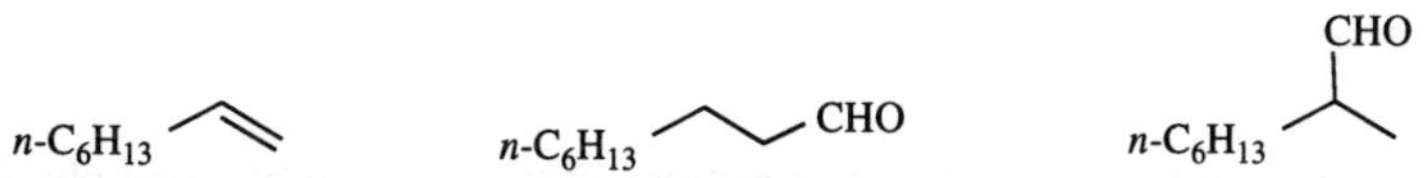

5.42 **5.43** **5.44**

The hydroformylation reaction of the alkene **5.42** with CO and H_2 (100 bar and 350 K) in a two phase alkene: 30% *N,N*-dimethylformamide and 70% water system with 0.03 mol % of catalyst **5.33** results in a quantitative conversion with 76% regioselectivity in favour of **5.43** over **5.44**, and a turnover of 3172. This indicates that **5.33** is > 150 times more efficient than is the conventional Rh^+ catalyst, and demonstrates the effect of complexation of **5.42** in the βCD annulus of **5.33**.

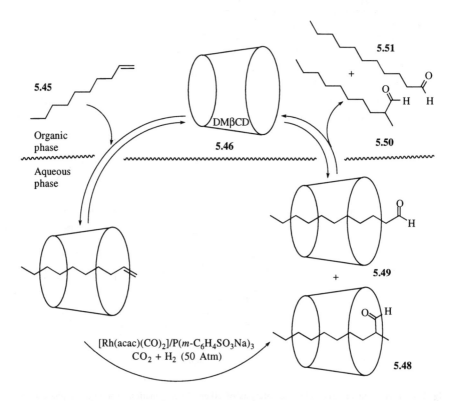

Fig. 5.5. Rhodium(I) catalysed hydroformylation of dec-1-ene **5.45** in a two phase undecane/water system where DMβCD (**5.46** where the 7 C(2) and 7 C(6) methoxy groups are not shown) acts as an inverse phase transfer catalyst.

A different approach to catalysing reactions of water-insoluble long chain alkenes involves transfer of dec-1-ene **5.45** from the organic phase (dec-1-ene/undecane in 20/1 mol ratio) by DMβCD **5.46** acting as an inverse phase transfer catalyst in the complex **5.47** to the aqueous phase where hydroformylation at 50 Atm of CO_2/H_2 and 350.2 K is catalysed by $[Rh(acac)(CO_2)_2]/P(m\text{-}C_6H_4SO_3Na)_3$ as shown in Fig. 5.5 [68]. This results in a 95% yield of the complexed *i*- and *n*-aldehydes **5.48** and **5.49** which transfer to the organic phase in a 9:1 **5.50**:**5.51** ratio. The 5% side products are mainly dec-2-ene, dec-3-ene and dec-4-ene. While some other water soluble modified CDs are quite effective as phase transfer catalysts, DMβCD has proven to the most effective of those so far tried [68-71]. The relative ineffectiveness of βCD is attributable to its poor organic-phase solubility. The oxidation of other olefins (C8-C16) to the corresponding ketones is also greatly enhanced (90% yields under oxygen at 353.2 K) through the use of DMβCD as the inverse phase transfer catalyst in a similar two-phase system where $PdSO_4/H_9PV_6Mo_6O_{40}/CuSO_4$ is the composite water soluble catalyst [72,73].

Due to the soft base nature of phosphorus donor atoms, CDs substituted with phosphine groups coordinate soft acid metals and represent a class of organometalloCDs which participate in different catalyses from those of the borderline hard to hard acid first row transition metal ions. This is illustrated by the catalysed hydrogenation and hydroformylation reactions discussed above. Nevertheless, only a few phosphine based metalloCDs, in addition to **5.32-5.36**, have been prepared and these are exemplified by βCD-based **5.52** and DMβCD-based **5.53** (methoxy groups at C(2) and C(6) are not shown), neither of which appear to have been deployed in catalytic studies. Thus, **5.52** incorporates a norbornadiene Rh^+ moiety bound through sulfur to a C(6) of βCD [74], and **5.53** has a ferrocenyl bisdiphenyl-phosphine moiety bound to DMβCD through an O(3) [75]. In the latter case, Pd^{2+} coordinates to the diphenylphosphine groups to form an organometalloCD which appears to aggregate in water.

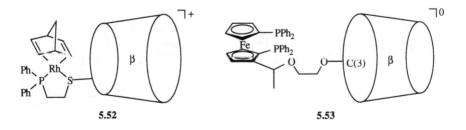

5.52   **5.53**

The ferrocenyl group itself plays an important mediating role in the electrochemical conversion of benzyl alcohol to benzaldehyde in acetonitrile as proposed in Fig. 5.6 [76]. The lipophilic organometalloCD **5.54** has one ferrocenyl group appended at the primary face of βCD where all of the remaining twenty hydroxy groups are acetylated. Electrochemical oxidation of the ferrocenyl moiety to ferrocinium ion causes the latter to move out of the βCD annulus to form **5.55** as indicated by circular dichroic spectroscopy. Complexation of benzyl alcohol in **5.56** and electron transfer produces **5.57** with the completion of the electrochemical cycle to give benzaldehyde. Thus, after 12 hours electrolysis time, 36.8% of the benzyl alcohol is converted to benzaldehyde with a turnover of 22.2 in the presence of **5.54** (0.0112 mol dm^{-3}, benzyl alcohol 0.337 mol dm^{-3}). In the presence of methylferrocene carbonate (0.0118 mol dm^{-3}, benzyl alcohol 0.381 mol dm^{-3}) only 0.07% is converted to benzaldehyde after 24 hours electrolysis time, and addition of hepta(2,3,6-O-acetyl)-βCD only increases this yield to 0.91% over 24 hours. This shows the importance of the complexation of benzyl alcohol by **5.54** in mediating the electrochemical oxidation. A similar situation prevails in the analogous oxidation of 1-naphthylmethanol to 1-naphthaldehyde.

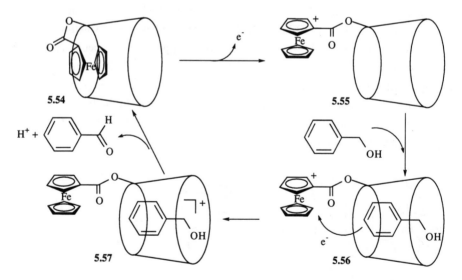

Fig. 5.6. The electrochemical redox cycle for the organometalloCD mediated oxidation of benzyl alcohol to benzaldehyde in acetonitrile.

The parallel orientation of the ferrocenyl axis with that of the βCD annulus in **5.54** shown in Fig. 5.6 is also the orientation observed in its analogues where the ferrocene carboxylate moiety is either bound directly to C(6) of βCD (**5.58**) or through a trimethylene link (**5.59**), as shown by circular dichroic spectroscopy [77]. When the annulus is enlarged there is an opportunity for this orientation to change and, although it has not proven possible to determine the orientation in the γCD **5.60**, it does change to an orientation where the ferrocenyl axis is at 90° to the γCD axis in **5.61**. The impact of the intramolecular complexation of the ferrocenyl moiety on the formation of ternary organometalloCDs is considerable. Thus, in 20% ethylene glycol aqueous solution at 298.2 K, K_{11} for the complexation of adamantan-1-ol is 1.45×10^4, 2.88×10^3 and 4.05×10^2 dm^3 mol^{-1} for βCD, **5.58** and **5.59**, respectively, are consistent with the complexation of adamantan-1-ol being progressively impeded as the ferrocenyl moiety penetrates deeper into the βCD annulus. The analogous $K_{11} = 5.16 \times 10^3$, 2.23×10^2 and 1.72×10^2 dm^3 mol^{-1} for γCD, **5.60** and **5.61**, respectively, consistent with a similar interpretation but reflecting the looser fit of the γCD annulus to adamantan-1-ol. Similar variations in K_{11} are found for the complexation of *l*-borneol, cyclododecanol, *l*-menthol and cyclohexanol.

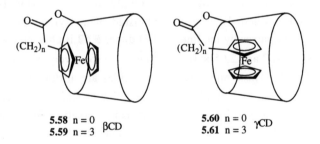

5.58 n = 0
5.59 n = 3 βCD

5.60 n = 0 γCD
5.61 n = 3

5.6. Catalytic Metallocyclodextrins: Multi-Cyclodextrin Systems

Breslow and co-workers have made extensive studies of the metalloCD **5.62**, where two βCDs are linked by a metallobipyridyl group, and its complexation and catalytic hydrolysis of a range of guest species exemplified by **5.63-5.70**

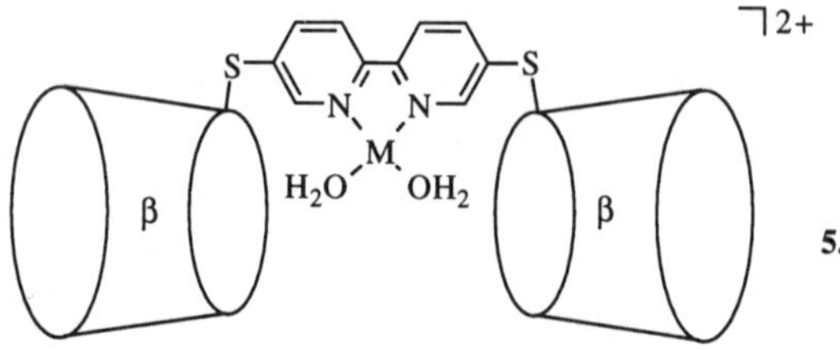

⌐2+

5.62

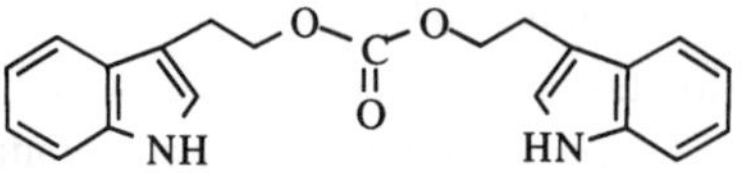

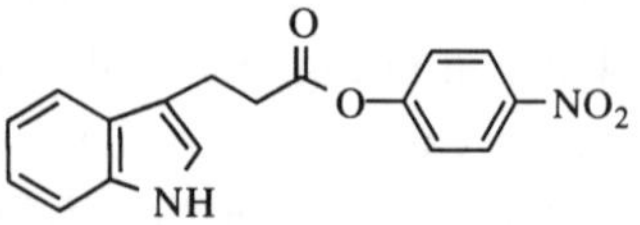

5.66

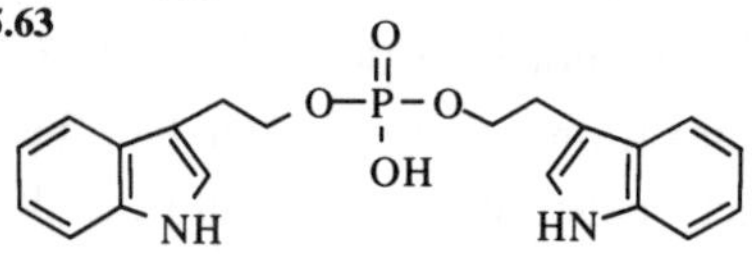

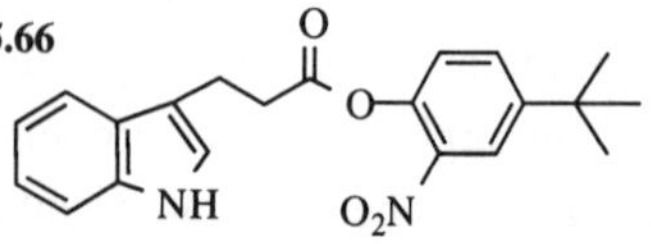

5.67

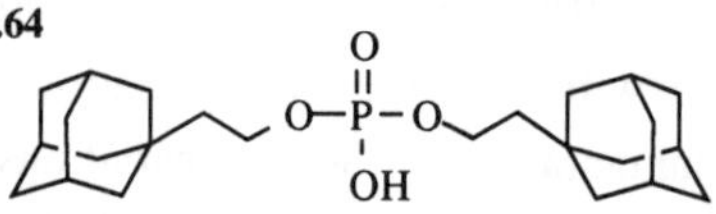

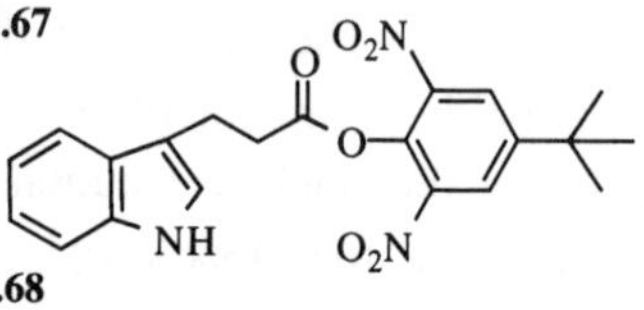

5.68

5.63

5.64

5.65

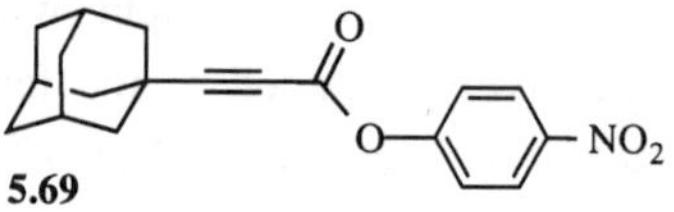

5.69

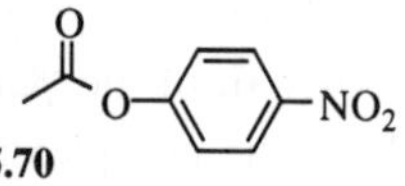

5.70

[49,78,79]. In the absence of M^{2+}, the complexation of bis(indol-3-ylethyl)-carbonate, bis(indol-3-ylethyl)phosphate and bis(adamant-1-ylethyl)phosphate **5.63-5.65** [78] in 1:1 complexes by the linked βCD dimer **5.62** is characterised by K_{11} = 1.43 × 10^6, 1.00 × 10^6 and 2.05 × 10^7 dm^3 mol^{-1}, respectively, at pH 7 in aqueous solution at 298.2 K. These high complex stabilities are attributable to the ditopic guests being complexed by both βCD annuli, and that of **5.65** in particular is further enhanced by the strong complexation of its adamantyl moieties. When M^{2+} = Zn^{2+}, K_{11} = 7.70 × 10^6, 5.47 × 10^7 and 1.02 × 10^9 dm^3 mol^{-1}, respectively, for the ternary metalloCDs formed by **5.63-5.65**. Thus, the presence of the Zn^{2+} centre increases K_{11} by 5, 55 and 50-fold, respectively, consistent with guest coordination by Zn^{2+}. The greater increase for the two phosphates is thought to reflect the dissociation of the phosphate proton and the tetrahedral coordination of the resulting anion by Zn^{2+}, while possibly only one carbonate oxygen is available for coordination of **5.63**. The increases in the strength of complexation of **5.64** and **5.65** in the presence of Zn^{2+} is similar to that expected of a metalloenzyme where substrate complexation in the transition state is stronger than in the ground state.

Such complexation of phosphate diesters is sometimes considered analogous to that in the tetrahedral transition states for ester and amide hydrolysis in enzymatic systems. Thus, it is anticipated that when $M^{2+} = Zn^{2+}$ **5.62** should act as a catalyst for ester hydrolysis, and this is found to be so not only for Zn^{2+} but also for other metal ions.

The variation of the observed rate constant, k_{obs}, for the **5.62** ($M^{2+} = Cu^{2+}$) catalysed ester hydrolysis of 4-nitrophenyl 3-indolepropionate **5.66** in the pH range 6.5 to 9.0 in aqueous 10^{-2} mol dm^{-3} HEPES buffer is given by:

$$k_{obs} = kK_aK_{11}(K_a + [H^+])^{-1}$$

where $pK_a = 7.15$, $K_{11} = 7.0 \times 10^4$ dm^3 mol^{-1} and $k = 2.05 \times 10^{-4}$ s^{-1} at 310.2 K in aqueous solution [79]. ($K_{11}^{-1} = K_M = 1.4 \times 10^{-5}$ mol dm^{-3}, the Michaelis constant for the equilibrium between **5.71**, **5.66** and **5.72**.) This is consistent with the mechanism shown in Fig. 5.7. The dissociation of a proton from Cu^{2+} coordinated water in **5.62** produces the hydroxo ligand nucleophile in **5.71** which

Table 5.5. Ester Hydrolysis Catalysed by **5.62** ($M^{2+} = Cu^{2+}$).[a]

Guest ester	pH	k_{obs} s^{-1}	k_u[b] s^{-1}	k_{obs}/k_u
5.66	7.0	5.49×10^{-4}	3.00×10^{-8}	18300
5.66	8.0	1.04×10^{-3}	1.00×10^{-3}	10400
5.67[c]	8.0	1.35×10^{-4}	1.50×10^{-7}	900
5.68[c]	8.0	1.74×10^{-4}	1.00×10^{-7}	1740
5.69	7.0	6.67×10^{-3}	3.00×10^{-8}	225000
5.69	8.0	1.20×10^{-2}	1.80×10^{-7}	66700
5.70	8.0	9.61×10^{-5}	1.42×10^{-5}	7

[a]In aqueous 10^{-2} mol dm^{-3} HEPES buffer at 310.2 K. All solutions are 1.0×10^{-4} mol dm^{-3} in **5.62** (no M^{2+}), 2.0×10^{-4} mol dm^{-3} in CuCl$_2$ and 6.0×10^{-4} mol dm^{-3} in guest ester. [b]Rate constant observed in the absence of **5.62** and Cu^{2+}. [c]In 60% aqueous 10^{-2} mol dm^{-3} HEPES buffer 40% dimethyl sulfoxide solution.

attacks the carbonyl carbon in the Michaelis-like complex **5.72**. Simultaneously Cu^{2+} coordinates the carbonyl oxygen of **5.66** whose aromatic moieties are bound in the βCD annuli. At pH 7.0 and 8.0, $k_{obs} = 5.49 \times 10^{-4}$ s^{-1} and 1.04×10^{-3} s^{-1}, respectively, which represents 1.83×10^4- and 1.04×10^4-fold increases over k_u. This indicates the great gain in catalytic effect achieved through the careful design of **5.62** to strongly complex and accurately position **5.66** for nucleophilic attack by the Cu^{2+} coordinated hydroxo nucleophile. At least 50 turnovers occur for the catalysis of the hydrolysis of **5.66** and **5.69** by **5.62** when $M^{2+} = Cu^{2+}$. (Much smaller catalytic effects occur in the presence of either Cu^{2+} or βCD alone or **5.62** in the absence of Cu^{2+}.)

Table 5.6. Ester Hydrolysis Catalysed by **5.63** and **5.73**.[a]

M^{m+}	Catalyst	Guest ester	k_{obs} s^{-1}	k_{obs}/k_u
none	none	**5.66**	3.00×10^{-8} $(k_u)^b$	1
Ni^{2+}	**5.62**	**5.66**	3.00×10^{-4}	10000
Cu^{2+}	**5.62**	**5.66**	5.50×10^{-4}	18300
Ni^{2+}	**5.73**	**5.66**	1.77×10^{-3}	59000
Cu^{2+}	**5.73**	**5.66**	2.70×10^{-4}	900
Co^{2+}	**5.73**	**5.66**	1.40×10^{-4}	4700
Zn^{2+}	**5.73**	**5.66**	8.89×10^{-3}	300000
Tb^{2+}	**5.73**	**5.66**	5.5×10^{-4}	18000
Eu^{2+}	**5.73**	**5.66**	6.2×10^{-4}	21000
none	none	**5.69**	3.00×10^{-8} $(k_u)^b$	1
Cu^{2+}	**5.62**	**5.69**	6.8×10^{-3}	220000
Zn^{2+}	**5.62**	**5.69**	1.2×10^{-3}	40000
Zn^{2+}	**5.73**	**5.69**	5.12×10^{-2}	1700000
Ni^{2+}	**5.73**	**5.69**	1.12×10^{-2}	371000

[a] In aqueous 10^{-2} mol dm^{-3} HEPES buffer at pH 7 and 310.2 K. All solutions are 1.0×10^{-4} mol dm^{-3} in **5.62** or **5.73** (no M^{m+}), 1.0×10^{-3} mol dm^{-3} in M^{m+} and 6.0×10^{-4} mol dm^{-3} in guest ester. [b] Rate constant observed in the absence of **5.62** or **5.73** and M^{m+}.

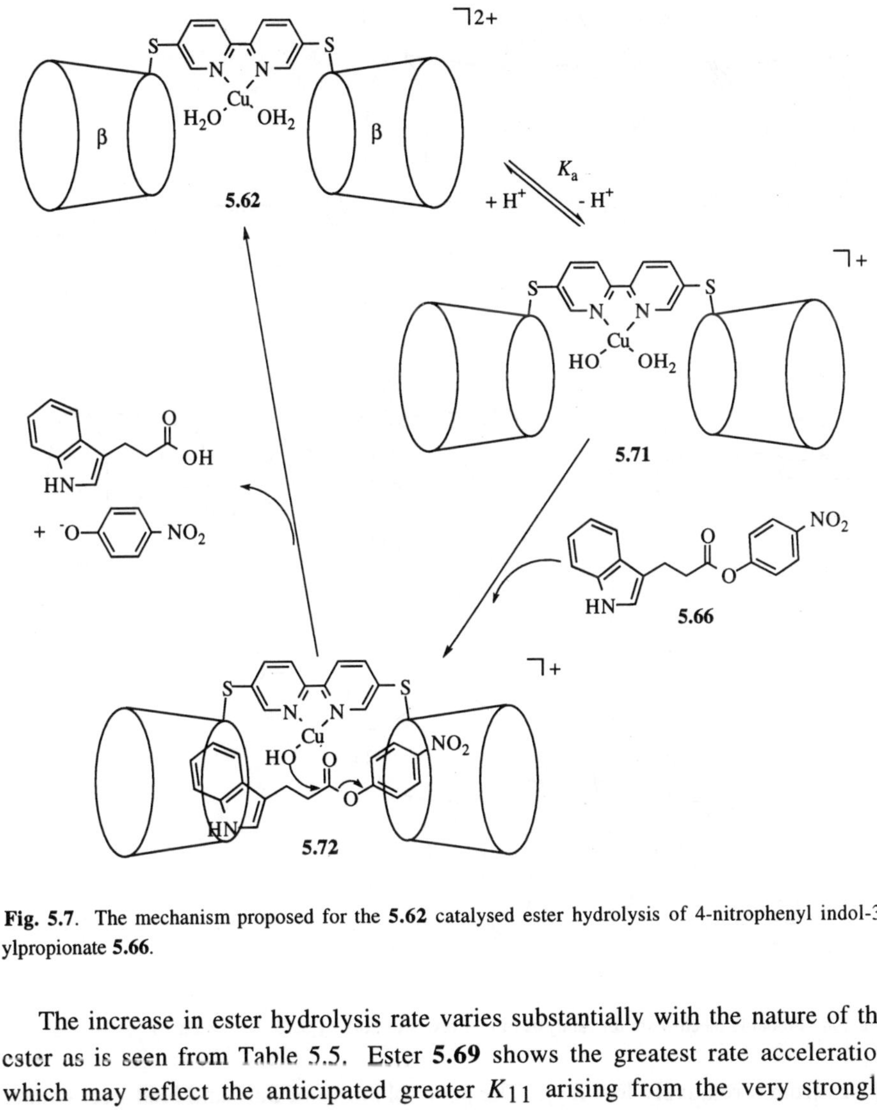

Fig. 5.7. The mechanism proposed for the **5.62** catalysed ester hydrolysis of 4-nitrophenyl indol-3-ylpropionate **5.66**.

The increase in ester hydrolysis rate varies substantially with the nature of the ester as is seen from Table 5.5. Ester **5.69** shows the greatest rate acceleration which may reflect the anticipated greater K_{11} arising from the very strongly complexing adamantyl group. This is also seen for the formation of the ternary metalloCD between **5.62** (M^{2+} = Zn^{2+}) and **5.65**. The small acceleration observed for 4-nitrophenyl acetate **5.70**, which has only one coordinating group, probably

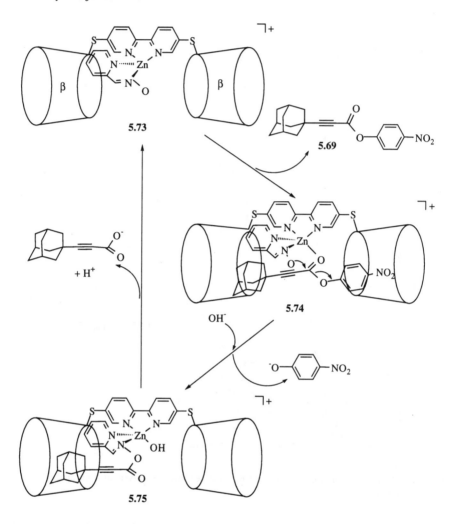

Fig. 5.8. Probable mechanism for catalysis of hydrolysis of **5.69** by **5.73** ($M^{2+} = Zn^{2+}$).

arises from a relatively low K_{11}. The poor solubility of **5.67** and **5.68** necessitates the use of aqueous 40% dimethyl sulfoxide solutions and K_{11} is lowered as a consequence.

The effectiveness of **5.62** as a catalyst varies with the nature of M^{m+} as is seen from Table 5.6 where Cu^{2+} is a more effective catalytic centre for the hydrolysis of ester **5.66** than is Ni^{2+}, and Cu^{2+} is a more effective catalytic centre for ester **5.69** than is Zn^{2+}. However, this may change with the occupancy of the first coordination sphere of M^{m+}. Such a change occurs when **5.62** coordinates the additional ligand, 2-pyridinecarbaldehyde oxime, to form **5.73** shown in Fig. 5.8. The catalysis of ester hydrolysis by **5.73** appears to occur through the quaternary metalloCD Michaelis-like intermediate **5.74**, where the additional ligand, 2-pyridinecarbaldehyde oxime, participates in the ester hydrolysis mechanism (Fig. 5.8). Thus, for the hydrolysis of **5.66** in the presence of **5.62** Cu^{2+} is a more effective centre than is Ni^{2+}, but this order of effectiveness is reversed for **5.73**. This may reflect the differing ligand field restraints placed on these ions by their different d orbital occupancies, and may partly explain why Zn^{2+}, which experiences no such constraints, is the most effective metal centre for **5.73**. To some extent this is borne out by the large catalytic effect of **5.73** with a Zn^{2+} centre in the hydrolysis of **5.69** (Table 5.6). An increase in the effectiveness of a metal centre in **5.73** over **5.62** is consistent with the ground state ternary metalloCD **5.74** more closely approaching the entatic state for ester hydrolysis.

Another example of a quaternary metalloCD being involved in catalysis is provided by the hydrolysis of bis(4-nitrophenyl)phosphate **5.76** to produce two moles of 4-nitrophenolate and inorganic phosphate in the presence of **5.62** (M^{m+} = La^{3+}) and H_2O_2 which is thought to proceed through the quaternary metalloCD **5.77** [80]. The oxidative hydrolysis of 6×10^{-5} mol dm^{-3} **5.76** is characterised by k_{obs} (= k_u) = 1.1×10^{-11} s^{-1} at 298.2 K in aqueous HEPES buffer at pH 7. This hydrolysis is greatly accelerated as shown by k_{obs} = 1.07×10^{-6} s^{-1} when the concentrations of **5.76**, La^{3+} and H_2O_2 = 6×10^{-5}, 1×10^{-4} and 4.8×10^{-2} mol dm^{-3}, respectively. When **5.62** is added as the free linked CD to 2×10^{-4} mol dm^{-3} concentration, k_{obs} increases to 3.37×10^{-5} s^{-1} which forms the basis for the postulation of intermediate **5.77**. In contrast, **5.78**, for which k_u = 4.9×10^{-7} s^{-1}, shows a very small acceleration in hydrolysis rate in the presence of either 1×10^{-3} mol dm^{-3} La^{3+} or 4.8×10^{-2} mol dm^{-3} H_2O_2, separately or together, probably because La^{3+} does not coordinate **5.78** to a significant extent. Neither is there acceleration in the presence of 1×10^{-3} mol dm^{-3} La^{3+} and 2×10^{-4} mol dm^{-3} **5.62** added as the free linked CD, but when 4.8×10^{-2} mol dm^{-3} H_2O_2 is added, k_{obs} = 1.88×10^{-4} s^{-1} for the hydrolysis of **5.78**, which suggests the formation of an intermediate similar to **5.77**.

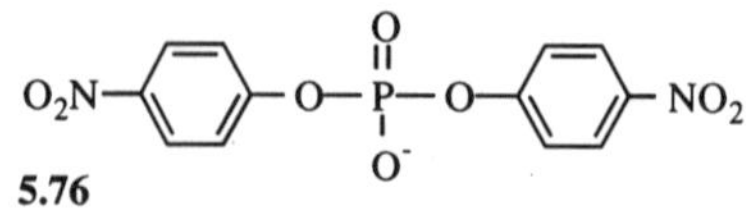

5.76

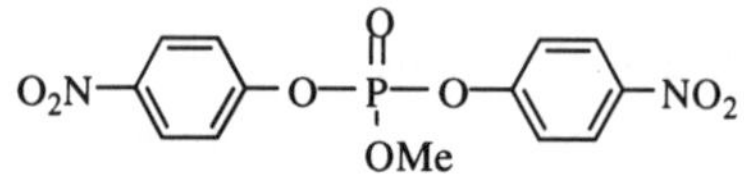

5.78

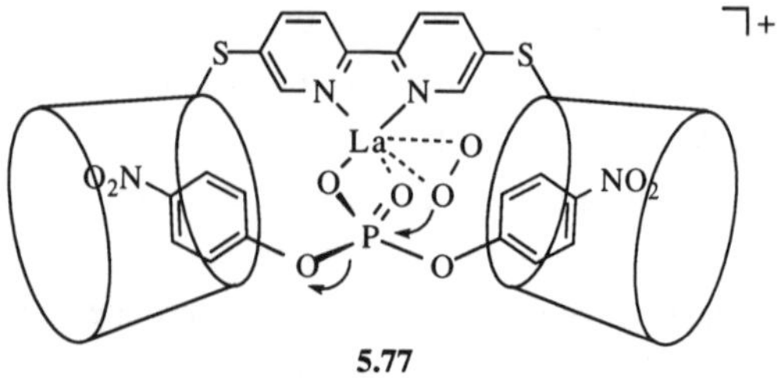

5.77

The βCD-based dimers **5.79-5.82** have longer linkers that incorporate the metal ion coordinating phenanthroline moiety which is more rigid than the bipyridyl moiety [81]. ROESY ^{1}H NMR studies indicate that the phenanthroline of **5.80** is sandwiched between the primary faces of the two βCD, and a modelled structure is consistent with this. In aqueous solution, both Cu^{2+} and Eu^{3+} are coordinated by **5.80** but Zn^{2+} and La^{3+} are weakly coordinated at best. The metalloCD formed by Cu^{2+} and **5.80** is quite effective in catalysing the hydrolysis of the phosphodiester bond of bis(4-nitrophenyl)phosphate in the presence of H_2O_2.

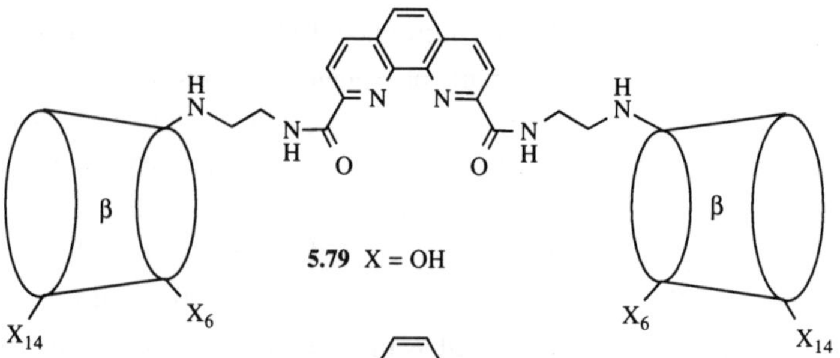

5.79 X = OH

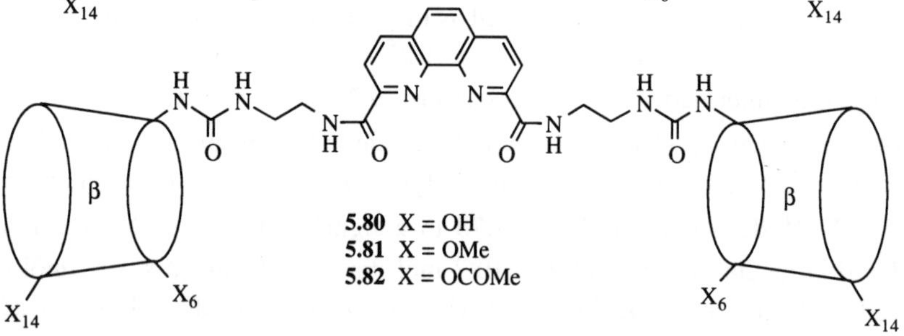

5.80 X = OH
5.81 X = OMe
5.82 X = OCOMe

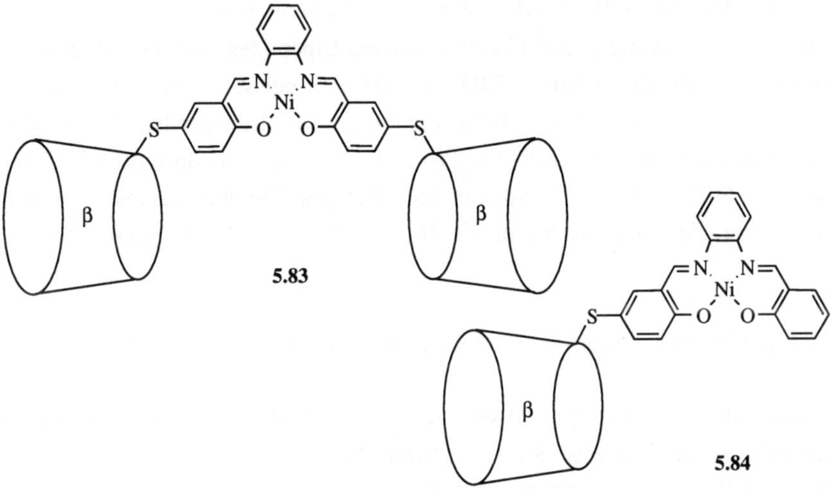

5.83

5.84

Table 5.7. Amino acid Sequences Selected by the MetalloβCDs **5.83** and **5.84** in the Assay of a Tripeptide Library.[a]

AA3	AA2	AA1	Frequency of % occurrence with:	
			5.83	**5.84**
L-Phe	D-Pro	X	36	46
X	L-Phe	D-Pro	16	8
D-Phe	L-Pro	X	28	31
X	D-Phe	L-Pro	20	0

[a]X, which represents the third amino acid of the tripeptide, was any of: Gly, D-Ala, L-Ala, D-Val, L-Val, D-Leu, L-Leu, D-Phe, L-Phe, D-Pro, L-Pro, D-Ser, L-Ser, D-Thr, L-Thr, D-Asp, L-Asp, D-Glu, L-Glu, D-Asn, L-Asn, D-Gln, L-Gln, D-His, L-His, D-Lys, L-Lys, D-Arg and L-Arg.

Although not directly related to catalytic studies, the use of orange **5.83** and **5.84** in screening a tripeptide library on hydrophilic poly(ethyleneglycol)polystyrene (TentaGel) beads represents a novel application of these metalloβCDs [82]. The principles of combinatorial chemistry were employed in screening the library for differences in peptide complexation which could not have been as rapidly achieved through conventional complexation studies. The library has the general structure

AA3-AA2-AA1-NH(CH$_2$)$_2$-TentaGel with 29 different amino acids at each site so that it contains maximally 29^3 (24389) different tripeptides. About 1 in 200 of the library beads show the colour of **5.83** and **5.84** after equilibration in water at pH 7, indicating complexation of an amino acid moiety. All of the beads selected by **5.83** contain either the sequence L-Phe-D-Pro or D-Phe-L-Pro, as do most of the beads selected by **5.84** (Table 5.7). None of the other possible phenylalanine-containing sequences are selected and neither are the D-Phe-D-Pro and L-Phe-L-Pro sequences.

5.7. Corrin and Porphyrin Metallocyclodextrins and Catalysis

Although vitamin B$_{12}$ is often encountered in its cyanocob(III)alamin and aquocob(III)alamin forms **5.85** and **5.86** (Fig. 5.9), respectively, it is as the adenosylcob(III)alamin coenzyme (coenzyme B$_{12}$ **5.87**) which combines with a high molecular weight protein or apoenzyme to form the fully functional holoenzyme. The first holoenzyme component determines the type of reaction which is catalysed and the second component determines the specificity and reaction rate of the catalysis. In one such catalytic cycle, homolysis of the cobalt(III)-carbon bond in coenzyme B$_{12}$ (**5.87**) produces B$_{12(r)}$ with cobalt in oxidation state (II) and an adenosyl radical in the first stage. Subsequently, hydrogen atom transfer from the holoenzyme-bound substrate to the adenosyl radical is followed by substrate rearrangement and hydrogen atom return to recreate the adenosyl radical which re-coordinates B$_{12(r)}$ to produce coenzyme B$_{12}$ and complete the catalytic cycle.

Several attempts have been made to mimic the vitamin B$_{12}$ catalytic process, one of which involves the displacement of CN$^-$ in **5.85** by βCD coordinated through a C(6) carbon to give the metalloCD **5.88** in Fig. 5.10 where it is shown with a guest in the annulus mimicking the holoenzyme bound substrate in the biological cycle [83,84]. This is in equilibrium with B$_{12(r)}$ (**5.89**) and the radical βCD complex **5.90** where homolysis of the C-I bond and transfer of I$^-$ produces the modified βCD complex **5.91**, containing the radical guest. When the guest is either benzyl- or *tert*-butylbenzyl iodide, the product **5.91** is obtained consistent with B$_{12(r)}$ (**5.89**) and the βCD radical **5.90** being produced as reaction intermediates. The βCD moiety of **5.88** is expected to complex benzyl- and *tert*-butylbenzyl iodide strongly and mimic the strong and selective substrate complexation by the apoenzyme in the B$_{12}$

holoenzyme. While the full catalytic B_{12} cycle is not achieved, this system clearly mimics several aspects of the holoenzyme including group transfer to form a substrate radical.

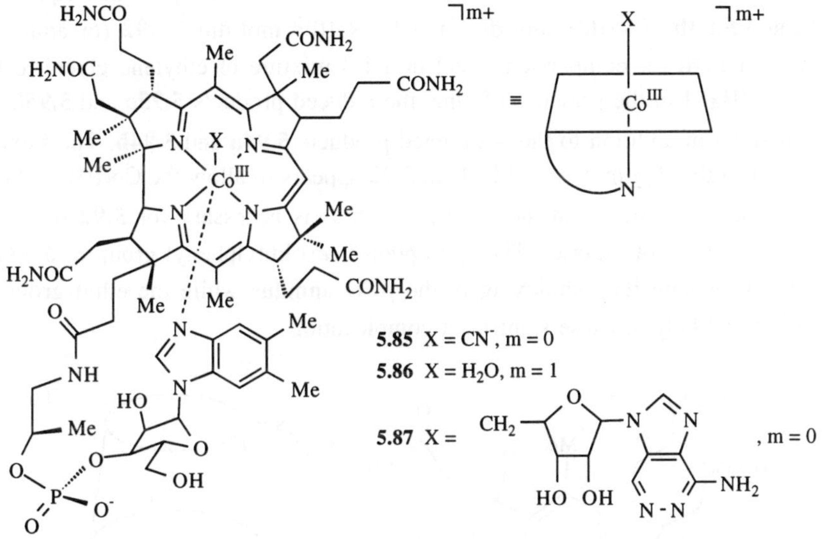

Fig. 5.9. The corrin ring based structures of vitamin B_{12} (**5.85**), aquocob(III)alamin (**5.86**) and vitamin B_{12} co-enzyme (**5.87**).

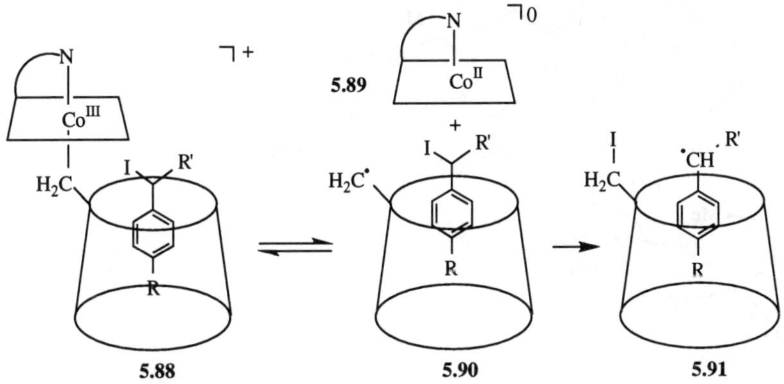

Fig. 5.10. Mode of action of a βCD-B_{12} enzyme mimic. In **5.88**, cobalt(III) is bound to a C(6) of βCD which has complexed a guest molecule to form a ternary metalloCD.

The metalloCD **5.92** arising from binding modified B_{12} to βCD at C(6) catalyses the rearrangement of **5.93a** to **5.94a** and **5.93b** to **5.94b** as shown in Fig. 5.11, and comes close to mimicking methylmalonyl-*CoA* mutase which catalyses highly stereoselective 1,2-shifts at saturated hydrocarbon centres [84]. When equimolar **5.93a** and **5.93b** (7×10^{-5} mol dm^{-3}) and 9×10^{-5} mol dm^{-3} **5.92** (or aquocob-(III)alamin **5.86** for comparison) react in a 1:4 mixture of ethylene glycol to 8% aqueous NH$_4$Cl in the presence of zinc, the reduced products **5.95a** and **5.95b** are also obtained in addition to the rearranged products **5.94a** and **5.94b**. The flexible link between the B_{12} moiety and βCD in **5.92** appears to allow the Co(III) centre to reach a guest complexed in the βCD annulus as is necessary for **5.92** to mimic methylmalonyl-*CoA* mutase. The hydrophobic *tert*-butylphenyl group of **5.93a** is expected to ensure its complexing in the βCD annulus while the ethyl group of **5.93b** is less likely to cause significant complexation.

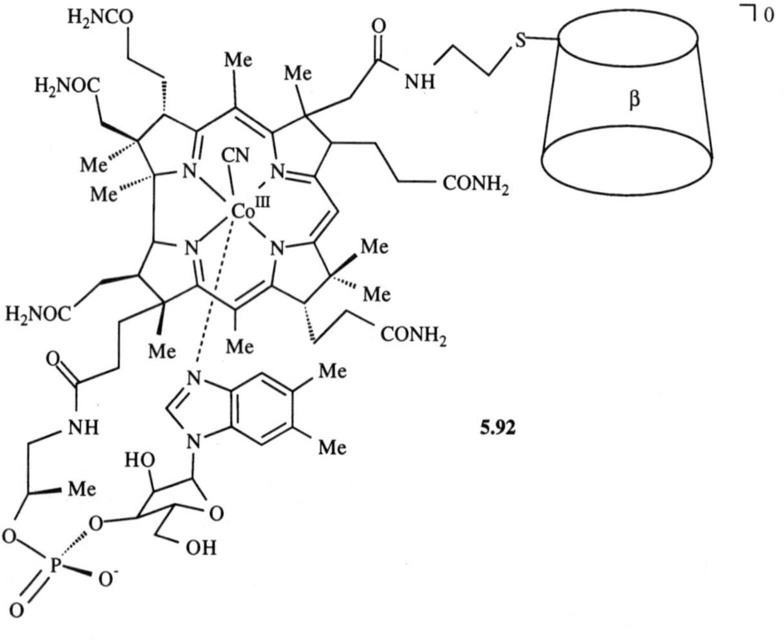

5.92

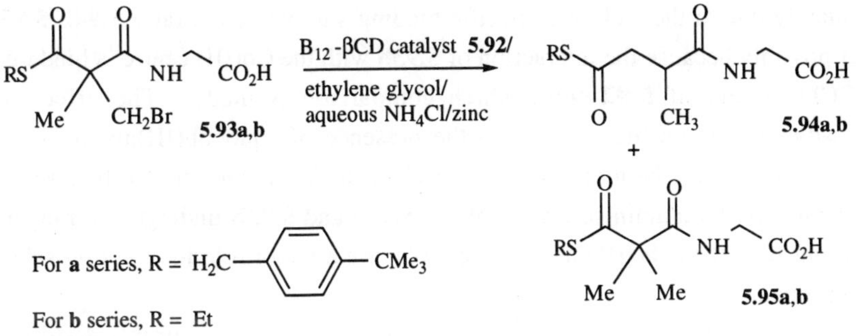

Fig. 5.11. Reaction of **5.93a,b** to **5.94a,b** and **5.95a,b** catalysed by B_{12}-βCD metalloCD **5.92**.

These expectations appear to be borne out by the data in Table 5.8. The ratios of the products **5.94a:5.95a** and **5.94b:5.95b** in the presence of **5.92** are quite different; 0.43 and 0.16, respectively. The ratio **5.94a:5.95a** decreases markedly in the presence of 1-aminoadamantane which occupies the βCD annulus of **5.92**, largely to the exclusion of **5.93a**, while **5.94b:5.95b** changes to a much smaller extent. The **5.94a:5.95a** ratio is smaller in the presence of aquocob(III)alamin **(5.86)**

Table 5.8. Selectivities of the Catalyst **5.92** and Aquocob(III)alamin.[a]

Catalyst	5.94a:5.95a[b]	5.94b:5.95b[b]	Preference[c]
5.92	0.43	0.16	2.73
5.92-1-aminoadamantane	0.16	0.21	0.77
aquocob(III)alamin	0.23	0.53	0.43
aquocob(III)alamin-1-aminoadamantane	0.22	0.46	0.48
5.92 : aquocob(III)alamin	1.89	0.30	6.34
5.92 : **5.92**-1-aminoadamantane	2.74	0.77	3.56
aquocob(III)alamin : aquocob(III)alamin-1-aminoadamantane	1.04	1.16	0.89

[a]Equimolar **5.93a** and **5.93b** (7×10^{-5} mol dm^{-3}) and 9×10^{-5} mol dm^{-3} **5.92** or aquocob(III)-alamin reacted in a 2:8 mixture of ethylene glycol to 8% aqueous NH$_4$Cl in the presence of zinc. [b]Ratio of products obtained [c]Ratio of values in second and third columns.

presumably due to the lack of a specific binding site, while the ratio **5.94b:5.95b** is larger possibly because the interaction of **5.93b** with the Co(III) centre is hindered by the βCD moiety of **5.92** with which comparison is made. The effect of 1-aminoadamantane on the reaction in the presence of aquocob(III)alamin is small consistent with there being no obvious mechanism for inhibition. In the absence of either aquocob(III)alamin or **5.92** < 5% of **5.93a** and **5.93b** undergo rearrangement consistent with the cob(III)alamin moiety playing a major role in directing substrate rearrangement.

Aqueous solutions of the Mn^{2+}, Mn^{3+} and Fe^{3+} metallo-tetrakis(4-sulfonato-phenyl)porphyrins show small changes in their uv-visible absorption spectra and increases in their water proton spin-lattice relaxation rates on the addition of αCD, βCD and γCD [85]. This is interpreted in terms of the formation of complexes where the metalloporphyrin is sandwiched between two CDs so that its plane is parallel to their annular faces. However, it is reported that Zn^{2+} and Fe^{3+} tetrakis(4-sulfonatophenyl)porphyrin form strong complexes with βCD where each 4-sulfonatophenyl group is complexed within a βCD annulus [86]. A third type of structure is also proposed for the complexation of tetrakis(4-(3-aminopropyloxy)-phenyl)porphyrin where the secondary faces of two heptakis(2,6-di-*O*-methyl)-βCDs almost touch as they each complex almost half each of the porphyrin [87]. While these observations are very interesting, most research in this area now concentrates on attaching CDs to the porphyrin structure to generate binary metalloCDs where the collective guest complexing properties of the strongly coordinated metal ion and the linked CD generate ternary metalloCDs with impressive catalytic properties.

The hemoprotein based cytochrome P_{450} acts as a selective Fe^{2+}/Fe^{3+} electron transfer catalyst in the hydroxylation and epoxidation reactions:

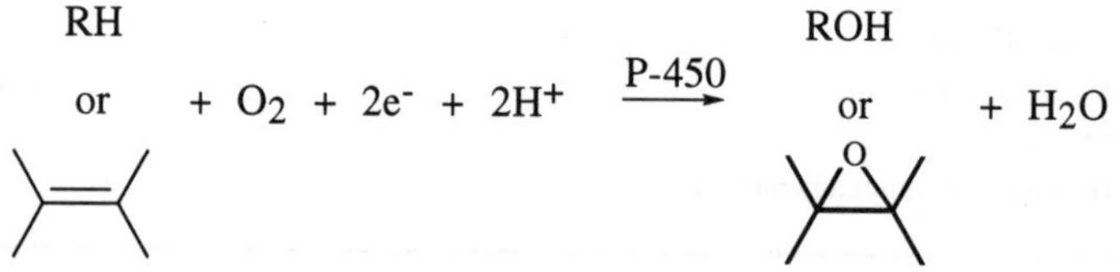

In an effort to mimic cyctochrome P_{450}, the metalloCD of heptakis-(2,6-di-*O*-methyl)-βCD (DMβCD) linked through C(2) to an Fe^{3+} porphyrin (**5.96**) (where the 6 and 7 methoxy groups at C(2) and C(6), respectively, are not shown) has been used as the catalyst in the enantioselective oxidation of (*S*)-α-pinene (**5.97S**) and (*R*)-α-pinene (**5.97R**). The catalytically active species is generated through irradiation with light of λ > 350 nm [88]. For a racemic mixture of **5.97S** and **5.97R** in benzene, the catalytic turnover is 144 and the mol % of **5.98** produced is 35 (5), **5.99** 18 (5), **5.100** 30 (6), **5.101** 14 (4), **5.102** 3 (10) and **5.103** 0, where the figures in parentheses refer to the modest enantiomeric excess of the (*S*)-enantiomer. The proportions of the products vary with the solvent, and in acetonitrile and acetone the enantiomeric excess of the (*S*)-products tends to be greater. This may be because these solvents cause a higher proportion of pinene to be complexed in the DMβCD annulus which results in a greater extent of enantioselectivity for (*S*)-pinene over (*R*)-pinene.

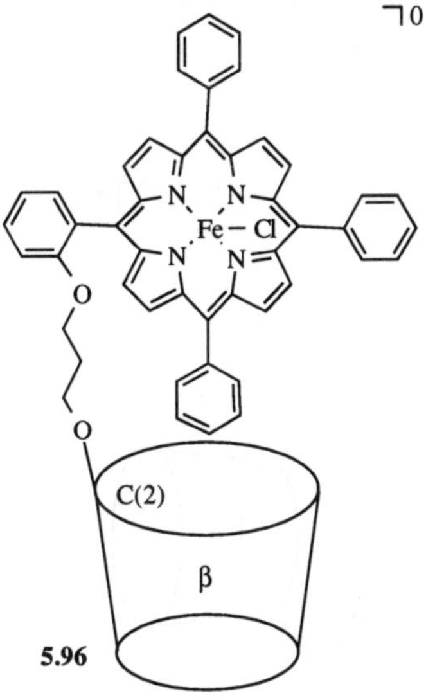

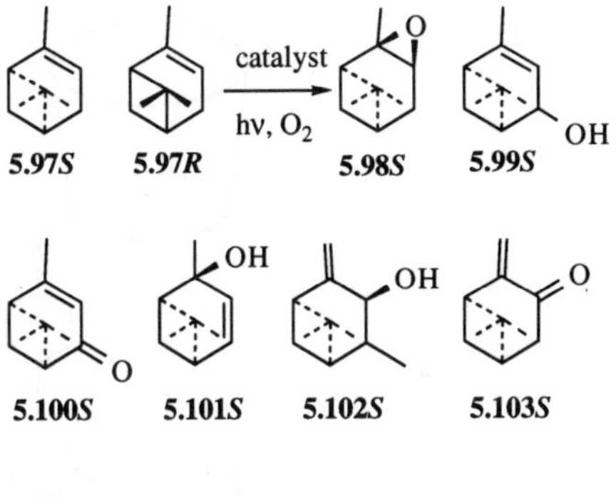

Another attempt to mimic cytochrome P_{450} involves two βCDs bound to an Fe^{2+} porphyrin to give the sandwiched structure **5.104**, in which a guest-binding site is positioned to either side of the porphyrin plane [89]. This latter aspect raises the possibility that **5.104** may be able to complex guests to form Michaelis complexes, as do cytochrome P_{450} and similar hemoproteins with substrates. In the presence of **5.104**, the epoxidation of cyclohexene occurs in 55% yield in aqueous phosphate buffer using iodosylbenzene as the oxygen source, while only a trace of epoxycyclohexane is detected when the simple tetrakis(*p*-sulfonatophenyl)porphyrin-atoiron(II) is used as the catalyst [90]. Under similar conditions **5.104** catalyses the epoxidations of norbornene, styrene and *p*-chlorostyrene in 14%, 18% and 29% yield.

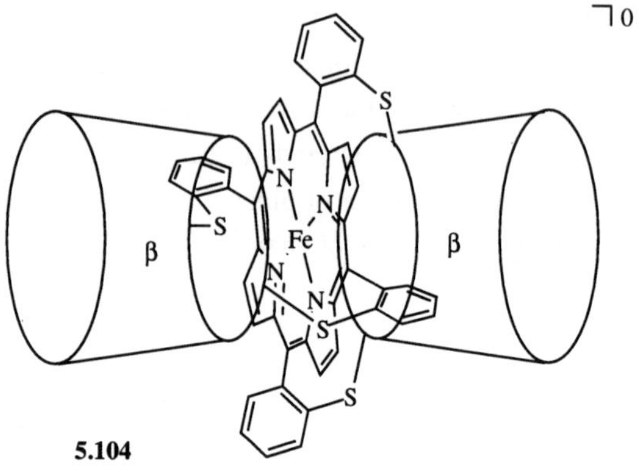

5.104

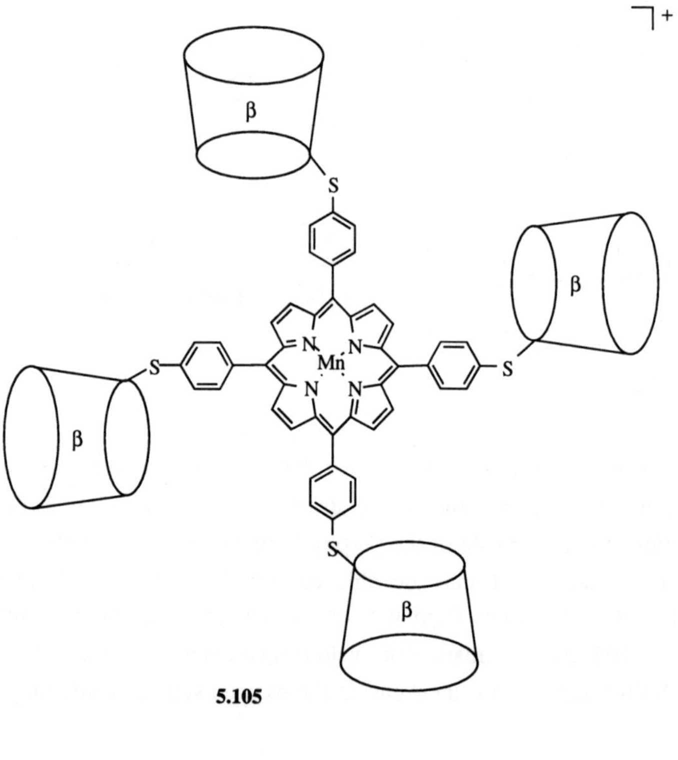

5.105

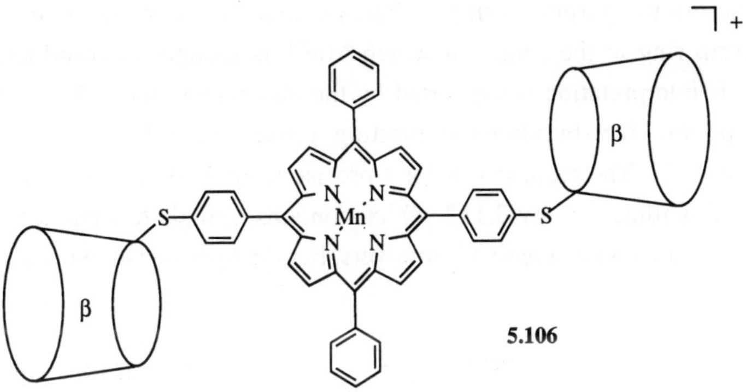

5.106

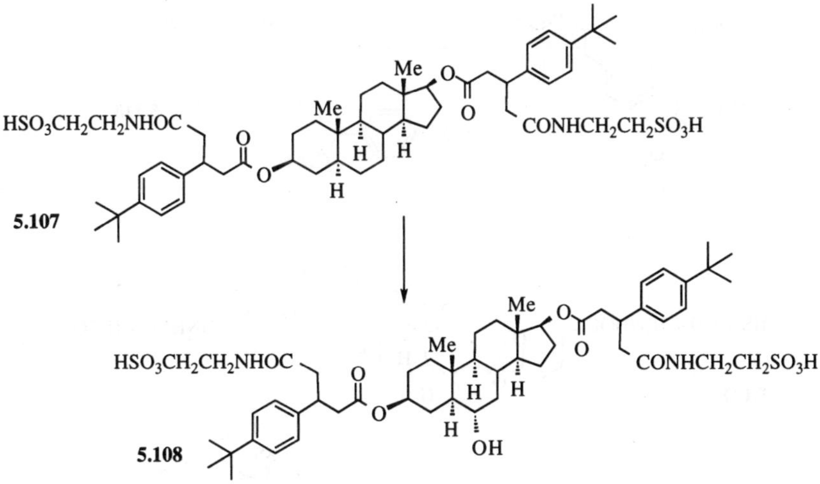

5.107

5.108

Through elegant design of the Mn^{3+} metalloCDs **5.105** and **5.106**, it has proved possible to mimic the P_{450} catalysed hydroxylations and epoxidations in larger molecules [91-93]. Thus, steroid **5.107** is hydroxylated to give the single product **5.108**, with at least 4 turnovers in the presence of **5.105** with iodosobenzene as the oxidant [92,93]. However, **5.105** is oxidised during the catalytic cycle. Under similar conditions **5.109** gives exclusively **5.110** with up to 650 turnovers at rates comparable to those for P_{450} enzymes [93,94]. The selectivity of these hydroxylations is consistent with **5.107** and **5.109** being bound to **5.105** by the complexation of their two *tert*-butylphenyl groups by βCD moieties in *trans* positions across the porphyrin ring. Thus, the hydroxylation site is directly above the porphyrin ring at the centre of which Mn^{3+} is thought to coordinate the oxo ligand. This interpretation is supported by the observation that **5.111**, which lacks the hydrophobic *tert*-butylphenyl binding groups of **5.107** and **5.109**, is not hydroxylated [92] The requirement for a precise fit to **5.105** of the compound to be hydroxylated is illustrated by **5.112**, which contains a single *tert*-butylphenyl group and is hydroxylated with at least 10 turnovers, but the hydroxylation occurs at several sites.

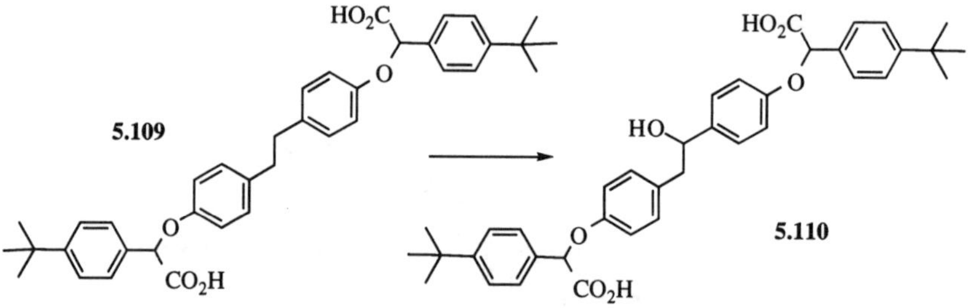

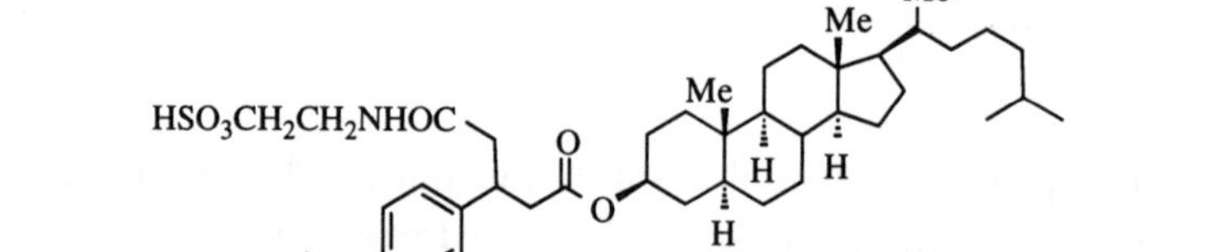

The tetrafluorination of each of the four phenyl rings of **5.105** stabilises it towards oxidation and 187 turnovers have been achieved with this catalyst in the conversion of **5.107** to **5.108** which compares very favourably with the 5 turnovers achieved by **5.105** [94].

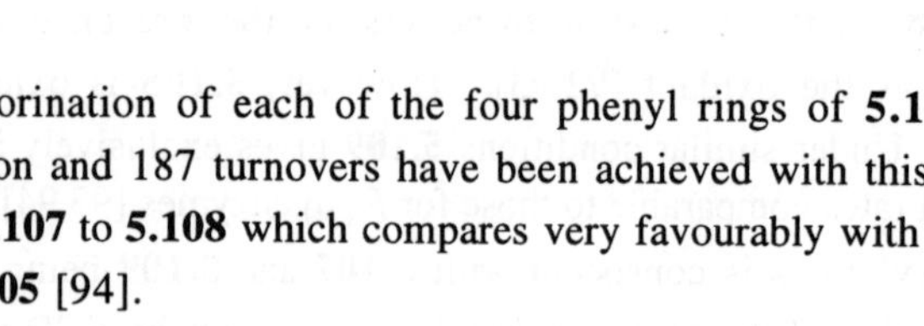

The epoxidation of the stilbene derivative **5.113** to give **5.114** is catalysed by both **5.105** and **5.106**, when iodosobenzene is the oxidant, 15 and 16 times more rapidly than **5.115** is converted to its corresponding epoxide [91-94]. For the first of these epoxidations the catalytic effectiveness of **5.105** is shown by the observation of up to 40 turnovers. The analogous epoxidation of **5.116** in the presence of **5.105** is 21 times faster than that of **5.115**. This variation in epoxidation rates is attributed to the hydrophobic nitrophenyl and *tert*-butylphenyl groups of **5.113** and **5.116** binding more strongly in complexes of **5.105** and **5.106** than do the hydrophilic end groups of **5.115**. Thus, under similar conditions the proportion of **5.113** and **5.116** complexed is greater than that of **5.115**. In the complexes of **5.105**, **5.113** and **5.116** are thought to be bound by the *trans* βCD moieties so that the double bond epoxidised is held above the porphyrin, as is a regiochemical necessity in the case of their complexes with **5.106**. The very poor selectivity between the stibene derivatives observed in the presence of the *cis* isomer of **5.106**, where such regiochemistry in the complex is much less probable, supports this inter-pretation.

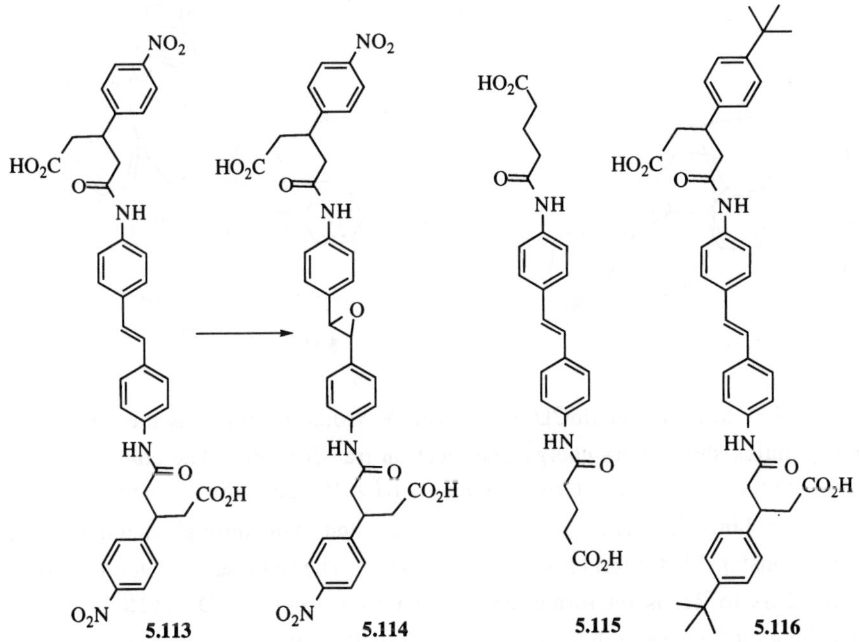

5.8. Energy and Electron Transfer in Metallocyclodextrins

Energy and electron transfer are intimately related processes and both have been observed in metalloCDs. An interesting example of energy transfer is seen with **5.117** where the dansyl group of 6^A-(5-dansylamino-3-azapentylamino)-6^A-deoxy-βCD self-complexes in the βCD annulus to give a strong fluorescence [95]. On coordination of Cu^{2+} by the three amine groups this fluorescence is quenched through the d^9 electronic manifold of Cu^{2+}. (It has been suggested that the sulfonamide group also coordinates, but such coordination would be very strained given the necessity to form a four-membered chelate ring.) Fe^{2+}, Ni^{2+} and Co^{2+}, all of which are potential quenchers, are much less effective than Cu^{2+} in quenching the dansyl fluorescence which may indicate that they coordinate less strongly due to steric restraints. Such restraint is suggested as the reason for Cu^{2+} being ineffective in quenching the fluorescence of the 6^A-(2-dansylaminoethylamino)-6^A-deoxy-βCD analogue of **5.117**.

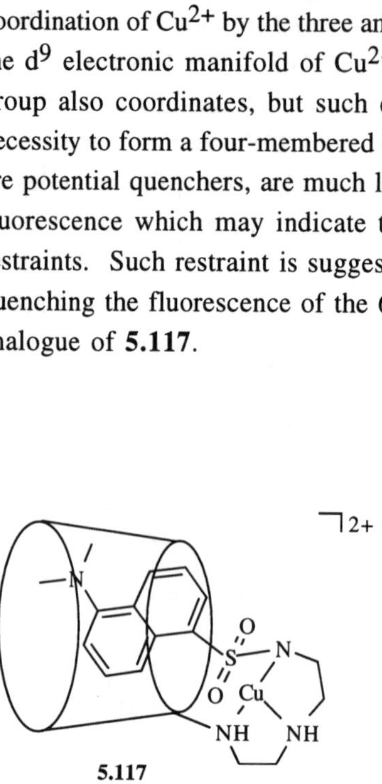

5.117 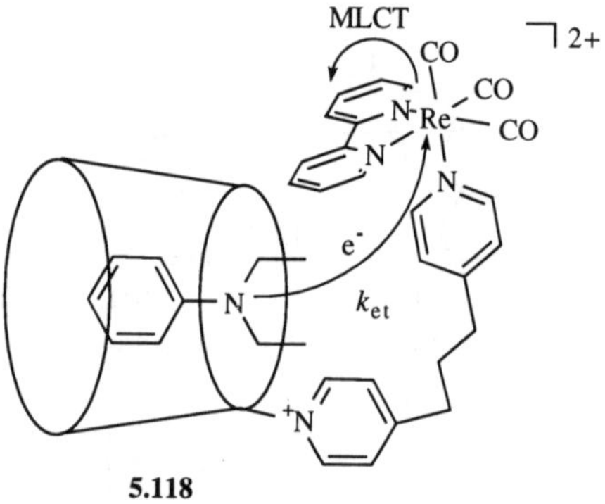   5.118

The Re^+ ternary metalloCD **5.118** with *N,N*-diethylaniline as the guest in the βCD annulus shows both energy and electron transfer [96]. The complexation of *N,N*-diethylaniline is characterised by $K_{11} = 1.1 \times 10^3$ dm^3 mol^{-1}. In the absence of *N,N*-diethylaniline, **5.118** is raised to an excited state through metal to ligand charge-transfer (MLCT) and fluoresces at 580 nm. The fluorescence lifetime shortens from 92 ns to 21 ns on formation of the ternary metalloCD **5.118** because of quenching through electron transfer from *N,N*-diethylaniline with an electron transfer rate constant $k_{et} = 3.7 \times 10^7$ s^{-1}.

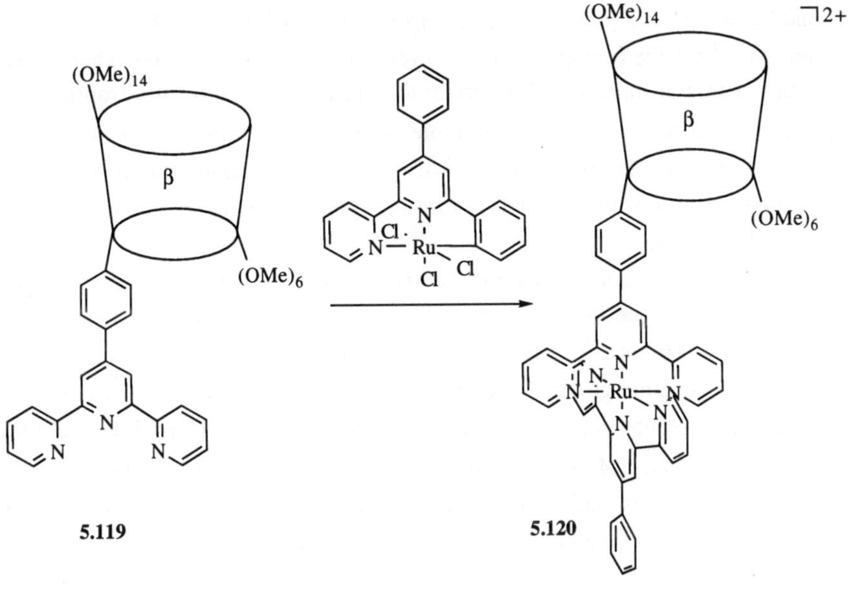

5.119   **5.120**

A similar quenching is seen when the tolyl-terpyridyl substituted TMβCD **5.119** coordinates Ru^{2+} to give the luminescent metalloCD **5.120** [97]. The formation of a ternary metalloCD by **5.120** with anthraquinone-2-carboxylic acid in water partially quenches the MLCT emission of Ru^{2+} probably through electron transfer from anthraquinone-2-carboxylic acid to the Ru^{2+} centre.

Some of the trivalent lanthanides possess luminescent properties which make them particularly interesting for coordinating to CDs and thereby producing light harvesting ternary metalloCDs [98-101]. The hard acid character of the trivalent lanthanides gives them a preference for coordinating to hard base oxygen donor CD substituents exemplified by the ether oxygens of the coronand substituents shown in **5.121** and **5.122** (formed through the 6^A- and $6^A,6^D$-substitution of βCD by 1,4,10,13-tetraoxa-7,16-diazacyclooctadecane [98,99,102]. While **5.121** and **5.122** are very stable, detailed formation studies have not been published and it is interesting to briefly consider the alkali metal ion analogues of **5.121**. The alkali metal ions resemble the trivalent lanthanides in their hard acid character, and also in size in the case of the heavier alkali metal ions. In *N,N*-dimethylformamide the complexation of 4-nitrophenolate by metal ion free **5.121** is characterised by $K_{11} = 7.5 \times 10^3$, $2.8 \times$

10^4 and 9.0×10^3 dm^3 mol^{-1}, which compare with 1.17×10^3, 4.0×10^2 and 7.2×10^2 dm^3 mol^{-1} for the formation of the corresponding βCD complexes where in both cases the successive values of K_{11} pertain to the Li$^+$, Na$^+$ and K$^+$ 4-nitrophenolates, respectively. It appears that the complexation of 4-nitrophenolate in the βCD annulus of **5.121** stabilises the coordination of the alkali metal ion by the diazacrown ether substituent which in turn provides an electrostatic attraction for 4-nitrophenolate [102].

Generally Eu^{3+} complexes exhibit strong red luminescence arising from transitions between the lowest energy $^5D_0$ excited state to the $^7F_0$ (580 nm), $^7F_1$ (592 nm), $^7F_2$ (616 nm), $^7F_3$ (650 nm), $^7F_4$ (700 nm), $^7F_5$ (750 nm), and $^7F_6$ (810 nm) components of the ground states manifold with the transitions to $^7F_1$, $^7F_2$, and $^7F_4$ accounting for 95% of the emission intensity. Both **5.121** and **5.122** and their parent complex ion, 1,4,10,13-tetraoxa-7,16-diazacyclooctadecaneeuropium(III), exhibit these emissions, and also dominant absorptions at 394 and 470 nm assigned to transitions $^5L_6 \leftarrow {}^7F_0$ and $^5D_2 \leftarrow {}^7F_0$, respectively.

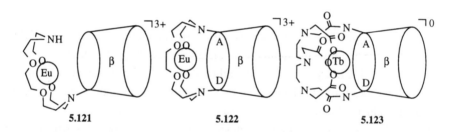

In acetonitrile solution the addition of benzene has little effect on the emission intensity of 1,4,10,13-tetraoxa-7,16-diazacyclooctadecaneeuropium(III) on excitation at the benzene 254 nm absorption frequency, but **5.121** shows a substantial increase in emission intensity under the same conditions [98]. This is attributed to absorption-energy-transfer-emission (AETE) occurring when benzene complexes in the βCD annulus of **5.121** and, in its excited state, acts as an energy donor to Eu^{3+} which is in close proximity. The absence of AETE for 1,4,10,13-tetraoxa-7,16-diazacyclooctadecaneeuropium(III) is attributed to its inability to bind benzene so that energy transfer from benzene can only occur through a bimolecular route which is very inefficient because of the short excited state lifetime of benzene.

In aqueous solution, picolinic and benzoic acid enhance the emission intensity of **5.121** much more strongly than does benzene [100]. This is probably because picolinate and benzoate both complex inside the βCD annulus and simultaneously coordinate to Eu^{3+} in the ternary metalloCD and thereby decrease the distance between the aromatic energy donor and Eu^{3+} to enhance the efficiency of AETE. In contrast, it appears that the 1,4,10,13-tetraoxa-7,16-diazacyclooctadecaneeuropium(III)

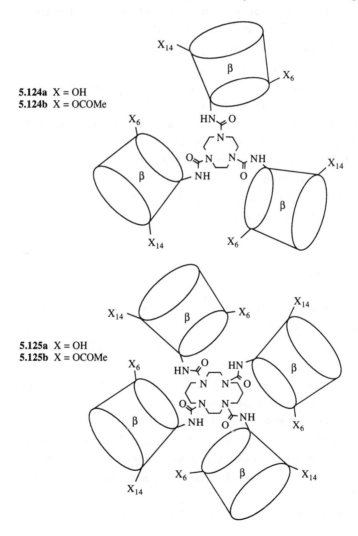

5.124a X = OH
5.124b X = OCOMe

5.125a X = OH
5.125b X = OCOMe

substituent swings away from the βCD moiety to which it is attached in the ternary metalloCD formed by benzene and thereby lowers the efficiency of AETE [100]. Addition of benzene to aqueous solutions of **5.122** where Eu^{3+} is tethered more closely to the βCD annulus causes little increase in Eu^{3+} luminescence. This is because the inclusion of benzene is weak ($K_{11} < 10$ dm^3 mol^{-1}) which is probably a consequence of the close proximity of Eu^{3+} decreasing the effective hydrophobicity of the βCD annulus. However, polar pyridine includes more strongly in **5.121** and **5.122** where $K_{11} = 1.05 \times 10^3$ and 3.48×10^2 dm^3 mol^{-1}, respectively. In both cases Eu^{3+} luminescence is strongly increased through AETE, the more so for **5.122**, probably because pyridine and Eu^{3+} are in closer proximity. Similarly, in the metalloCD **5.123** where Tb^{3+} is coordinated by the amino and carboxylate groups of a 6^A- and 6^D-substituted βCD, Tb^{3+} emits a strong luminescence at 544 nm when either naphthalene or 1,2,4,5-tetramethylbenzene complexed in the βCD annulus is excited at 275 or 278 nm, respectively [100,101].

The substitution of βCD on each nitrogen of 1,4,7-triazanonane and 1,4,7,10-tetraazadodecane to produce **5.124a**, **5.124b**, **5.125a** and **5.125b** has also been reported [103]. The Eu^{3+} complex of **5.125a** has luminescence lifetimes of 0.66 ms and 3.03 ms in methanol and D_2O, respectively.

5.9. References

1.   D. R. Alston, T. H. Lilley and J. F. Stoddart, *J. Chem. Soc., Chem. Commun.* (1985) 1600.

2.   D. R. Alston, A. M. Z. Slawin, J. F. Stoddart and D. J. Williams, *J. Chem. Soc., Chem. Commun.* (1985) 1602.

3.   D. R. Alston, A. M. Z. Slawin, J. F. Stoddart and D. J. Williams, *Angew. Chem., Int. Ed. Engl.* **24** (1985) 786.

4.   H. Yokoi, M. Satoh and M. Iwaizumi, *J. Am. Chem. Soc.* **113** (1991) 1530.

5.   K. Yamanari, M. Nakamichi and Y. Shimura, *Inorg. Chem.* **28** (1989) 248.

6.   Y. Wang, S. Mendoza and A. E. Kaifer, *Inorg. Chem.* **37** (1998) 317.

7.   T. Matsue, D. H. Evans, T. Osa and N. Kobayashi, *J. Am. Chem. Soc.* **107** (1985) 3411.

8.   K. Mochida and Y. Matsui, *Chem. Lett.* (1976) 963.

9. M. McNamara and N. R. Russell, *J. Inclusion Phenom. Mol. Recogn. Chem.* **13** (1992) 145.

10. R. Fuchs, N. Habermann and P Klüfers, *Angew. Chem., Int. Ed. Engl.* **32** (1993) 852.

11. P. Klüfers and J. Schuhmacher, *Angew. Chem., Int. Ed. Engl.* **33** (1994) 1863.

12. P. Klüfers, H. Piotrowski and J. Uhlendorf, *Chem. Eur. J.* **3** (1997) 601.

13. S. E. Brown, J. H. Coates, C. J. Easton, S. J. van Eyk, S. F. Lincoln, B. L. May, M. A. Stile, C. B. Whalland and M. J. Williams, *J. Chem. Soc., Chem. Commun.* (1994) 47.

14. S. E. Brown, J. H. Coates, C. J. Easton and S. F. Lincoln, *J. Chem. Soc., Faraday Trans.* **90** (1994) 739.

15. C. A. Haskard, C. J. Easton, B. L. May and S. F. Lincoln, *Inorg. Chem.* **35** (1996) 1059.

16. R. M. Smith and A. E. Martell, in *Critical Stability Constants* (Plenum Press, New York, 1975) Vol. 1.

17. V. Cucinotta, F. D'Alessandro, G. Impellizzeri, G. Maccarrone, E. Rizzarelli and G. Vecchio, *J. Chem. Soc., Perkin Trans. 2* (1996) 1785.

18. S. E. Brown, C. A. Haskard, C. J. Easton and S. F. Lincoln, *J. Chem. Soc., Faraday Trans.* **91** (1995) 1013, 4335.

19. V. Cucinotta, A. Mangano, G. Nobile, A. M. Santoro and G. Vecchio, *J. Inorg. Biochem.* **52** (1993) 183.

20. V. Cucinotta, G. Grasso, S. Pedotti, E. Rizzarelli, G. Vecchio, B. Di Blasio, C. Isernia, M. Saviano and C. Pedone, *Inorg. Chem.* **35** (1996) 7535.

21. K. Matsumoto, Y. Noguchi and N. Yoshida, *Inorg. Chim. Acta* **272** (1998) 162.

22. R. Deschenaux, M. M. Harding and T. Ruch, *J. Chem. Soc., Perkin Trans. 2* (1993) 1251.

23. R. Deschenaux, T Ruch, P.-F. Deschenaux, A. Juris and R. Zeissel, *Helv. Chim. Acta* **78** (1995) 619.

24. R. Deschenaux, A. Greppi, T. Ruch, H.-P. Kriemler, F. Raschdorf and R. Ziessel, *Tetrahedron Lett.* **35** (1994) 2165.

25. A. W. Coleman, C.-C. Ling and M. Miocque, *Angew. Chem., Int. Ed. Engl.* **31** (1992) 1381.

26. C.-C. Ling, M. Miocque and A. W. Coleman, *J. Coord. Chem.* **28** (1993) 313.

27. A. W. Coleman, C.-C. Ling and M. Miocque. *J. Coord. Chem.* **26** (1992) 137.

28. R. Bonomo, V. Cucinotta, F. D'Alessandro, G. Impellizzeri, G. Maccarrone, G. Vecchio and E. Rizzarelli, *Inorg. Chem.* **30** (1991) 2708.

29. G. Impellizzeri, G. Maccarrone, E. Rizzarelli, G. Vecchio, R. Corradini and R. Marchelli, *Angew. Chem., Int. Ed. Engl.* **30** (1991) 1348.

30. R. Corridini, A. Dossena, G. Impellizzeri, G. Maccarrone, R. Marchelli, E. Rizzarelli, G. Sartor and G. Vecchio, *J. Am. Chem. Soc.* **116** (1994) 10267.

31. R. P. Bonomo, B. Di Blasio, G. Maccarrone, V. Pavone, C. Pedone, E. Rizzarelli, M. Saviano and G. Vecchio, *Inorg. Chem.* **35** (1996) 4497.

32. V. Cucinotta, F. D'Alessandro, G. Impellizzeri and G. Vecchio, *J. Chem. Soc., Chem. Commun.* (1992) 1743.

33. R. P. Bonomo, V. Cucinotta, F. D'Alessandro, G. Impellizzeri, G. Maccarrone and E. Rizzarelli, *J. Inclusion Phenom. Mol. Recogn. Chem.* **15** (1993) 167.

34. R. P. Bonomo, S. Pedotti, G. Vecchio and E. Rizzarelli, *Inorg. Chem.* **35** (1996) 6873.

35. V. Cucinotta, G. Grasso, S. Pedotti, E. Rizzarelli, G. Vecchio, B. Di Blasio, C. Isernia, M. Saviano and C. Pedone, *Inorg. Chem.* **35** (1996) 7535.

36. R. P. Bonomo, G. Impellizzeri, G. Pappalardo, E. Rizzarelli and G. Vecchio, *Gazz. Chim. Ital.* **123** (1993) 593.

37. V. Cucinotta, F. D'Alessandro, G. Impellizzeri, G. Pappalardo, E. Rizzarelli and G. Vecchio, *J. Chem. Soc., Chem. Commun.* (1991) 293.

38. B. Di Blasio, V. Pavone, F. Nastri, C. Isernia, M. Saviano, C. Pedone, V. Cucinotta, G. Impellizzeri, E. Rizzarelli and G. Vecchio, *Proc. Natl. Acad. Sci. USA* **89** (1992) 7218.

39. T. J. Wenzel, M. S. Bogyo and E. L. Lebeau, *J. Am. Chem. Soc.* **116** (1994) 4858.

40. I. Tabushi, *Acc. Chem. Res.* **15** (1982) 66.

41. R. Breslow, *Recl. Trav. Chim. Pays-Bas* **113** (1994) 493.

42. R. Breslow, *Pure Appl. Chem.* **66** (1994) 1573.

43. R. Breslow, *Acc. Chem. Res.* **28** (1995) 146.

44. A. J. Kirby, *Angew. Chem., Int. Ed. Engl.* **35** (1996) 707.

45. Y. Murakami, J. Kikuchi, Y. Hisaeda and O. Hayashida, *Chem. Rev.* **96** (1996) 721.

46. T. Osa and I. Suzuki, *Reactivity of Included Guests*, in *Comprehensive Supramolecular Chemistry*, eds. J. L. Atwood, J. E. D. Davies, D. D. MacNicol, F. Vögtle and J.-M. Lehn, Vol. 3, eds. J. Szejtli and T. Osa (Pergamon, Oxford, 1996), p. 367.

47. M. Komiyama and H. Shigekawa, *Cyclodextrins as Enzyme Models*, in *Comprehensive Supramolecular Chemistry*, eds. J. L. Atwood, J. E. D. Davies, D. D. MacNicol, F. Vögtle and J.-M. Lehn, Vol. 3, eds. J. Szejtli and T. Osa (Pergamon, Oxford, 1996), p. 401.

48. S. F. Lincoln, *Coord. Chem. Rev.* **166** (1997) 255.

49. R. Breslow and S. D. Dong, *Chem. Rev.* **98** (1998) 1997.

50. R. Breslow and L. E. Overman, *J. Am. Chem. Soc.* **92** (1970) 1075.

51. E. U. Akkaya and A. W. Czarnik, *J. Am. Chem. Soc.* **110** (1988) 8553.

52. E. U. Akkaya and A. W. Czarnik, *J. Phys. Org. Chem.* **5** (1992) 540.

53. M. I. Rosenthal and A. W. Czarnik, *J. Inclusion Phenom. Mol. Recogn. Chem.* **10** (1991) 119.

54. R. Breslow and S. Singh, *Bioorganic Chem.* **16** (1988) 408.

55. M. Komiyama and Y. Matsumoto, *Chem. Lett.* (1989) 719.

56. I. Tabushi, N. Shimizu, T. Sugimoto, M. Shiozuka and K. Yamamura, *J. Am. Chem. Soc.* **99** (1977) 7100.

57. I. Tabushi, Y. Kuroda and A. Mochizuki, *J. Am. Chem. Soc.* **102** (1980) 1152.

58. I. Tabushi and Y. Kuroda, *J. Am. Chem. Soc.* **106** (1984) 4580.

59. V. Cucinotta, F. D'Alessandro, G. Impellizzeri and G. Vecchio, *Carbohydr. Res.* **224** (1992) 95.

60. R. P. Bonomo, E. Conte, G. De Guidi, G. Maccarrone, E. Rizzarelli and G. Vecchio, *J. Chem. Soc., Dalton Trans.* (1996) 4351.

61. G. Condorelli, L. L. Constanzo, G. De Guidi, S. Giuffrida, E. Rizzarelli and G. Vecchio, *J. Inorg. Biochem.* **54** (1994) 257.

62. R. Fornasier, E. Scarpa, P. Scrimin, P. Tecilla and U. Tonellato, *J. Inclusion Phenom. Mol. Recogn. Chem.* **14** (1992) 205.

63. Y. Matsui, T. Yokoi and K. Mochida, *Chem. Lett.* (1976) 1037.

64. H.-J. Schneider and F. Xiao, *J. Chem. Soc., Perkin Trans. 2* (1992) 387.

65.   M. Bonchio, T. Carofiglio, F. Di Furia and R. Fornasier, *J. Org. Chem.* **60** (1995) 5986.

66.   M. Yashiro, T. Takarada, S. Miyama and M. Komiyama, *J. Chem. Soc., Chem. Commun.* (1994) 1757.

67.   M. T. Reetz and S. R. Waldvogel, *Angew. Chem., Int. Ed. Engl.* **36** (1997) 865.

68.   E. Monflier, G. Fremy, Y. Castanet and A. Mortreux, *Angew. Chem., Int. Ed. Engl.* **34** (1995) 2269.

69.   E. Monflier, S. Tilloy, G. Fremy, Y. Castanet and A. Mortreux, *Tetrahedron Lett.* **36** (1995) 9481.

70.   E. Monflier, S. Tilloy, F. Bertoux, Y. Castanet and A. Mortreux, *New. J. Chem.* (1997) 857.

71.   E. Monflier, S. Tilloy, Y. Castanet and A. Mortreux, *Tetrahedron Lett.* **39** (1998) 2959.

72.   E. Monflier, E. Blouet, Y. Barbaux and A. Mortreux, *Angew. Chem., Int. Ed. Engl.* **33** (1994) 2100.

73.   E. Monflier, S. Tilloy, G. Fremy, Y. Barbaux and A. Mortreux, *Tetrahedron Lett.* **36** (1995) 387.

74.   M. T. Reetz and J. Rudolph, *Tetrahedron Asym.* **4** (1993) 2405.

75.   M. Sawamura, K. Kitayama and Y. Ito, *Tetrahedron Asym.* **4** (1993) 1829.

76.   I. Suzuki, Q. Chen, Y. Kashiwagi, T. Osa and A. Ueno, *Chem. Lett.* (1993) 1719.

77.   I. Suzuki, Q. Chen, A. Ueno and T. Osa, *Bull. Chem. Soc. Jpn.* **66** (1993) 1472.

78.   R. Breslow and B. Zhang, *J. Am. Chem. Soc.* **114** (1992) 5882.

79.   B. Zhang and R. Breslow, *J. Am. Chem. Soc.* **119** (1997) 1676; **120** (1998) 5854.

80.   R. Breslow and B. Zhang, *J. Am. Chem. Soc.* **116** (1994) 7893.

81.   F. Sallas, A. Marsura, V. Petot, I. Pintér, J. Kovács and L. Jicsinszky, *Helv. Chim. Acta* **81** (1998) 632.

82.   M. Maletic, H. Wennemers, D. Q. McDonald, R. Breslow and W. C. Still, *Angew. Chem., Int. Ed. Engl.* **35** (1996) 1490.

83.   R. Breslow, P. J. Duggan and J. P. Light, *J. Am. Chem. Soc.* **114** (1992) 3982.

84. M. Rezac and R. Breslow, *Tetrahedron Lett.* **38** (1997) 5763.
85. S. K. Sur and R. G. Bryant, *J. Phys. Chem.* **99** (1995) 4900.
86. S. Mosseri, J. C. Mialocq, B. Perly and P. Hambright, *J. Phys. Chem.* **95** (1991) 2196.
87. J. S. Manka and D. S. Lawrence, *J. Am. Chem. Soc.* **112** (1990) 2440.
88. L. Weber, I. Imiolczyk, G. Haufe, D. Rehorek and H. Hennig, *J. Chem. Soc., Chem. Commun.* (1992) 301.
89. Y. Kuroda, T. Hiroshige, T. Sera, Y. Shiroiwa, H. Tanaka and H. Ogoshi, *J. Am. Chem. Soc.* **111** (1989) 1912.
90. Y. Kuroda, T. Hiroshige and H. Ogoshi, *J. Chem. Soc., Chem. Commun.* (1990) 1594.
91. R. Breslow, X. Zhang and Y. Huang, *J. Am. Chem. Soc.* **119** (1997) 4535.
92. R. Breslow, Y. Huang, X. Zhang and J. Yang, *Proc. Nat. Acad. Sci. USA.* **94** (1997) 11156.
93. R. Breslow, B. Gabriele and J. Yang, *Tetrahedron Lett.* **39** (1998) 2887.
94. R. Breslow, X. Zhang, R. Xu, M. Maletic and R. Merger, *J. Am. Chem. Soc.* **118** (1996) 11678.
95. R. Corradini, A. Dossena, G. Galaverna, R. Marchelli, A. Panagia and G. Sartor, *J. Org. Chem.* **62** (1997) 6283.
96. A. Nakamura, S. Okutsu, Y. Oda, A. Ueno and F. Toda, *Tetrahedron Lett.* **35** (1994) 7241.
97. S. Weidner and Z. Pikramenou, *J. Chem. Soc., Chem. Commun.* (1998) 1473.
98. Z. Pikramenou and D. G. Nocera, *Inorg. Chem.* **31** (1992) 532.
99. Z. Pikramenou, K. M. Johnson and D. G. Nocera, *Tetrahedron Lett.* **34** (1993) 3531.
100. Z. Pikramenou, J. Yu, R. B. Lessard, A. Ponce, P. A. Wong and D. G. Nocera, *Coord. Chem. Rev.* **132** (1994) 181.
101. C. M. Rudzinski, W. K. Hartmann and D. G. Nocera, *Coord. Chem. Rev.* **171** (1998) 115.
102. I. Willner and Z. Goren, *J. Chem. Soc., Chem. Commun.* (1983) 1469.
103. F. Charbonnier, T. Humbert and A. Marsura, *Tetrahedron Lett.* **39** (1998) 3481.

CHAPTER 6
MODIFIED CYCLODEXTRINS

MULTISITE COMPLEXATION OF GUESTS AND LINKED CYCLODEXTRINS

6.1. Multiple Complexation of Guest Species by Cyclodextrins

A major thrust in the modification of natural CDs has been the introduction of additional guest complexing sites. Such modifications can have fascinating consequences both for the complexes formed and the complexed guest. It is instructive to introduce this area of CD chemistry through a brief examination of the multiple complexing of guest species by natural CDs which occurs quite widely. This is exemplified by the complexation of o-, p- and α-fluoro-*trans*-cinnamate and o,p- and α,p-difluoro-*trans*-cinnamate by αCD where both 1:1 and 2:1 complexes are formed in the sequential equilibria:

$$\alpha CD + \text{cinnamate} \xrightleftharpoons{K_{11}} \alpha CD.\text{cinnamate}$$

$$\alpha CD + \alpha CD.\text{cinnamate} \xrightleftharpoons{K_{21}} (\alpha CD)_2.\text{cinnamate}$$

The 1:1 and 2:1 complexes are characterised by different ^{19}F NMR chemical shifts and as consequence a biphasic variation of the ^{19}F NMR chemical shift of the cinnamate is observed with increase in αCD concentration as the cinnamate exchanges rapidly between the different magnetic environments of the two complexes

[1].   For *p*-fluoro-*trans*-cinnamate, K_{11} and K_{21} = 109 and 35 dm^3 mol^{-1}, respectively, in aqueous 0.1 mol dm^{-3} NaCl at 294.0 K.

When two different guests are present, the possibility of i) forming two different 1:1 complexes with a single CD arises together with the possibility of forming ii) dimeric 2:1 complexes with one or other of the guests, iii) dimeric 2:2 complexes where two of the same guest are complexed, and iv) dimeric 2:1:1 complexes where one of each guest is complexed. Possibilities i), iii) and iv) are realised in the 6^A-*O*-α-D-glucosyl-βCD (G$_1$-βCD, **6.1**) complexation of 4-(dimethylamino)benzonitrile (DMABN) in the presence of either benzonitrile or anisole as shown in Fig. 6.1 [2]. Changes in the fluorescence of DMABN with changes in solution composition are consistent with the formation of G$_1$-βCD.DMABN (**6.2**) and a homodimer (G$_1$-βCD)$_2$.(DMABN)$_2$ (**6.4**), together with the analogous complexes **6.3** and **6.6** where DMABN is replaced by either benzonitrile or anisole.   A heterodimer, (G$_1$-βCD)$_2$.DMABN.benzonitrile (**6.5**) or its anisole analogue is also formed.   Both the homo- and heterodimer dimer complexes are probably stabilised by

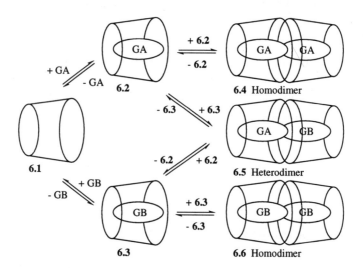

Fig. 6.1. The formation of homo- and heterodimers by 6^A-*O*-α-D-glucosyl-βCD, G$_1$-βCD (**6.1**) with two guests GA and GB where GA is 4-(dimethylamino)benzonitrile and GB is either benzonitrile or anisole. The orientations shown for the complexes are representative only.

interactions between the guests and simultaneous interactions of each guest with both G_1-βCD hosts. In 9:1 ratio by volume water:acetonitrile solvent at 298.2 K, K_{11} = 158 and 86 mol dm^{-3} for G_1-βCD.DMABN and G_1-βCD.benzonitrile, respectively, and the dimerisation of these complexes to form $(G_1$-βCD$)_2$.(DMABN)$_2$ and $(G_1$-βCD$)_2$.DMABN.benzonitrile is characterised by $K_D = 4.1 \times 10^2$ mol dm^{-3} and 5.0×10^1 mol dm^{-3}, respectively.

The stability of dimer complexes can be quite high as is exemplified by the complexes formed between heptakis(2,6-di-O-methyl)-βCD (**6.7**, DMβCD) and the tetraaminoporphyrin (**6.8**) shown in Fig. 6.2 [3]. The spectrophotometrically determined $K_{11} = 7.7 \times 10^4$ mol dm^{-3} and $K_{21} = 5.9 \times 10^4$ mol dm^{-3}, respectively,

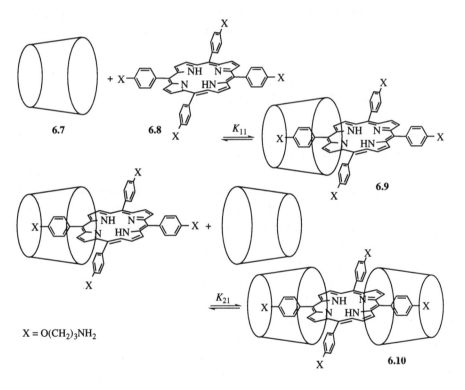

$X = O(CH_2)_3NH_2$

Fig. 6.2. The complexation by DMβCD (**6.7**) of the tetraaminoporphyrin **6.8** to form the 1:1 and 2:1 complexes **6.9** and **6.10**, respectively.

at 333.2 K in aqueous succinate buffer at pH 5.0. These high stabilities are attributed to the complexation of both the phenyl group and part of the hydrophobic periphery of the porphyrin macrocycle in the DMβCD annulus. The alignment of the secondary faces of each DMβCD over the porphyrin macrocycle is deduced from ^{1}H NOESY NMR spectroscopy. When the two porphyrin macrocycle ring protons of **6.8** are displaced by the coordination of Fe^{3+}, 1:1 and 2:1 complexes analogous to **6.9** and **6.10** are formed with $K_{11} = 3.5 \times 10^4$ mol dm^{-3} and $K_{21} = 9.0 \times 10^2$ mol dm^{-3}, respectively, at 298.2 K in aqueous citrate buffer at pH 3.0. The marked decrease in the relative stability of the Fe^{3+} dimer complex is attributed to the introduction of a positive charge into the hydrophobic microenvironment created when two DMβCDs complex the tetraaminoporphyrin simultaneously. Other complexations of porphyrins and their use as substituents in modified CDs are discussed in several of the sections that follow.

6.2. Dimerisation of Modified Cyclodextrins

While the natural CDs show little tendency to dimerise in solution, modified CDs form dimers if they either possess opposite charges or are substituted by groups which are complexed by another CD. An example of a dimer formed between CDs of opposite charge is provided through the interaction of the two modified βCDs produced through substitution at all seven C(6) sites by either $-NH_2$ (βCD(NH$_2$)$_7$) or $-SCH_2CO_2H$ (βCD(SCH$_2$CO$_2^-$)$_7$(H$^+$)$_7$) [4] . This creates 7 positively charged and 7 negatively charged modified βCDs, respectively, at low and high pH so that a solution of both modified βCDs contains a range of opposite and highly charged species at intermediate pH values. The formation of at least five electrostatically bound heterodimers, [βCD(NH$_2$)$_7$.βCD(SCH$_2$CO$_2^-$)$_7$(H$^+$)$_{14-n}$]$^{(7-n)+}$, where n ranges from 5 to 9, occurs in the equilibrium:

$$[βCD(NH_2)_7(H^+)_{(14-i-n)}]^{(14-i-n)+} + [βCD(SCH_2CO_2^-)_7(H^+)_i]^{(7-i)-} \underset{}{\overset{P_n}{\rightleftharpoons}}$$
$$[βCD(NH_2)_7.βCD(SCH_2CO_2^-)_7(H^+)_{(14-n)}]^{(7-n)+}$$

where for the heterodimers the number of protons varies in integers (i) from 7 to 14 and the corresponding variation in charge ranges from 0 to 7. The potentiometrically determined phenomenological stability constants, P_n, encompass the range: log P_n =

7.5, 8.35, 10.25, 8.6, and 6.6 as n increases from 5 to 9 in aqueous solution at 298.2 K.

The photochemically generated radical cations of 6^A-deoxy-6^A-(1'-hexyl-4,4'-bipyridin-1-yl)-βCD (βCDC$_n$V$^{\cdot+}$, **6.11**) and their heptyl- and octyl- analogues form homodimers, as shown in Fig. 6.3 [5]. The homodimer (**6.12**) formation results from the complexation of the alkyl tails of adjacent βCDC$_n$V$^{\cdot+}$ monomers, and is disrupted by either an amphiphile such as *n*-octyl sulfate being competitively complexed by βCDC$_n$V$^{\cdot+}$ to form a 1:1 complex (**6.13**) or βCD competing for complexation of the tail to form a heterodimer (**6.14**). The homodimer formation is characterised by the spectrophotometrically determined $K_D = 1.0 \times 10^2$, 4.0×10^4, 8.9×10^5, and 6.8×10^6 dm^3 mol^{-1} where the alkyl tail is methyl, hexyl, heptyl, and octyl, respectively, in aqueous 0.1 mol dm^{-3} NaCl solutions at 298.2 K. This variation of K_D demonstrates the systematic variation of complex stability with tail length.

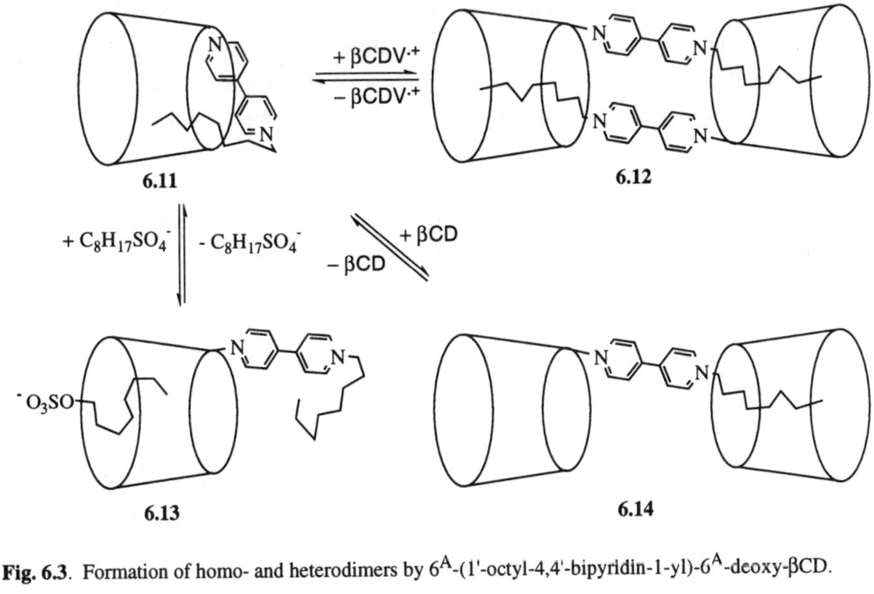

Fig. 6.3. Formation of homo- and heterodimers by 6^A-(1'-octyl-4,4'-bipyridin-1-yl)-6^A-deoxy-βCD.

Such dimerisation is quite common, and a second example is provided by a spectrophotometric and circular dichroic study of the modified βCD substituted at C(6) by the amino group of 5-(4-aminophenyl)-10,15,20-tris(4-sulfonato)phenyl)-porphyrin. It is found that a head-to-tail dimer forms where the porphyrin moiety of one modified βCD enters the secondary face of another, while the head-to-head dimer analogous to **6.12** forms to a lesser extent [6].

In the solid state the proximity of one CD to another is inevitably close, nevertheless, it is of interest to note that complexation of the tail of one modified CD by another occurs here also. X-ray crystallography shows that the alkyl tail of each 6^A-(6-aminohexylamino)-6^A-deoxy-βCD enters the secondary face of the annulus of an adjacent CD molecule and protrudes from the primary face in the formation of polymer-like columns [7]. Similarly, head-to-tail arrangements of 6^A-azido-6^A-deoxy-αCD and of 2^A-*O*-allyl-αCD form helical columns where the azido and allyl tails, respectively, enter the annuli of adjacent CD molecules [8].

As in the case of 1:1 CD.guest complexes, the relative sizes and orientations of the CD annulus and the guest are important factors affecting the stability of CD dimers. Thus, the pyrene moiety linked to γCD through amide and ester bonds at O(6) in **6.15a** and **6.15b** enters the γCD annulus to form the dimers **6.16a** and **6.16b** for which spectrophotometrically determined K_D values of 1.74×10^5 and 1.53×10^4 mol dm^{-3}, respectively, are obtained in aqueous 10% dimethyl sulfoxide at 298.2 K [9]. Fluorescence studies are consistent with the pyrene moieties in **6.16a** and **6.16b** being sufficiently close to form an eximer. However, the isomers of **6.15b** where the pyrene moiety is linked to γCD through an ester bond at either O(2) or O(3) do not form dimers to a detectable extent consistent with dimer stability being critically dependent on the orientation of the pyrene moieties. The βCD analogue of **6.15b** does not form a dimer probably because of the smaller size of the βCD annulus.

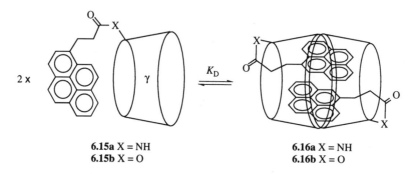

2 x K_D

6.15a X = NH   **6.16a** X = NH
6.15b X = O   **6.16b** X = O

6.3. Covalently Linked Cyclodextrin Dimers

The complexation of free and appended guest species in the dimer formations considered in the preceding sections prompts the obvious query as to what extent cooperativity between covalently linked CDs might enhance guest complexation. This has resulted in the synthesis of a wide range of linked CDs. Thus, sulfide [10], disulfide [11-21], dithioether [14,15,17-19,22-27], diether [28,29], diamine [30,31], diester [13,21,16,32], diamide [26,33-39], imidazole [18,21], benzimidazole [40], porphyrin [22,25,26,41] and urea [39,42] linked CDs have been synthesised and their complexing properties studied. The linking of two CDs may be achieved by substituting either a primary hydroxy group in each CD, or a secondary hydroxy group in each CD, or a primary hydroxy group in one CD and a secondary hydroxy group in the other CD. In some cases two hydroxy groups are substituted on each CD, and in others different combinations of αCD, βCD and γCD have been linked together.

If the two CD moieties of a linked dimer complex the guest simultaneously this may increase the complex stability to a substantially greater extent than that expected from the statistical effect, under which circumstances a cooperative effect arises. This is often observed, as is exemplified by the complexation of the three dyes **6.17**, **6.18** and **6.19** by the βCD dimers **6.20**, **6.21a-6.21d**, and **6.22a** and **6.22b**. Because 6-(*p*-toluidinyl)naphthalene-2-sulfonate (TNS⁻, **6.17**) fluoresces weakly in water but strongly in a hydrophobic environment, its fluorescence increases greatly on complexation in the annuli of **6.20**, **6.21a-6.21d**, and **6.22a** and **6.22b**. The change in TNS⁻ fluorescence with increasing linked CD concentration, in aqueous 0.10 mol dm⁻³ phosphate buffer at pH 7.0 and 298.2 K, yields $K_{11} = 4.5 \times 10^4$, 3.3×10^4, 1.1×10^4, 1.7×10^4, and 1.3×10^4 dm³ mol⁻¹, respectively, for the complexes formed with **6.20** and **6.21a-6.21d** [38]. For the TNS⁻ complexes formed by **6.22a** and **6.22b**, K_{11} = 1.05×10^4 and 6.7×10^3 dm³ mol⁻¹, respectively [35]. The general increase in stability as the linker length decreases in this group of complexes is consistent with an optimisation of the hydrophobic interaction between both TNS⁻ aromatic moieties and the two βCD annuli. The decrease in the stability of the **6.22b**.TNS⁻ complex may indicate a secondary effect of a stereochemical constraint on stability, however, the similar stabilities of **6.21c**.TNS⁻ and **6.22a**.TNS⁻ show that the change in βCD orientation in these complexes has little effect on stability.

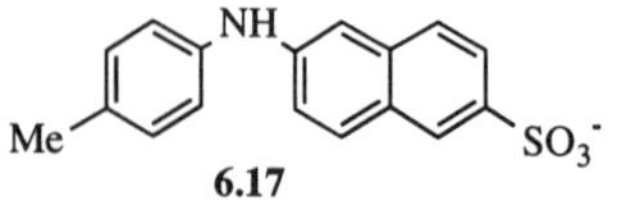

6.17

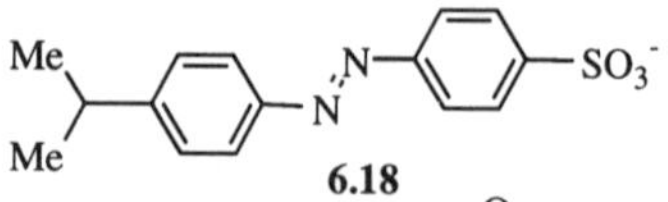

6.18

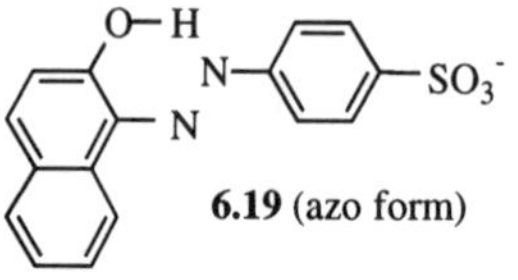

6.19 (azo form)

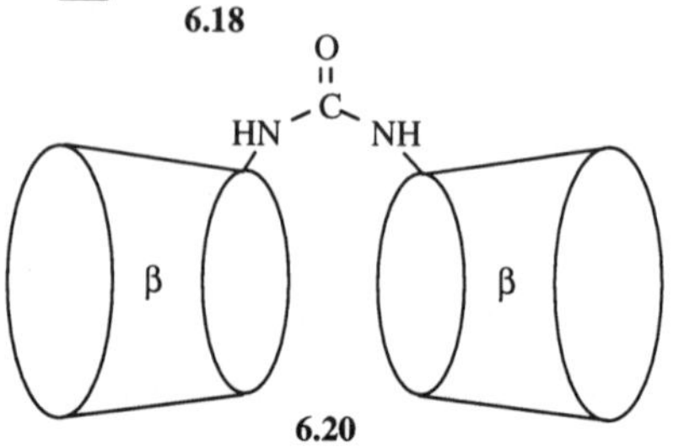

6.20

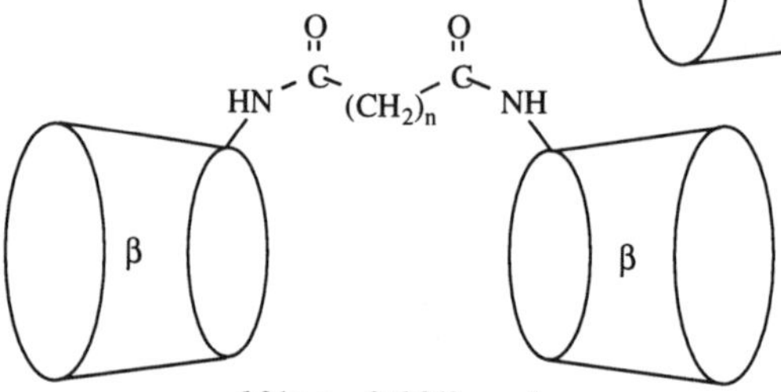

6.21a n = 0 **6.21b** n = 1
6.21c n = 2 **6.21d** n = 3

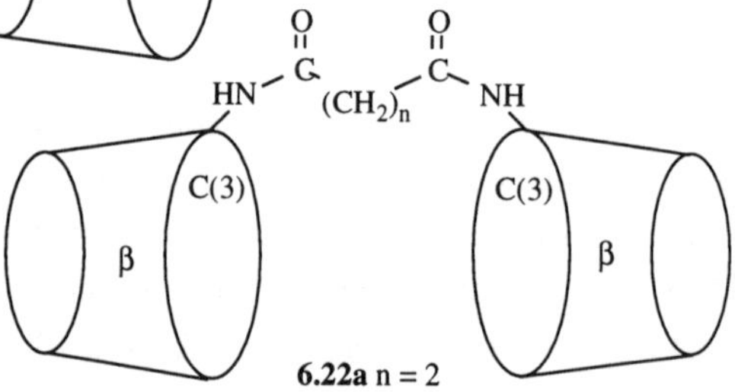

6.22a n = 2
6.22b n = 8

The complexation of TNS⁻ by βCD has generated considerable interest, and under the conditions of the above studies may either be fitted to a model where βCD.TNS⁻ alone forms and $K_{11} = 1.85 \times 10^3$ dm³ mol⁻¹ or to a model where both βCD.TNS⁻ and (βCD)$_2$.TNS⁻ form and $K_{11} = 3.14 \times 10^3$ dm³ mol⁻¹ and $K_{21} = 86$ dm³ mol⁻¹ [38]. It is seen from comparison with these data that there is substantial cooperativity in the complexation of TNS⁻ by **6.20**, **6.21a-6.21d**, and **6.22a** and **6.22b**, and that the lengthening of the linker in **6.22b** decreases the cooperativity. In the latter case, it is possible that some of the decrease in stability of the **6.22b** complex is because the linker itself partially complexes and competes with TNS⁻ for complexation in the βCD annulus, as has been reported for an analogue of **6.22b** in which one of the βCDs is replaced by an αCD [36]. (The ¹H NMR spectra of **6.22b** and its analogue, where a βCD is replaced by αCD, show more anomeric ¹H resonances than anticipated from the formal magnetic equivalences of the anomeric hydrogens which is consistent with the octamethylene linker being partially complexed inside a CD annulus [37].) Similar cooperativities are found for the complexation of Methyl Orange and Tropaeolin by **6.20**, **6.21a** and **6.21c** [34]. An analogue of **6.22a**, where one βCD is replaced by αCD, shows cooperative and site-specific binding of isoamyl *p*-dimethylaminobenzoate, where the isoamyl group complexes in the βCD annulus and the 4-dimethylaminobenzoate moiety partially enters the αCD annulus [34].

A spectrophotometric study finds that Methyl Orange (**6.18**) is also complexed cooperatively by **6.20**, **6.21a** and **6.21c** as shown by $K_{11} = 1.05 \times 10^5$, 1.92×10^5 and 2.5×10^4 dm^3 mol^{-1}, respectively, in aqueous 0.10 mol dm^{-3} phosphate buffer at pH 9.0 and 298.2 K which compare with $K_{11} = 2.16 \times 10^3$ dm^3 mol^{-1} for the βCD complex [43]. Tropaeolin (**6.19**) similarly shows cooperative complexing by **6.20**, **6.21a** and **6.21c** as shown by $K_{11} = 1.39 \times 10^4$, 7.4×10^3 and 4.6×10^3 dm^3 mol^{-1}, respectively, in aqueous 0.10 mol dm^{-3} phosphate buffer at pH 5.5.0 and 298.2 K which compare with $K_{11} = 7.1 \times 10^2$ dm^3 mol^{-1} for the βCD complex. These data show the larger Tropaeolin (**6.19**) to be less strongly complexed probably because of its greater rigidity and more angular structure by comparison with Methyl Orange.

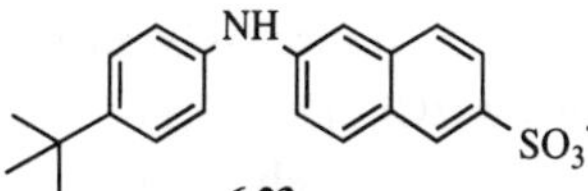

6.23

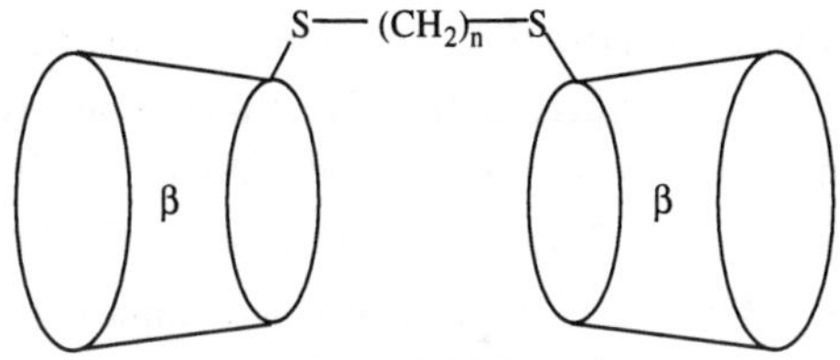

6.24a n = 0 **6.24b** n = 2
6.24c n = 3 **6.24d** n = 4
6.24e n = 5

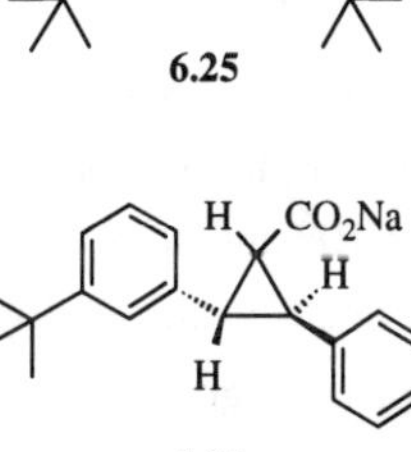

6.25

6.26

6.27

A series of 1:1 complexes of 6-(4-*tert*-butylanilino)-naphthalene-2-sulfonate (**6.23**, BNS⁻) with the βCD dimers **6.24a-6.24e**, linked by substitution of an OH(6) by a sulfur of -S(CH₂)ₙS-, shows a steady decrease in stability as the linker lengthens from n = 2 to n = 6, so that K_{11} decreases from 8.2 x 10⁶ to 1.5×10^5 dm³ mol⁻¹ [19]. However, when n = 0, K_{11} drops to 7.9×10^4 dm³ mol⁻¹, and this is attributed to a destabilising decrease in the stereochemical match of the hydrophobic areas of the linked βCD and BNS⁻. When the nature of the aromatic guest is varied, K_{11} for the complexes formed with **6.24a** varies over a range $< 3 \times 10^3$ to 3.5×10^8 dm³ mol⁻¹ in water at 298.2 K where the guests forming the least and most stable complexes are the *cis*-stilbene **6.25** and the cyclopropane **6.26** [13]. Under similar conditions, the hydrophobic nature of cholesterol (**6.27**) causes it to form a strong complex with the analogue of **6.24a**, where the disulfide linker is replaced by a sulphide linker, and K_{11} = 3.3-5.54 × 10⁶ dm³ mol⁻¹, although it contains no aromatic moiety [10]. For the analogous BNS⁻ complex, $K_{11} = 6.37$-7.40×10^5 dm³ mol⁻¹.

Changing the size of the CD annulus causes substantial changes in the stability of 1:1 complexes which depend on the relative sizes of the CD and the guest as discussed in preceding chapters. Such changes also affect the stabilities of 1:1 complexes formed by linked CDs. Thus, **6.24a** complexes Methyl Orange and Ethyl Orange 22.4 and 108.6 times more strongly in 1:1 complexes, respectively, than does its analogue where both βCD moieties are replaced by αCD [11]. (For Methyl Orange $K_{11} = 5.83 \times 10^5$ and 2.60×10^4 dm³ mol⁻¹, respectively, and for Ethyl Orange $K_{11} = 2.03 \times 10^6$ and 1.87×10^4 dm³ mol⁻¹, respectively.) Nevertheless, both **6.24a** and its αCD analogue complex these dyes much more strongly than do βCD and αCD. In the analogue of **6.24a** where both βCD moieties are replaced by γCD, the larger annular size results in the dominant 1:2 complex accommodating two dye molecules and β_{12} ($K_{11} \times K_{12}$) = 1.06×10^{11} and 3.60×10^{10} dm⁶ mol⁻² for Methyl Orange and Ethyl Orange, respectively, at pH 10.6 and 298.2 K [12]. These values are consistent with a high degree of cooperativity in complexing the two dye molecules simultaneously as a comparison with $\beta_{12} = 8.67 \times 10^6$ and 4.36×10^7 dm⁶ mol⁻² for the analogous γCD.dye₂ complexes shows. A head-to-tail analogue of **6.24a** where one βCD is linked through C(3) and the other through C(6) has been prepared, but no studies of the effect of this linkage variation on complexation appear to have been reported [20].

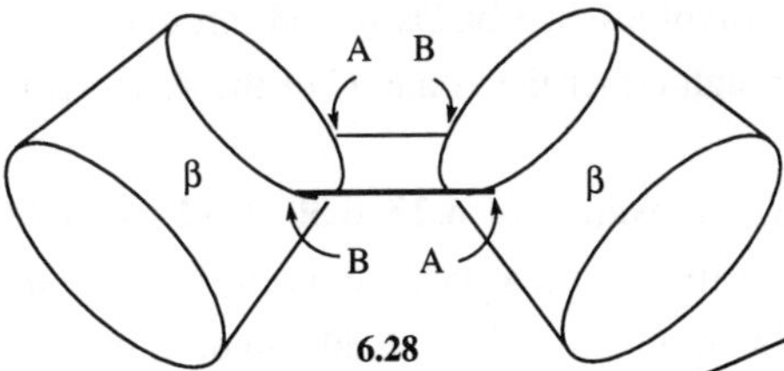

6.28

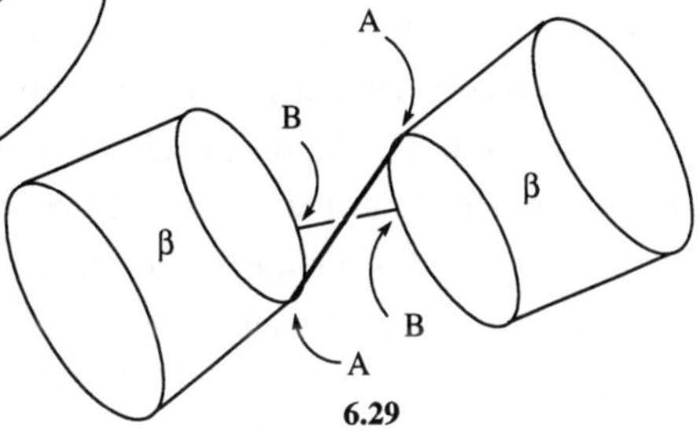

where

$\underline{\qquad}$ = S—S

6.29

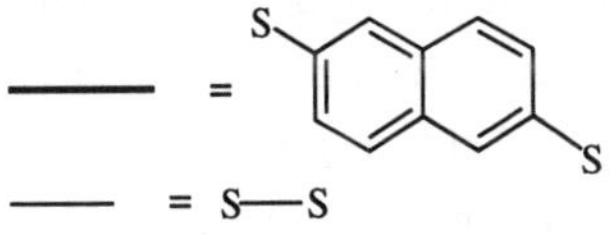

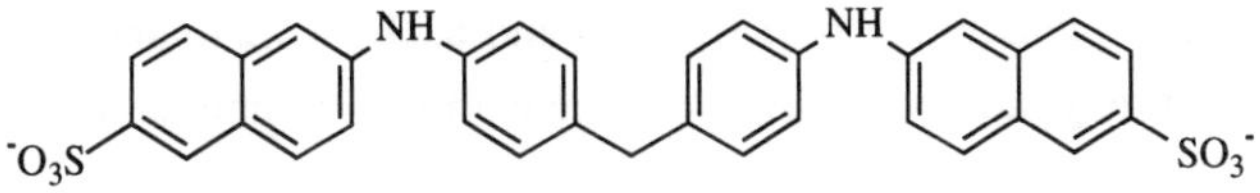

6.30

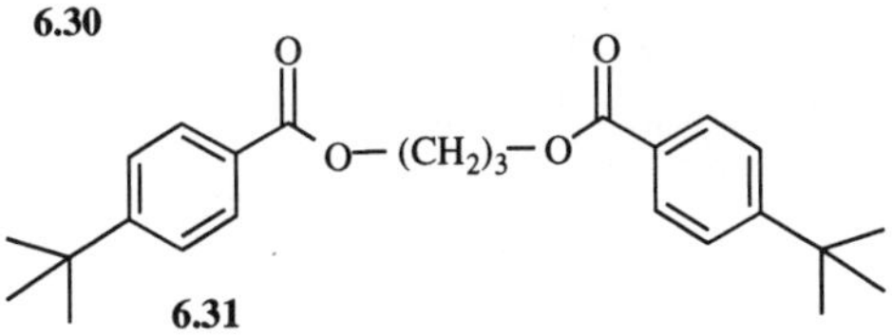

6.31

The stereochemical aspects of the formation of linked CD complexes have been studied through the double linking of βCD in the occlusive or 'clamshell' isomer **6.28**, and the aversive or 'loveseat' isomer **6.29** [14,18,44]. The occlusive **6.28** closes on ditopic guests like a clamshell, leading to strong complexation, while the aversive **6.29** shows no cooperative complexation of ditopic guests because the two βCD annuli are directed away from each other. Thus, the large $K_{11} = 4 \times 10^6$ dm^3 mol^{-1} for the complexation of **6.23** by **6.28** is attributed to cooperative complexation while $K_{11} = 2 \times 10^5$ dm^3 mol^{-1} for the complexation of **6.23** by **6.29** is not much greater than that for complexation by βCD. Longer ditopic guests fit the steric requirements for complexation by **6.28** more closely and are complexed much more strongly as illustrated by $K_{11} = 4 \times 10^8$ and 1×10^{10} dm^3 mol^{-1}, respectively, for the complexing of **6.30** and **6.31**.

Another doubly linked βCD system involves two βCDs linked through two 1,2-diaminoethane moieties substituting through either the A and C or the A and D C(6) sites [30].

Calorimetric studies show that the complexation of **6.23, 6.30, 6.32** and **6.33** by βCD is dominantly enthalpy driven and the cooperativity between the two linked βCD moieties in **6.24a, 6.34** and **6.35** in complexing **6.23, 6.30** and **6.33** is due to a much greater ΔH^o than that observed for the complexing of these guests by βCD (Table 6.1) [17]. This contrasts with the observation that hydrophobic interactions [45] and the formation of chelated metal complexes tend to be entropy driven [46]. The linear relationship between $T\Delta S^o$ and ΔH^o in Table 6.1 is consistent with an enthalpy/entropy compensation which probably mainly arises through hydration changes accompanying complexation [16,17]. The decreases in heat capacity, ΔC_p^o, arising from the complexation of **6.31** by βCD and **6.31** by **6.24a** are -400 and -657 J mol^{-1} K^{-1}, respectively, and typify hydrophobic complexing interactions [45,47,48].

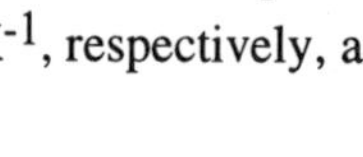

6.32 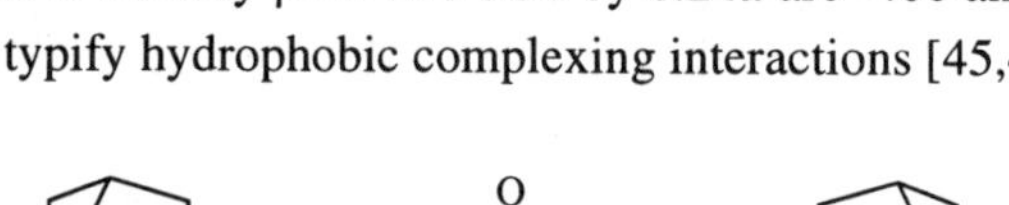　　　　　　　　　　　　　**6.33**

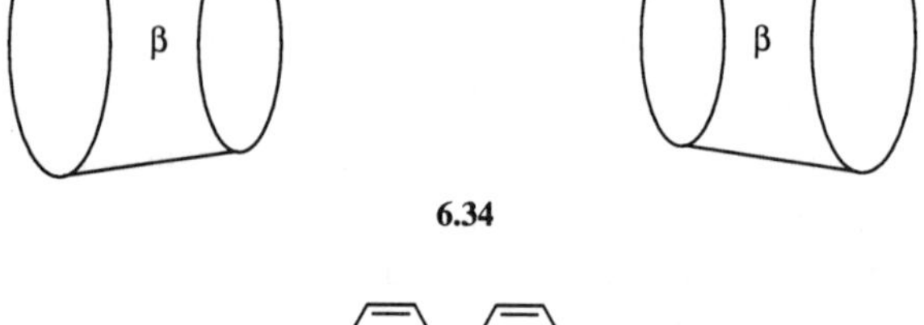

6.34

6.35

Table 6.1. Parameters for Guest Complexation by βCD and Linked βCDs.[a]

CD	Guest	K_{11} or K_{12} $dm^3 \, mol^{-1}$	ΔG^o kJ mol^{-1}	ΔH^o kJ mol^{-1}	$T\Delta S^o$ kJ mol^{-1}
βCD[b]	**6.32**	3.95×10^4	-26.2	-21.8	4.44
βCD[b]	**6.33**	2.26×10^5	-30.5	-29.3	1.26
βCD[c]	**6.33**	4.39×10^3	-20.8	-16.1	4.73
6.24a[b]	**6.33**	1.79×10^7	-41.4	-67.6	-26.2
6.34[b]	**6.33**	1.13×10^7	-38.7	-60.5	-20.2
6.35[b]	**6.33**	2.14×10^6	-36.1	-62.3	-26.2
βCD[b]	**6.23**	5.57×10^4	-27.1	-25.3	1.76
6.24a[b]	**6.23**	3.67×10^6	-37.4	-65.5	-28.1
βCD[b]	**6.30**	8.05×10^4	-28.0	-18.5	9.50
βCD[c]	**6.30**	2.34×10^3	-19.2	-16.2	3.01
6.34[b]	**6.30**	3.50×10^7	-43.1	-89.5	-46.48

[a] In aqueous 0.020 mol dm^{-3} HEPES buffer solution at 298.2 K. [b] Complexation of first guest. [c] Complexation of second guest.

The bipyridyl component in the linker of **6.34** chelates metal ions and this greatly enhances its ability to complex guests possessing two hydrophobic moieties and also metal ion coordinating groups. This is illustrated by the complexation of **6.36** by **6.34** for which $K_{11} = 2.5 \times 10^6$ dm^3 mol^{-1} which is raised 10-fold in the presence of Zn^{2+} due to the simultaneous coordination of Zn^{2+} by the tryptophan groups of **6.36** and the bipyridyl nitrogens of **6.34** to produce a ternary metalloCD [24]. Such metal ion coordination affords an opportunity for a hydroxo ligand to make a nucleophilic attack on a guest in a ternary metalloCD as is proposed for Cu^{2+} in **6.37** [18,44,49,50]. Thus, while for the uncatalysed hydrolysis of the esters **6.38** and **6.39** the rate constant, k_u (310.2 K) = 3 × 10^{-8} s^{-1} in each case at pH 7.0, the Cu^{2+} metalloCD **6.37** catalyses their hydrolyses by several orders of magnitude under the same conditions, as shown by the respective rate constants, k_{cat} (310.2 K) = 6.8 × 10^{-3} and 5.5 × 10^{-4} s^{-1} [24]. The catalysis occurs through a nucleophilic attack by a hydroxo ligand (the pK_a of its conjugate acid aqua ligand is 7.15) on the carbonyl carbon as shown in **6.37**. With an excess concentration of the ester **6.38** at least 50 turnovers are observed for the hydrolysis. The linked CD **6.34** is also the basis for an impressive catalyst for cleavage of the phosphate esters **6.40a** and **6.40b** in the presence of La^{3+} and H_2O_2 [51]. This, together with the catalytic and other characteristics of metalloCDs, are discussed in detail in Chapter 5.

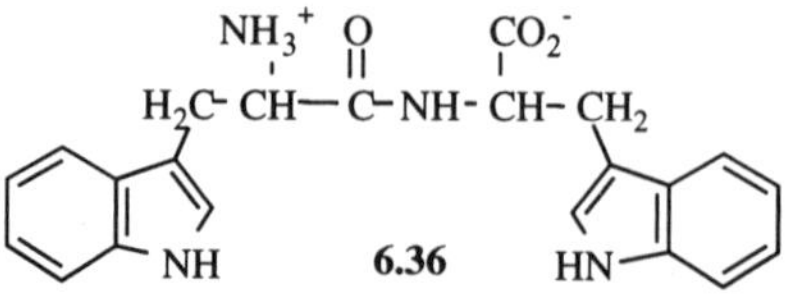

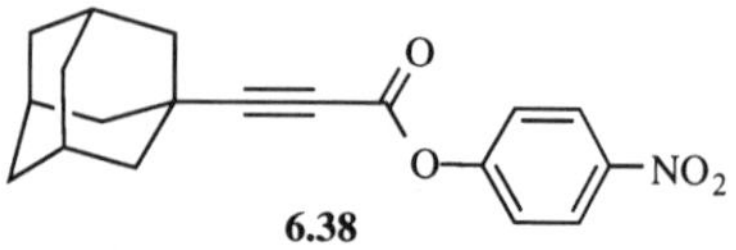

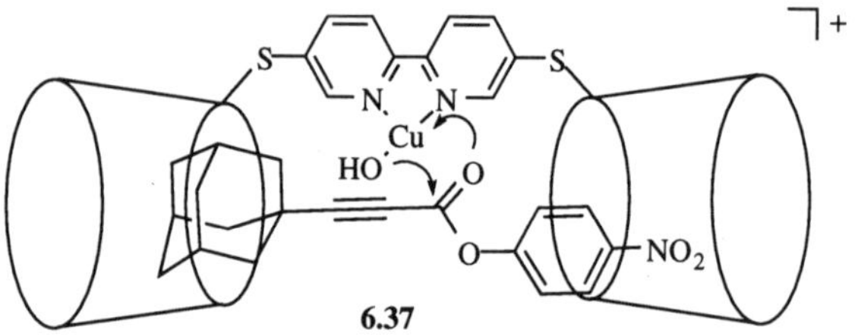

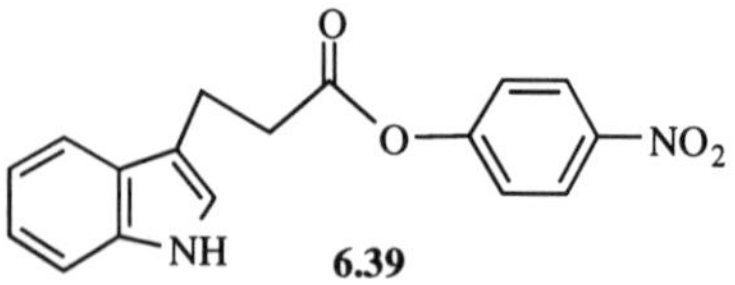

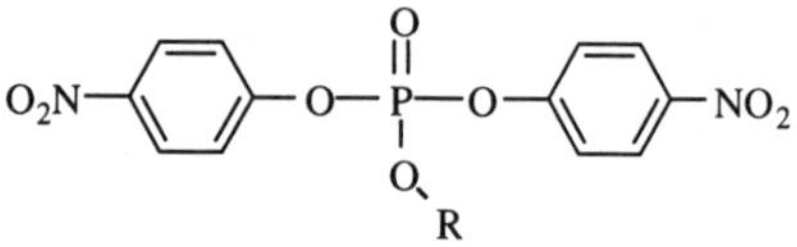

When two CDs are linked an opportunity arises not only to tailor the separation of the CDs to selectively complex a guest, but also to attach a catalytic group to the linker to produce selectivity in catalysing reaction of a guest. This approach has been adopted in a study of the catalysis of the hydrolysis of the 4-nitrophenyl alkanoates **6.41a-6.41d** by the linked βCD **6.42** where the histidine moiety is the catalytic group [52]. The catalysed hydrolysis follows Michaelis-Menten kinetics and the catalytic rate shows a significant dependence on the alkyl chain length in the **6.41a-6.41d** series as seen from Table 6.2. The positioning of the guests **6.41a-6.41d** within their **6.42** complexes appears to have a major influence on the magnitudes of k_{cat} and K_M, and the ratio k_{cat}/k_u. While, in terms of the ratio k_{cat}/k_u, **6.42** is not as effective a catalyst as **6.43** it does show a greater catalytic discrimination between guests. When either one of the βCD moieties of **6.42** is replaced by αCD, two isomers are possible and these have been synthesised [53]. Neither isomer is as effective a catalyst as is **6.42** in hydrolysing **6.41a**, **6.41b** and **6.41c**.

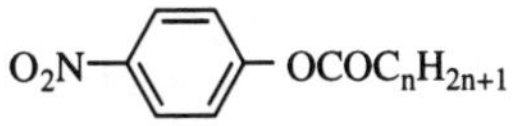

$$O_2N-\!\!\!\underset{}{\bigcirc}\!\!\!-OCOC_nH_{2n+1}$$

6.41a n = 1
6.41b n = 2
6.41c n = 5
6.41d n = 11

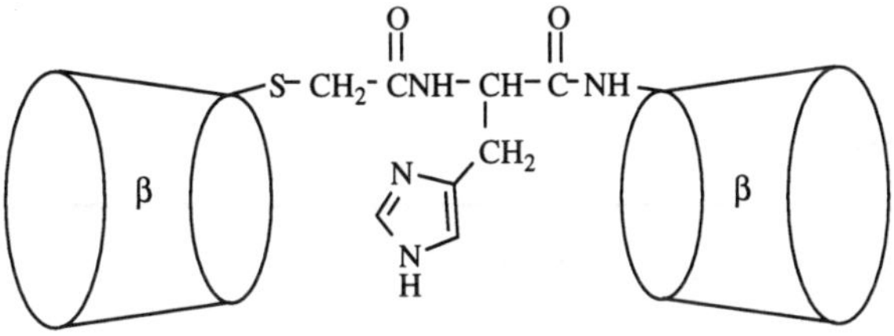

6.42

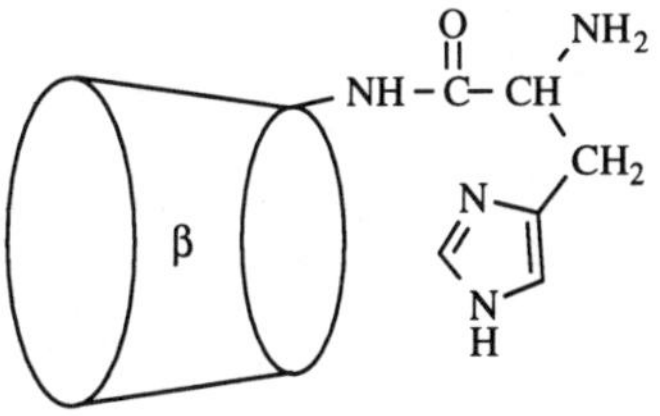

6.43

When a pendant group attached to a linker becomes sufficiently long it may be complexed by an adjacent linked CD and so on to from an aggregate as appears to be the case for **6.44** where the 2,3,6-termethylated αCDs are linked through C(6) [54]. In acid and basic aqueous solution the spectrum of the azophenol and azophenolate moieties of **6.44** undergo absorption maximum changes from 380 to 390 nm, and from 520 to 540 nm, respectively, over a two hour period at room temperature, consistent with the complexation of the pendant groups by the CD annuli of adjacent **6.44** to form aggregates.

Table 6.2. Parameters for Hydrolysis of the 4-Nitrophenyl Alkanoates **6.41a-6.41d** by the Modified Cyclodextrins **6.42** and **6.43**.[a]

CD	Guest	$k_{cat} \times 10^6$ s^{-1}	$K_M \times 10^6$ $dm^3\ mol^{-1}$	k_{cat}/K_M $dm^3\ mol^{-1}\ s^{-1}$	k_{cat}/k_u
6.42	**6.41a**	2690	3700	0.726	134
6.42	**6.41b**	2830	3170	0.893	191
6.42	**6.41c**	191	6.73	28.4	10.7
6.42	**6.41d**	86.2	12.5	6.90	12.2
6.43	**6.41a**	4700	6730	0.698	234
6.43	**6.41b**	4220	5080	0.830	285
6.43	**6.41c**	1630	1970	0.827	91.9

[a]In aqueous phosphate buffer at pH 7.8 and 298.2 K.

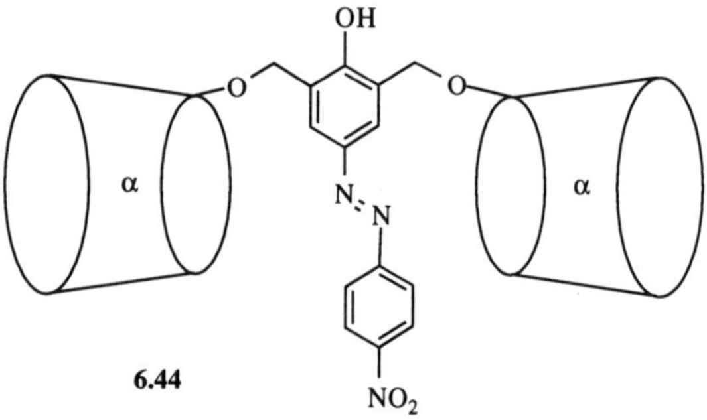

6.44

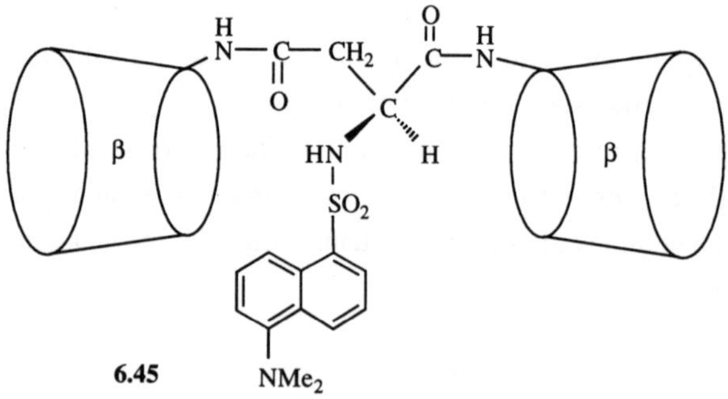

6.45

Intramolecular complexation of the dansyl pendant group of the βCD dimer **6.45**, where linking occurs through C(6), renders it fluorescent [55]. This fluorescence is markedly decreased when the steroids **6.46a** and **6.46b** are present consistent with the dansyl moiety being displaced from the βCD annulus through the formation of complexes characterised by $K_{11} = 6.9 \times 10^3$ and 1.3×10^3 dm^3 mol^{-1}, respectively, in aqueous solution at 298.2 K. However, steroid **6.46c** causes a small increase in the fluorescence of **6.45** which may be due to it being complexed without displacement of the dansyl moiety, while **6.46d** causes no change in fluorescence consistent with little if any complexation.

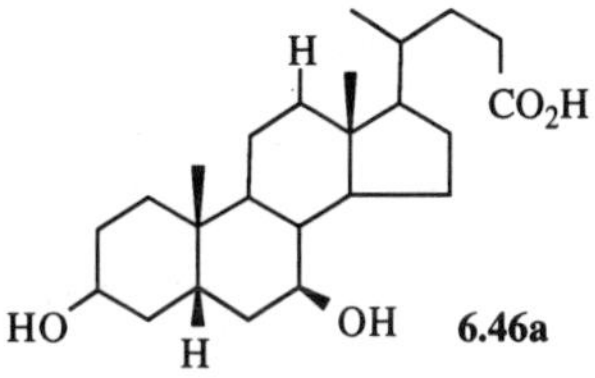

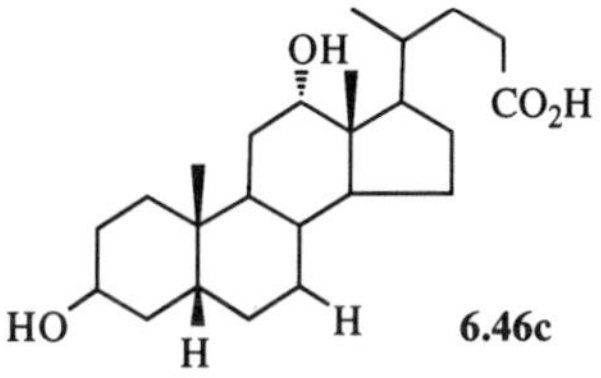

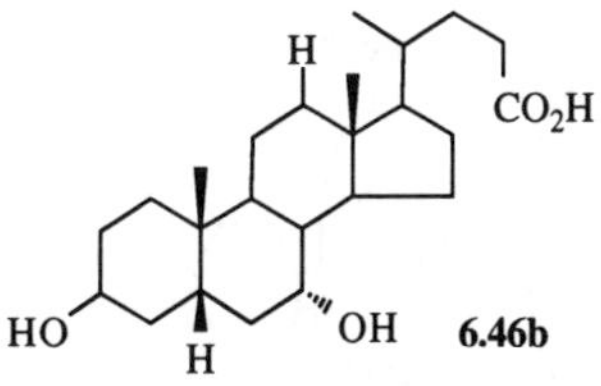

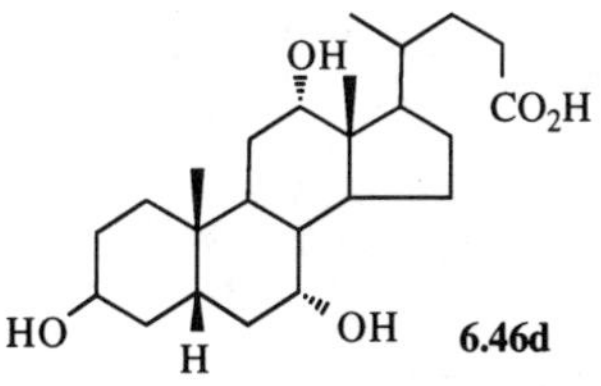

In principle, the complexation of two reactive guest molecules, one in each annulus of a linked CD dimer, might be expected to influence the direction of the reaction between them where more than one reaction path is possible. The first report of this effect is for the competitive formation of $\Delta^{2,2'}$-biindoline-3,3'-dione (indigo) **6.47** and $\Delta^{2,3'}$-biindoline-2',3-dione (indirubin) **6.48** from the oxidative dimerisation of 1*H*-indol-3-olate **6.49** and its condensation with 1*H*-indoline-2,3-dione **6.50**, respectively [56]. Under the same conditions, the ratio of the percentage yield of **6.47** to the percentage yield of **6.48** in the absence of any CD is 16:13, in the presence of βCD is 2.5:2.5, and in the presence of the linked βCD dimers **6.20**, **6.21a** and **6.21c** is 0.03:1.0, 0.2:0.6 and 0.5:0.7, respectively, after 16 hours at pH 10.0. In the presence of the CDs the yields of **6.47** and **6.48** are substantially decreased, probably because their complexation favours their unimolecular reactions over their bimolecular reactions which produce **6.47** and **6.48**. However, in the presence of the βCD dimers the product ratio **6.47**:**6.48** is increased markedly in favour of **6.48**, particularly for the linked dimer **6.20**. This change in product ratio is consistent with the simultaneous complexation of **6.49** and **6.50** in the orientation shown in **6.51** which leads to the formation of **6.48** in **6.52**, rather than **6.47**.

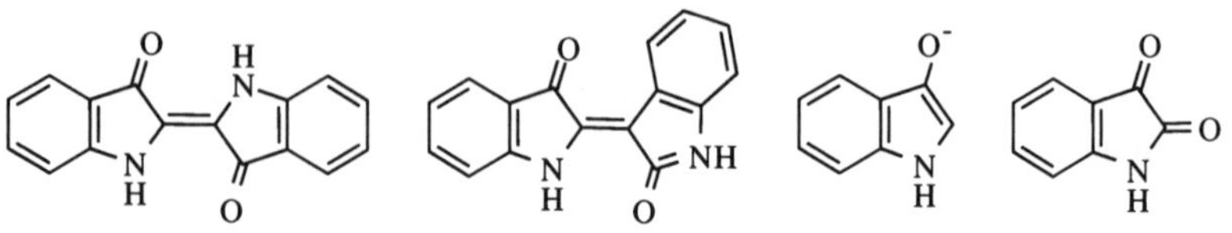

| 6.47 | 6.48 | 6.49 | 6.50 |

| 6.51 | 6.52 |

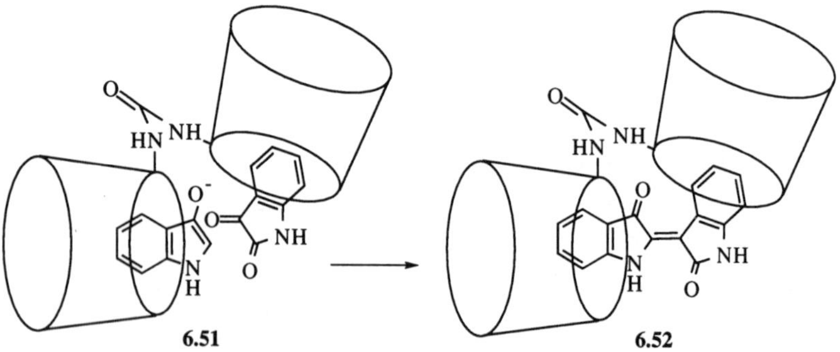

6.4. Porphyrin Complexing, Porphyrin Substituted Cyclodextrins and Related Systems

The porphyrins have attracted attention either as guest species in linked CD dimer complexes or as a component of the linker itself. This interest arises largely from a desire to gain further insight into the role of the porphyrin moiety in photosynthesis and in heme proteins. The complexation of the porphyrins **6.53a** and **6.53b** by βCD is characterised by $K_{11} = 1.4 \times 10^3$ and 1.7×10^3 dm^3 mol^{-1} at pH 7.0 and 298.2 K. Their complexation by the linked CD dimer **6.22a** is characterised by $K_{11} = 8 \times 10^5$ and 1.9×10^6 dm^3 mol^{-1}, respectively, and for complexation by **6.22b** $K_{11} = 4 \times 10^5$

and 9×10^5 d m^3 mol^{-1}, respectively, which demonstrates strong cooperativity between the two βCD moieties in the linked CD dimers [57]. A similarly strong cooperativity arises in the complexation of **6.53b** by **6.54** where $K_{11} > 5 \times 10^7$ dm^3 mol^{-1}. Isomerism can arise in such porphyrin complexation as is shown by the complex formed between **6.22a** and **6.53a**, which has a *syn* stereochemistry where the two βCD annuli complex adjacent aromatic groups, while the complex formed between **6.22b** and **6.53a** exists both as the *syn* isomer and the *anti* isomer where the two βCD annuli complex alternate aromatic groups. The more rigid **6.54** appears to form a 2:2 complex with **6.53a**.

6.53a R = SO$_3^-$
6.53b R = CO$_2^-$

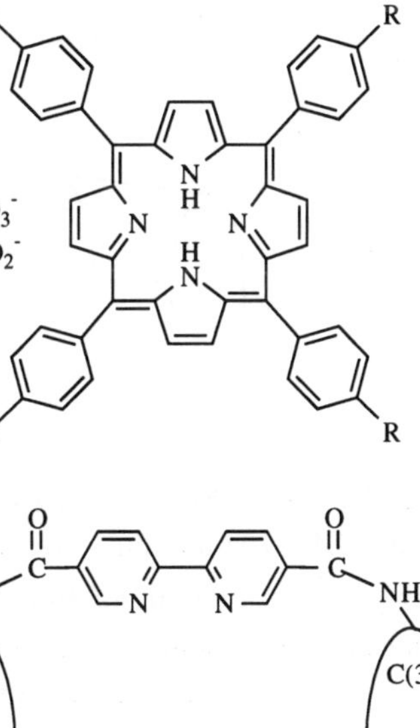

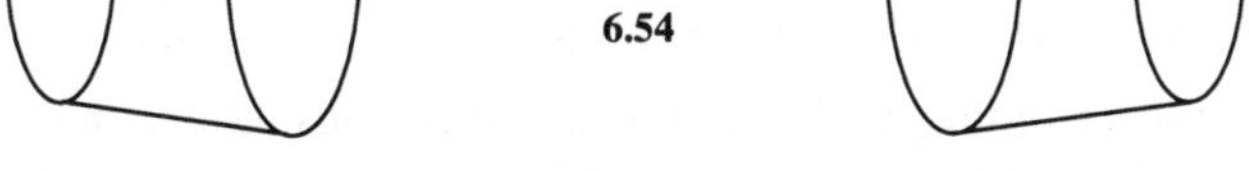

6.54

Metalloporphyrins are able to coordinate to metal binding sites in the linker as is shown by the complexation of the porphyrin **6.55a** and the metalloporphyrins **6.55b-6.55d** by the linked βCD dimer **6.56** [22,58]. Here, $K_{11} = 2.5 \times 10^4$, 3.4×10^6, 7.6×10^6 and 1.7×10^8 dm^3 mol^{-1} for the complexation of **6.55a-655d**, respectively, by the linked βCD dimer **6.56** at pH 7.0 and 298.2 K. The stronger complexation of the metalloporphyrins is attributed to coordination of the pyridine nitrogen of the host by the metal centre. This is consistent with the observation that the closely related linked βCD dimer **6.57**, which lacks a nitrogen in the linker, shows no enhanced complexation of the metalloporphyrins as shown by $K_{11} = 1.7 \times 10^4$, 1.9×10^4, 1.0×10^4 and 1.3×10^4 dm^3 mol^{-1}, respectively, for the complexation of **6.55a-6.55d**.

6.55a M = H,H   6.55b M = Zn^{2+}
6.55c M = Mn^{3+} 6.55d M = Co^{3+}

The metal complexes **6.58a** and **6.58b** are also strongly complexed by **6.56** as shown by $K_{11} = 2.1 \times 10^4$ and 2.4×10^4 mol dm^{-3}, respectively, while the lower K_{11} $= 1.8 \times 10^3$ mol dm^{-3} for the complexation of **6.58b** by **6.57** is consistent with metal ion coordination by **6.56** enhancing the stability of its complex with **6.58b**. A similar trend in stability is seen in the complexes of **6.56** with **6.59a** to **6.59d** where $K_{11} = 6.3 \times 10^4$, 1.3×10^5, 1.5×10^5 and 3.3×10^5 mol dm^{-3}, respectively, and those of **6.57** with **6.59a** and **6.59d** where $K_{11} = 5.6 \times 10^4$ and 5.0×10^4 mol dm^{-3}, respectively. In contrast, the presence of the pyridine nitrogen in **6.56** adds little to the stability of the complexes formed by **6.56** with **6.60a** and **6.60b** where $K_{11} = 1.4 \times 10^4$ and 1.7×10^4 mol dm^{-3}, respectively, by comparison with those formed by **6.57** with **6.60a** and **6.60b** where $K_{11} = 1.3 \times 10^4$ and 1.2×10^4 mol dm^{-3}, respectively. This difference probably arises because, unlike the borderline hard acid divalent first row transition metal ions, hard acid Ba^{2+}, La^{3+} and Eu^{3+} are unlikely to strongly coordinate the borderline soft base pyridine nitrogen of **6.56**. The complexes of **6.56** with **6.61a**, **6.61b** and **6.62** are characterised by $K_{11} = 2.2 \times 10^4$, 2.6×10^4 and 6.9×10^4 mol dm^{-3}, respectively. A variety of dyes are also complexed by **6.56** among which Methyl Orange (**6.18**) forms a complex characterised by $K_{11} = 1.6 \times 10^7$ mol dm^{-3} which appears to be the highest K_{11} reported for a CD complex of this dye [22].

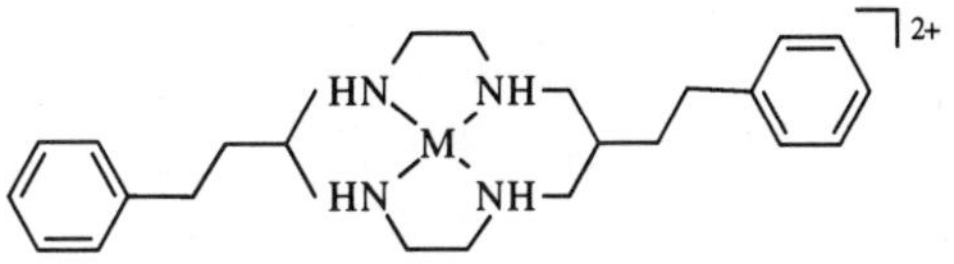

6.58a M = Cu^{2+}
6.58b M = Ni^{2+}

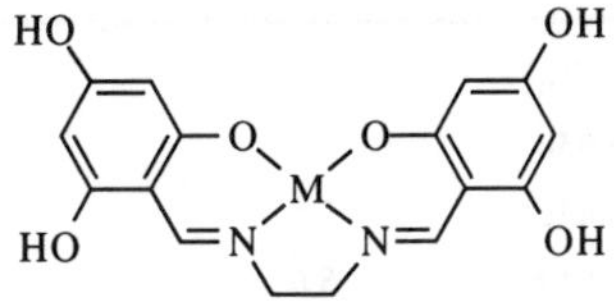

6.59a M = H,H **6.59b** = Ni^{2+}
6.59c M = Zn^{2+} **6.59d** = Co^{2+}

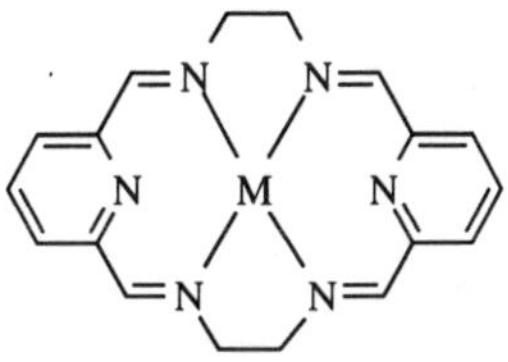

6.60a M = Ba^{2+}
6.60b M = La^{3+}

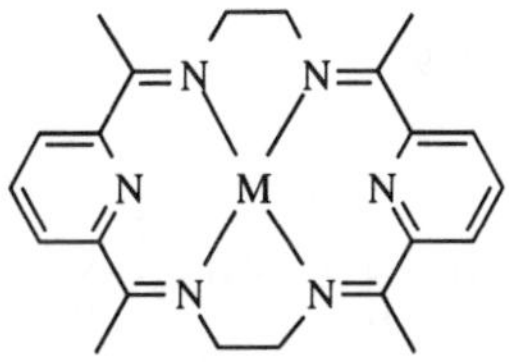

6.61a M = Eu^{3+}
6.61b M = La^{3+}

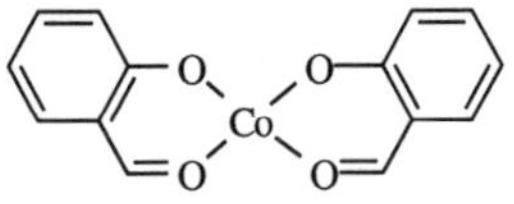

6.62

A competitive fluorescence study, where TNS$^-$ acts as the reference fluorescent guest, shows that the linked βCD dimers **6.63a-6.63d** and **6.64a** and **6.64b** form stable 1:1 complexes with the porphyrinoid guests **6.65-6.67** [31]. Generally, K_{11} increases with decrease in the linker length as shown in Table 6.3. The linked βCD dimers **6.63a-6.63d** and **6.64a** and **6.64b** were designed as potential carriers for porphyrinoid photosensitisers in photodynamic cancer therapy.

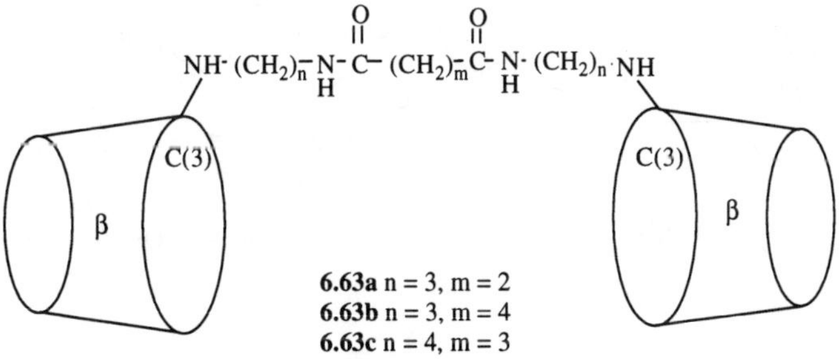

6.63a n = 3, m = 2
6.63b n = 3, m = 4
6.63c n = 4, m = 3
6.63d n = 4, m = 4

Table 6.3. Stability Constants for Linked βCD Complexation of Porphyrinoid Guests.[a]

CD	Guest = TNS⁻ $10^{-3}K_{11}$ mol dm⁻³	Guest = 6.65 $10^{-5}K_{11}$ mol dm⁻³	Guest = 6.66 $10^{-5}K_{11}$ mol dm⁻³	Guest = 6.67 $10^{-5}K_{11}$ mol dm⁻³
6.63a	3.6	4.5	3.3	2.1
6.63b	3.8	3.1	1.3	6.3
6.63c	5.5	1.3	150	8.4
6.63d	5.6	1.4	41	8.7
6.64a	8.3	0.58	40	5.0
6.64b	5.8		14	2.4

[a]In 0.1 mol dm⁻³ carbonate buffer at pH 9.

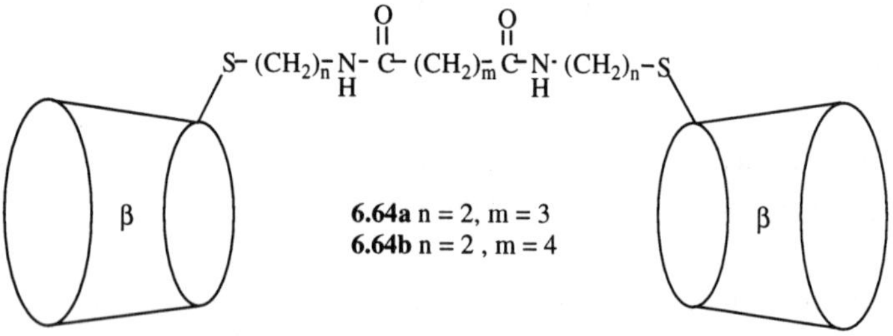

6.64a n = 2, m = 3
6.64b n = 2 , m = 4

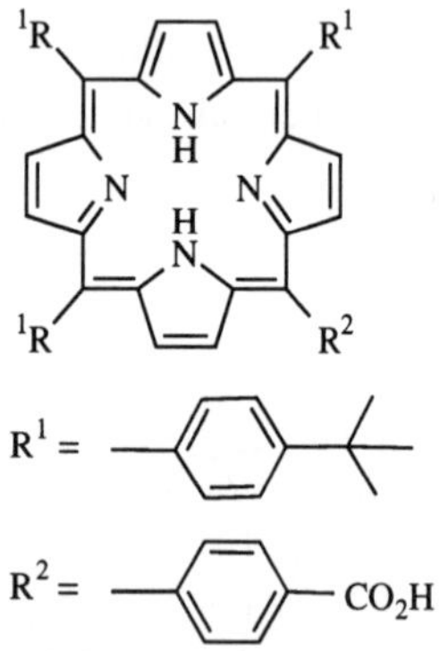

6.65

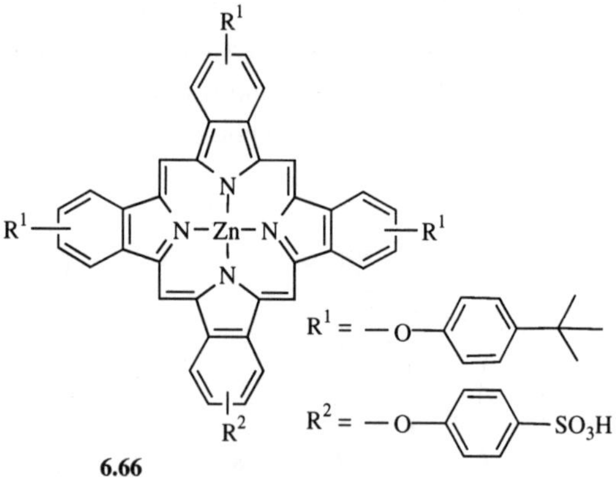

6.66

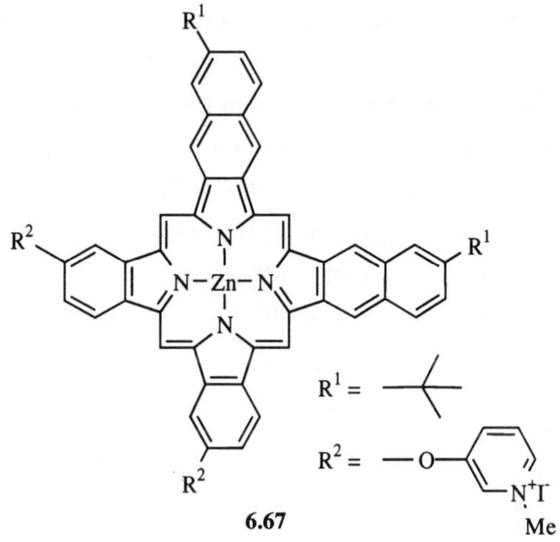

$R^1 =$

$R^2 = $

6.67

Me

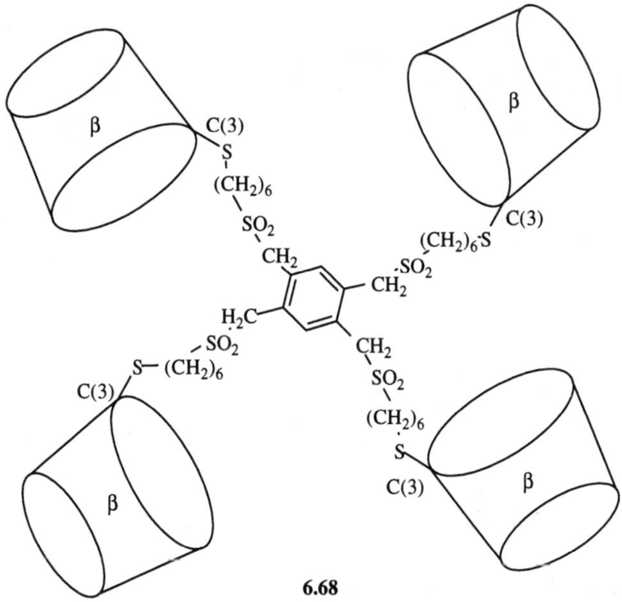

6.68

An interesting extension of linked CD chemistry is exemplified by the βCD tetramer **6.68** where each βCD linkage occurs at C(3) [59]. Both tetraarylporphyrins and metalloporphyrins are bound in 1:1 complexes with K_{11} values of up to 10^8 dm^3

mol^{-1}, but the number of porphyrin aryl substituents simultaneously complexed is uncertain.

In the linked βCD dimer **6.69** the porphyrin linker is bound to each βCD through sulfur at the A and D C(6) sites. This dimer forms the complexes **6.70** with the guests **6.71-6.73** (Fig. 6.4) where $K_{11} = 7.4 \times 10^3$, 2.2×10^4, and $> 5 \times 10^5$ dm^3 mol^{-1}, respectively, in aqueous solution at pH 9.0 and 296.2 K. The rate of electron transfer from the porphyrin linker of **6.69** to the guests in the complex **6.70** is characterised by $k_{et} = 2 \times 10^9$, 10^9, and 10^9 s^{-1}, respectively, as measured from the quenching of the porphyrin fluorescence in the presence of the guests **6.71-6.73** [25].

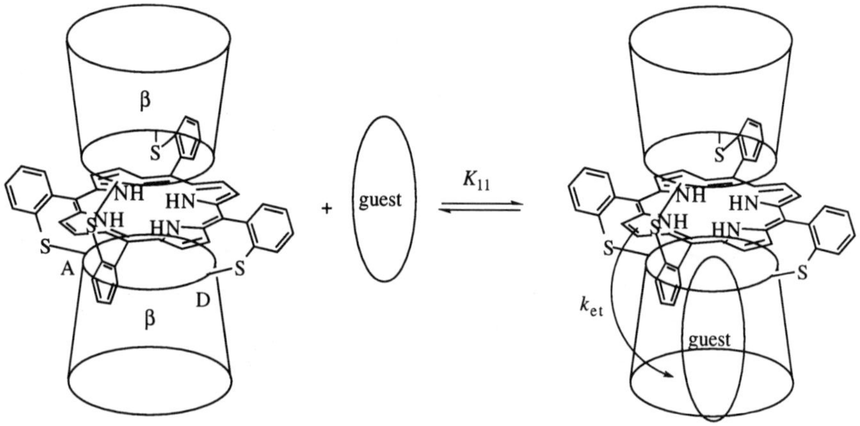

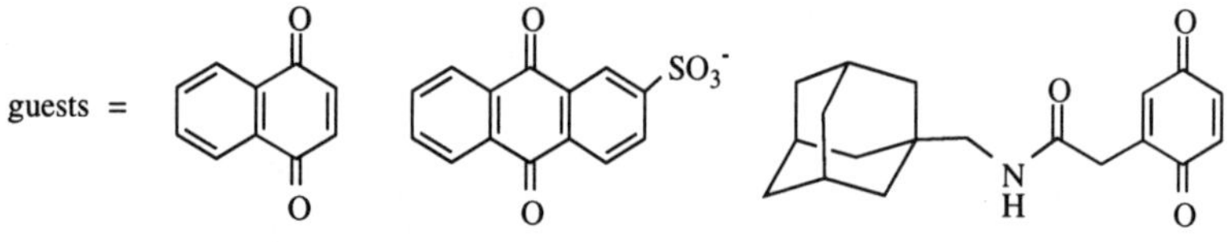

Fig. 6.4. Complexation of guests by **6.69** and electron transfer in the 1:1 complex.

The analogue of **6.69**, where each S link between the porphyrin and βCD is replaced by an amide link and each hydroxy group is replaced by a methoxy group, shows potential for similar electron transfer studies [26]. It is not necessary for the porphyrin to link two CDs for electron transfer between the porphyrin and the guest to occur. Thus, the modified βCD where a 5-(4-carboxyphenyl)-10,15,20-tris(4-methyl)phenyl)porphyrin is substituted onto C(6) of βCD through the carboxy group, shows light induced electron transfer from the porphyrin to complexed guest quinones and nitrobenzenes as detected by EPR spectroscopy [60]. Extensive studies of metalloCDs formed by porphyrin substituted βCDs have been reported [61] and are discussed in Chapter 5.

6.5. Calixarene Substituted Cyclodextrins

The substitution of the primary amine group of 6^A-(2-aminoethylamino)-6^A-deoxy-βCD by tetrakis(hydroxycarbonylmethoxy)calix[4]arene produces the water soluble modified CD **6.74**, where the calixa[4]arene substituent acts as a second hydrophobic binding group in addition to the βCD moiety [62]. While no complexation studies have been reported for **6.74**, such studies have been reported for the isomeric modified βCDs **6.75** and **6.76**, substituted at one OH(2) by the

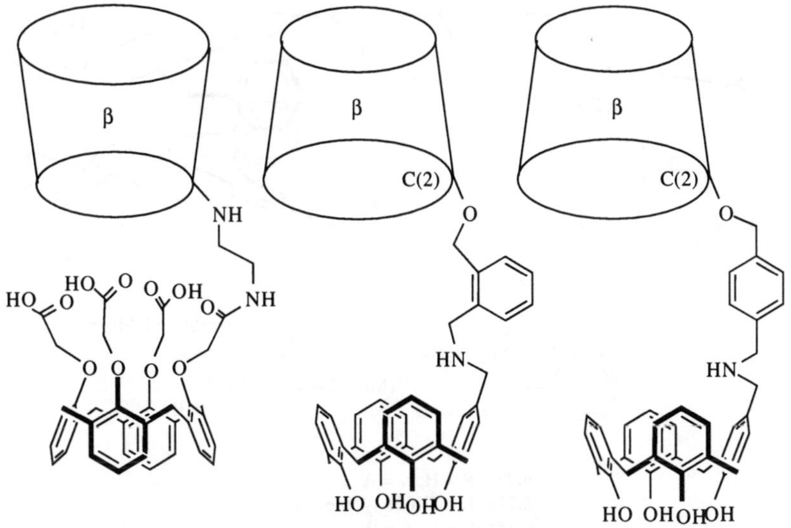

6.74   6.75   6.76

calix[4]arene moiety, and by OMe at the remaining six OH(2) sites [63,64]. The formation of 1:1 complexes by **6.75** and **6.76** with TNS⁻ (**6.17**) is characterised by $K_{11} = 3.05 \times 10^4$ and 1.53×10^5 dm³ mol⁻¹ in water at pH 7.0, respectively, while under the same conditions βCD forms a 1:1 complex with TNS⁻ where $K_{11} = 2.0 \times 10^3$ dm³ mol⁻¹. The greater K_{11} values observed for the complexes formed by **6.75** and **6.76** are attributed to the simultaneous complexing of the toluidinyl and naphthyl groups of TNS⁻. The lower stability of the complex formed by **6.75** is attributed to partial complexation of the calix[4]arene substituent in the βCD annulus competing with the complexation of TNS⁻. A similar effect is seen for the binding of 1-anilino-8-naphthalenesulfonate, ANS⁻, by **6.75** and **6.76** where $K_{11} = 2.3 \times 10^3$ and 2.48×10^4 dm³ mol⁻¹, respectively. (The analogues of **6.75** and **6.76** where each primary hydroxy group is replaced by an *O-tert*-butyldimethylsilyl group have also been prepared [64].)

The complexation of TNS⁻ and ANS⁻ by **6.75** and **6.76** is detected by the increased fluorescence of both guests. This fluorescence decreases when other guests such as steroids and terpenes compete in the complexation process, and the decrease

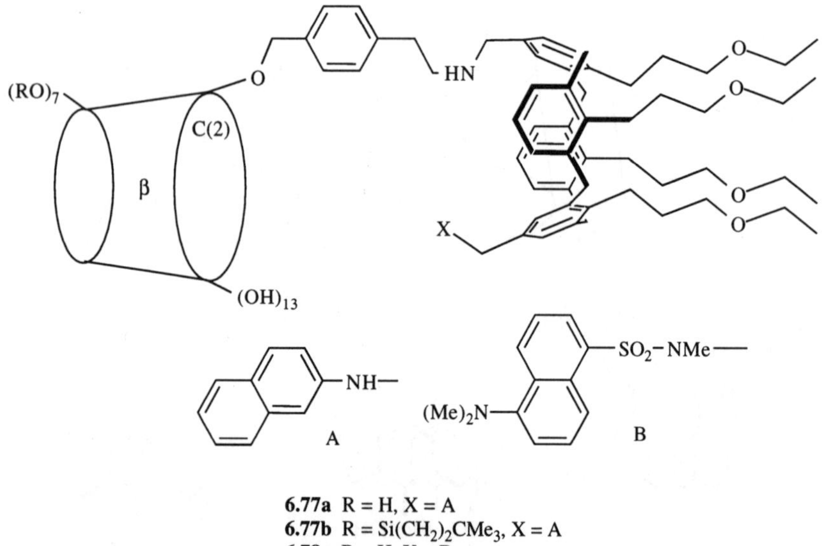

6.77a R = H, X = A
6.77b R = Si(CH₂)₂CMe₃, X = A
6.78a R = H, X = B
6.78b R = Si(CH₂)₂CMe₃, X = B

in fluorescence is an indication of the complexation of the steroids and terpenes [62]. A logical extension of this observation is to develop a sensor system which incorporates an intramolecular complexing fluorophore into the modified CD as exemplified by **6.77a** and **6.78a** (**6.77b** and **6.78b** are precursors in the preparation of **6.77a** and **6.78a**) [65]. While **6.77a** shows substantial fluorescence decreases in the presence of steroids and terpenes in aqueous solution ($K_{11} = 1.03 \times 10^{-4}$ dm^3 mol^{-1} in phosphate buffer at pH = 7 and $I = 0.02$ mol dm^{-3} for the complex of **6.77a** and norethindrone), consistent with the 2-naphthylamine fluorophore being exposed to the aqueous environment in the steroid and terpene complexes, **6.78a** shows no fluorescence change in the presence of either class of potential guests. It appears that the dansyl fluorophore is so strongly intramolecularly complexed that steroids and terpenes do not effectively compete for its place within the hydrophobic interior of **6.78a**.

Another attachment of a second complexing group is exemplified by the mono- and bis-substitution of 1,4,10,13-tetraoxa-7,16-diazacyclooctadecane through one and two C(6) sites of a single βCD, respectively [66,67]. Such attachment of second complexing groups to CDs has been used extensively in the metalloCD chemistry discussed in Chapter 5.

6.6. Polymeric Cyclodextrins

A logical development of the linked CD dimers is to increase the number of linkages to produce polymers, and this has resulted in some interesting systems. Early examples are the polyacryloyl-αCD **6.79** and poly-*N*-acryloyl-6-aminocaproyl-αCD **6.80** and their βCD analogues where the CD moieties are attached mainly through substitution at C(3) to the polymer by single linkers [68-70]. Polyacryloyl-βCD catalyses the hydrolysis of 4-nitrophenyl acetate and 4-nitrophenyl 4-nitrobenzoate through a mechanism which appears to involve cooperative complexing of these guest molecules by adjacent βCDs attached to the polymer. In the latter hydrolysis polyacryloyl-βCD is a 3.3-fold more effective catalyst than βCD, but poly-*N*-acryloyl-6-aminocaproyl-βCD is only 1.8-fold as effective consistent with increasing distance between the βCD moieties decreasing catalytic effectiveness. (Polyacryloyl-βCD also binds TNS⁻ (**6.17**) through simultaneous complexation by adjacent βCD moieties.) Another polymer, where βCD moieties are linked to a polyethylenimine backbone, catalyses the hydrolysis of two 4-*tert*-butylphenyl esters where the ethylenimine nitrogen acts as the nucleophile [71]. Michaelis-Menten kinetics are observed for the hydrolysis consistent with the guest ester being complexed by the polymer βCD moieties.

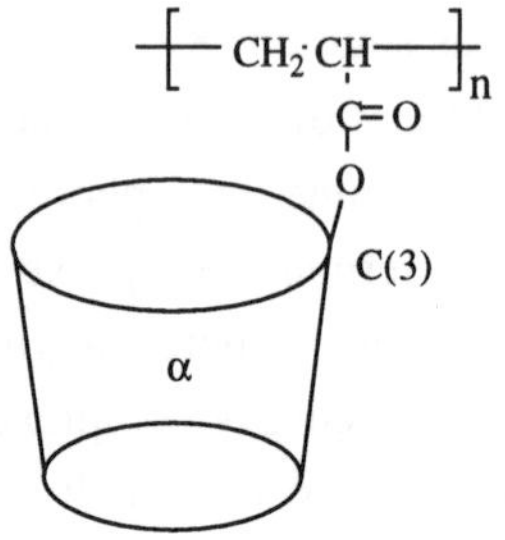

6.79

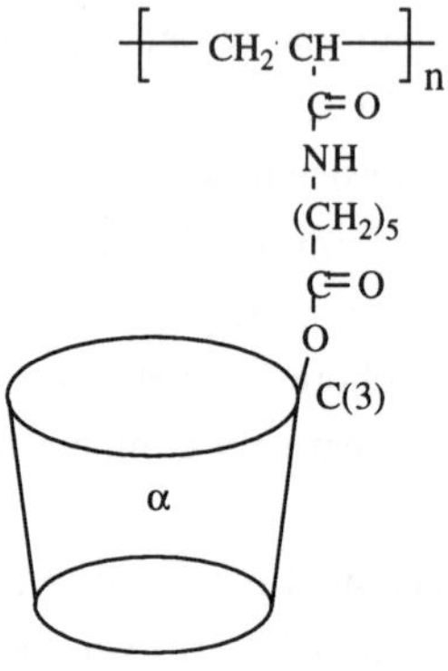

6.80

The reaction of βCD with epichlorohydrin produces a polymer where βCDs are linked mainly through their C(6) sites and are part of the polymer backbone (**6.81**) [72], while reaction of poly(allylamine) with monotosylated βCD produces a polymer where each βCD is attached to the polymer through a short linker, mainly through C(6) sites (**6.82**) [73]. Several examples of each polymer have been prepared where the average n varied. (It should be appreciated that a degree of heterogeneity exists in each polymer chain when the average n is referred to below.)

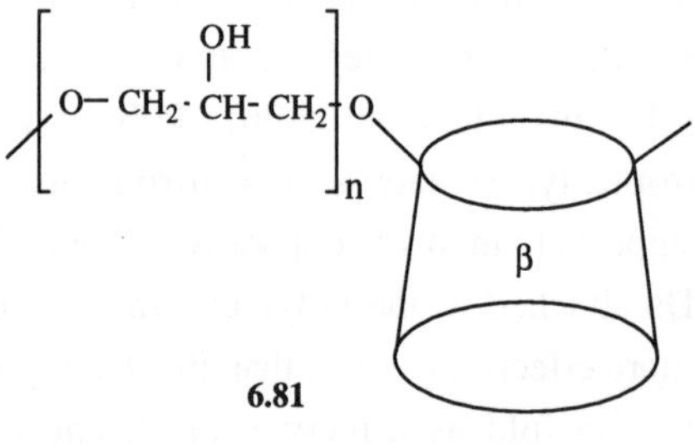

6.81

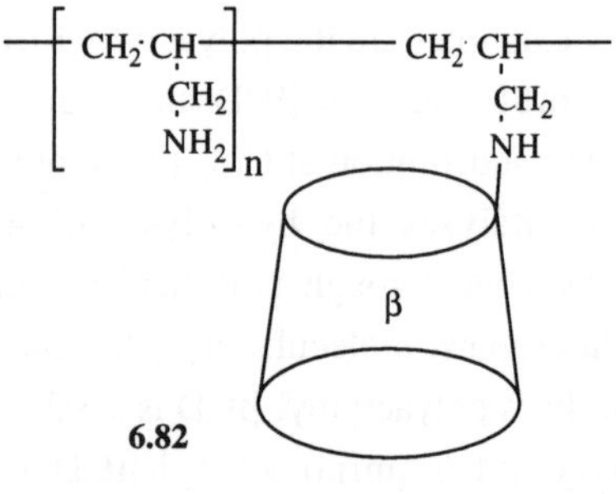

6.82

A very detailed luminescence study shows that **6.81** (shown as the polymer segment **6.83**) complexes pyrene (**6.84**) more strongly than does βCD provided that the distance between linked βCD moieties is sufficiently large [72]. This is attributed to cooperative complexation proceeding through **6.85**, where one end of a pyrene is complexed by one βCD moiety of **6.86** and the other end is complexed by another βCD moiety in the polymer chain. The distance of the second βCD moiety from the first has a considerable impact on pyrene complexation. This distance is controlled by the number of units (n) in the $-O(CH_2CH(OH)CH_2O)_n-$ linker joining the two βCD moieties at C(6). The apparent pyrene complexation constant, K_{app}, is given by:

$$K_{app} = (K_{int} + 1)K_{11}$$

When n = 1, $K_{app} = 1.2 \times 10^2$ dm^3 mol^{-1} and $K_{int} = 2.4$; when n = 2, $K_{app} = 1.8 \times 10^2$ dm^3 mol^{-1} and $K_{int} = 4$; when n = 3, $K_{app} = 5.1 \times 10^2$ dm^3 mol^{-1} and $K_{int} = 14$; when n = 4, $K_{app} = 1.11 \times 10^3$ dm^3 mol^{-1} and $K_{int} = 31$; and when n = 8, $K_{app} = 4.2 \times 10^3$ dm^3 mol^{-1} and $K_{int} = 1.2 \times 10^2$ in water at room temperature. This increase in K_{app} with n is attributed to the increasing ease of simultaneous complexation of pyrene by two βCD moieties in the same polymer chain. The higher the value of K_{int} the greater is the proportion of total complexed pyrene which is simultaneous complexation by two βCD moieties. Molecular modelling shows that when n = 1 such simultaneous complexation by adjacent βCDs is not possible and the third or fourth βCD along the polymer chain combines with the first in simultaneous complexation instead. A similar situation may exist when n = 2. For the initial complexation step for each polymer K_{11} is estimated as 35 dm^3 mol^{-1} which

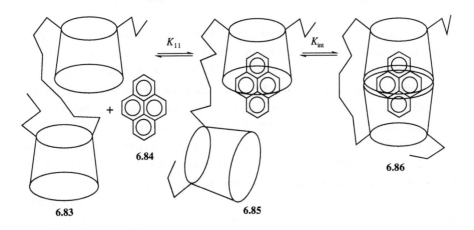

6.83 6.84 6.85 6.86

compares with $K_{app} = K_{11} = 1.4 \times 10^2$ dm^3 mol^{-1} for the βCD.pyrene complex. Thus, when n = 8 for the polymer K_{app} is 30 times greater than that for βCD.

Stronger complexation by epichlorohydrin-generated polymers of αCD, βCD and γCD, by comparison with that of the parent CDs, occurs for several guest molecules and is mainly attributed to cooperative complexing of the guests by adjacent CD moieties in the polymer [74,75]. It appears that the linker units may also serve to hold the CD moieties in optimal orientations for guest complexation and also increase the hydrophobicity of the polymer such that guest complexation is reinforced [75].

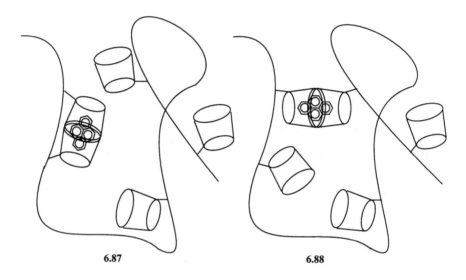

6.87   6.88

Luminescence studies of the complexation of pyrene by polymer **6.82** show that as the degree of substitution of βCD onto the polymer backbone increases from 0.5% to 23% of the amine groups being substituted by βCD (and n decreases) so K_{app} for the complexation of pyrene increases to 3.5×10^2 times that for pyrene complexation by βCD [73]. In principle, pyrene may be simultaneously complexed by either two adjacent appended βCD moieties (**6.87**) or by two βCDs separated by several polymer segments (**6.88**). If the latter complexation is significant complex stability should increase with polymer length, but this is not observed and it appears that complexation of pyrene by adjacent βCDs dominates.

The effect of the polymer structure on guest complexation may be exploited in molecular imprinting where a CD.guest complex is incorporated into a polymer such that subsequent removal of the guest leaves the guest imprint on the polymer so that it retains a strong selectivity for complexation of that guest. This principle of molecular imprinting has been applied to produce a cross-linked polymer which complexes cholesterol with a high degree of specificity as shown in Fig. 6.5 [76]. The βCD.cholesterol 3:1 complex **6.89** is linked and cross-linked with hexamethylene-diisocyanate to produce the polymer **6.90**, from which the templating cholesterol is then removed to produce **6.91**. This solid polymer is 15 times more effective in absorbing cholesterol from solution than is its analogue prepared in the absence of the cholesterol template. The nature of the linker is critical for the selective absorbing character of the polymer as is shown by the toluene-2,4-disocyanate linked analogue of **6.91** which, although 4.7 times more effective in absorbing cholesterol than **6.91**, is

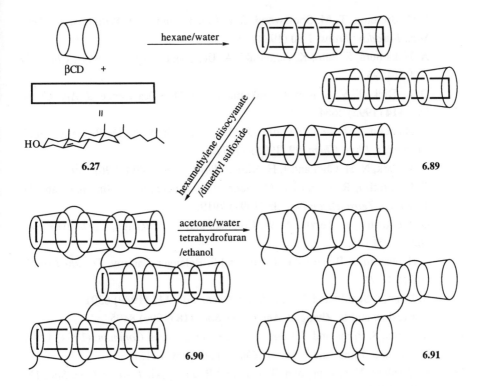

Fig. 6.5. Synthesis of a molecular imprinted polymer for the molecular recognition of cholesterol.

only 3.7 times more effective than its analogue prepared in the absence of the cholesterol template. The epichlorohydrin-linked polymer is much less effective than **6.91** in absorbing cholesterol.

Solid CD polymers have attracted attention as chromatographic materials but so far do not appear to have proved either as durable or versatile as the chromatographic materials generated through the binding of CDs to silica supports [77]. The latter chromatographic materials are capable of separating structural isomers, enantiomers and diastereomers [78-81].

6.7.   References

1.   I. M. Brereton, T. M. Spotswood, S. F. Lincoln and E. H. Williams, *J. Chem. Soc.*, *Faraday Trans. 1* **80** (1984) 3147.

2.   A. Nakamura, S. Sato, K. Hamasaki, A. Ueno and F. Toda, *J. Phys. Chem.* **99** (1995) 10952.

3.   D. L. Dick, T. V. S. Rao, D. Sukumaran and D. S. Lawrence, *J. Am. Chem. Soc.* **114** (1992) 2664.

4.   B. Hamelin, L. Jullien, F. Guillo, J.-M. Lehn, A. Jardy, L. De Robertis and H. Driguez, *J. Phys. Chem.* **99** (1995) 17877 .

5.   J. W. Park, N. H. Choi and J. H. Kim, *J. Phys. Chem.* **100** (1996) 769.

6.   T. Carofiglio, R. Fornasier, G. Gennari, V. Lucchini, L. Simonato and U. Tonellato, *Tetrahedron Lett.* **38** (1997) 7919.

7.   D. Mentzafos, A. Terzis, A. W. Coleman and C. de Rango, *Carbohydr. Res.* **282** (1996) 125.

8.   S. Hanessian, A Benalil, M. Simard and F. Bélanger-Gariépy, *Tetrahedron* **51** (1995) 10149.

9.   A. Ueno, I. Suzuki and T. Osa, *J. Am. Chem. Soc.* **111** (1989) 6391.

10.   R. Breslow and B. Zhang, *J. Am. Chem. Soc.* **118** (1996) 8495.

11.   K. Fujita, S. Ejima and T. Imoto, *J. Chem. Soc.*, *Chem. Commun.* (1984) 1277.

12.   K. Fujita, S. Ejima and T. Imoto, *Chem. Lett.* (1985) 11.

13.   R. Breslow, N. Greenspoon, T. Guo and R. Zarzycki, *J. Am. Chem. Soc.* **111** (1989) 8296.

14.   R. Breslow and S. Chung, *J. Am. Chem. Soc.* **112** (1990) 9659.

15. R. C. Petter, C. T. Sikorski and D. H. Waldeck, *J. Am. Chem. Soc.* **113** (1991) 2325.

16. R. Breslow, *Isr. J. Chem.* **32** (1992) 23.

17. B. Zhang and R. Breslow, *J. Am. Chem. Soc.* **115** (1993) 9353.

18. R. Breslow, *Rec. Trav. Chim. Pays-Bas* **113** (1994) 493.

19. C. T. Sikorski and R. C. Petter, *Tetrahedron Lett.* **35** (1994) 4275.

20. Y. Okabe, H. Yamamura, K. Obe, K. Ohta, M. Kawai and K. Fujita, *J. Chem. Soc., Chem. Commun.* (1995) 581.

21. R. Breslow, S. Halfon and B. Zhang, *Tetrahedron* **51** (1995) 377.

22. T. Jiang, D. K. Sukumaran, S.-D. Soni and D. S. Lawrence, *J. Org. Chem.* **59** (1994) 5149.

23. Y. Kuroda, T. Hiroshige, T. Sera, Y. Shiroiwa, H. Tanaka and H. Ogoshi, *J. Am. Chem. Soc.* **111** (1989) 1912.

24. R. Breslow and B. Zhang, *J. Am. Chem. Soc.* **114** (1992) 5882.

25. Y. Kuroda, M. Ito, T. Sera and H. Ogoshi, *J. Am. Chem. Soc.* **115** (1993) 7003.

26. Y. Kuroda, Y. Egawa, H. Seshimo and H. Ogoshi, *Chem. Lett.* (1994) 2361.

27. T. Jiang, D. K. Sukumaran, S.-D. Soni and D. S. Lawrence, *J. Org. Chem.* **59** (1994) 5149.

28. R. Deschenaux, A. Greppi, T. Ruch, H.-P. Kriemler, F. Raschdorf and R. Ziessel, *Tetrahedron Lett.* **35** (1994) 2165.

29. Y. Ishimura, T. Masuda and T. Iida, *Tetrahedron Lett.* **38** (1997) 3743.

30. I. Tabushi, Y. Kuroda and K. Shimokawa, *J. Am. Chem. Soc.* **101** (1979) 1614.

31. A. Ruebner, D. Kirsch, S. Andrees, W. Decker, B. Roeder, B. Spengler, R. Kaufmann and J. G. Moser, *J. Inclusion Phenom. Mol. Recogn. Chem.* **27** (1997) 69.

32. A. Harada, M. Furue and S. Nozakura, *Polymer J.* **12** (1980) 29.

33. J. H. Coates, C. J. Easton, S. J. van Eyk, S. F. Lincoln, B. L. May, C. B. Whalland and M. L. Williams, *J. Chem. Soc., Perkin Trans. 1* (1990) 2619.

34. Y. Wang, A. Ueno and F. Toda, *Chem. Lett.* (1994) 167.

35. F. Venema, C. M. Baselier, E. van Dienst, B. H. M. Ruël, M. C. Feiters, J. F. J. Engbersen, D. N. Reinhoudt and R. J. M. Nolte, *Tetrahedron Lett.* **35** (1994) 1773.

36. F. Venema, C. M. Baselier, M. C. Feiters and R. J. M. Nolte, *Tetrahedron Lett.* **35** (1994) 8661.

37. N. Birlirakis, B. Henry, P. Berthault, F. Venema and R. J. M. Nolte, *Tetrahedron* **54** (1998) 3513.

38. C. A. Haskard, C. J. Easton, B. L. May and S. F. Lincoln, *J. Phys. Chem.* **100** (1996) 14457.

39. C. J. Easton, S. J. van Eyk, S. F. Lincoln, B. L. May, J. Papageorgiou and M. L. Williams, *Aust. J. Chem.* **50** (1997) 9.

40. D.-Q. Yuan, K. Fujita, H. Mizaushima and M. Yamaguchi, *J. Chem. Soc., Perkin Trans. 1* (1997) 3135.

41. Y. Kuroda and H. Ogoshi, *Synlett.* (1994) 319.

42. F. Sallas, J. Kovács, I. Pintér, L. Jicsinszky and A. Marsura, *Tetrahedron Lett.* **37** (1996) 4011.

43. C. A. Haskard, B. L. May, T. Kurucsev, S. F. Lincoln and C. J. Easton, *J. Chem. Soc., Faraday Trans.* **93** (1997) 279.

44. R. Breslow, *Pure Appl. Chem.* **66** (1994) 1573.

45. C. Tanford, in *The Hydrophobic Effect: Formation of Micelles and Biological Membranes* (Wiley, New York, 1980) 2nd Edn.

46. D. F. Shriver, P. W. Atkins and C. H. Langford, in *Inorganic Chemistry* (Oxford, 1994) 2nd Edn.

47. P. L. Privalov and S. J. Gill, *Pure Appl. Chem.* **61** (1989) 1097.

48. R. Varadarajan, P. R. Connelly, J. M. Sturtevant and F. M. Richards, *Biochemistry* **31** (1992) 1421.

49. R. Breslow, *Acc. Chem. Res.* **28** (1995) 146.

50. Y. Murakami, J. Kikuchi, Y. Hisaeda and O. Hayashida, *Chem. Rev.* **96** (1996) 721.

51. R. Breslow and B. Zhang, *J. Am. Chem. Soc.* **116** (1994) 7893.

52. T. Akiike, Y. Nagano, Y. Yamamoto, A. Nakamura, H. Ikeda, A. Ueno and F. Toda, *Chem. Lett.* (1994) 1089.

53. H. Ikeda, S. Nishikawa, J. Takaoka, T. Akiike, Y. Yamamoto, A. Ueno and F. Toda, *J. Inclusion Phenom. Mol. Recogn. Chem.* **25** (1996) 133.

54. J. Hwa Jung, C. Takehisa, Y. Sakata and T. Kaneda, *Chem. Lett.* (1996) 147.

55. M. Nakamura, T. Ikeda, A. Nakamura, H. Ikeda, A. Ueno and F. Toda, *Chem. Lett.* (1995) 343.

56. C. J. Easton, J. B. Harper and S. F. Lincoln, *New J. Chem.* (1998) 1163.

57. F. Venema, A. E. Rowan and R. J. M. Nolte, *J. Am. Chem. Soc.* **110** (1996) 257.
58. T. Jiang and D. S. Lawrence, *J. Am. Chem. Soc.* **117** (1995) 1857.
59. T. Jiang, M. Li and D. S. Lawrence, *J. Org. Chem.* **60** (1995) 7293.
60. M. C. Gonzalez, A. R. McIntosh, J. R. Bolton and A. C. Weedon, *J. Chem. Soc., Chem. Commun.* (1984) 1139.
61. R. Breslow, X. Zhang, R. Xu, M. Maletic and R. Merger, *J. Am. Chem. Soc.* **110** (1996) 11678.
62. F. D'Alessandro, F. B. Gulino, G. Impellizzeri, G. Pappalardo, E. Rizzarelli, D. Sciotto and G. Vecchio, *Tetrahedron Lett.* **35** (1994) 629.
63. E. van Dienst, B. H. M. Snellink, I. von Piekartz, J. F. J. Engbersen and D. N. Reinhoudt, *J. Chem. Soc., Chem. Commun.* (1995) 1151.
64. E. van Dienst, B. H. M. Snellink, I. von Piekartz, M. H. B. Grote Gansey, F. Venema, M. C. Feiters, R. J. M. Nolte, J. F. J. Engbersen and D. N. Reinhoudt, *J. Org. Chem.* **60** (1995) 6537.
65. J. Bügler, J. F. J. Engbersen and D. N. Reinhoudt, *J. Org. Chem.* **63** (1998) 5339.
66. I. Willner and Z. Goren, *J. Chem. Soc., Chem. Commun.* (1983) 1469.
67. Z. Pikramenou, K. M. Johnson and D. G. Nocera, *Tetrahedron Lett.* **34** (1993) 3531.
68. A. Harada, M. Furue and S. Nozakura, *Macromolecules* **9** (1976) 701.
69. A. Harada, M. Furue and S. Nozakura, *Macromolecules* **9** (1976) 705.
70. A. Harada, M. Furue and S. Nozakura, *Macromolecules* **10** (1977) 676.
71. J. Suh, S. H. Lee and K. D. Zoh, *J. Am. Chem. Soc.* **114** (1992) 7916.
72. W. Xu, J. N. Demas, B. A. DeGraff and M. Whaley, *J. Phys. Chem.* **97** (1993) 6546.
73. M. Hollas, M.-A. Chung and J. Adams, *J. Phys. Chem. B* **102** (1998) 2947.
74. A. Harada, M. Furue and S. Nozakura, *Polymer J.* **13** (1981) 777.
75. T. C. Werner and I. M. Warner, *J. Inclusion Phenom. Mol. Recogn. Chem.* **18** (1994) 385.
76. H. Asanuma, M. Kakazu, M. Shibata, T. Hishiya and M. Komiyama, *J. Chem. Soc., Chem. Commun.* (1997) 1971.
77. S. Li and W. C. Purdy, *Chem. Rev.* **92** (1992) 1457.

78.   D. W. Armstrong, W. DeMond, A. Alak, W. L. Hinze, T. Riehl and K. H. Bui, *Anal. Chem.* **57** (1985) 234.

79.   W. L. Hinze, T. Riehl, D. W. Armstrong, W. DeMond, A. Alak and T. Ward, *Anal. Chem.* **57** (1985) 237.

80.   D. W. Armstrong, W. DeMond and B. P. Czech, *Anal. Chem.* **57** (1985) 481.

81.   D. W. Armstrong, T. Ward, R. D. Armstrong and T. E. Beesley, *Science* **232** (1986) 1132.

CHAPTER 7

MODIFIED CYCLODEXTRINS

CYCLODEXTRIN ROTAXANES AND CATENANES

7.1. Introduction

Rotaxanes and catenanes are unusual molecular species whose major components are held together mechanically. The name rotaxane is derived from the Latin *rota* for wheel and *axis* for axle in recognition that the macrocycle component of a rotaxane may be viewed as a wheel on an axle represented by its linear or threading molecular component. Similarly, the name catenane is derived from the Latin for chain, *catena*, in recognition that the interpenetrating macrocycles of the catenane may be likened to the links of a chain. A simple rotaxane consists of a macrocycle threaded onto a long molecule where large blocking end groups prevent the macrocycle from dethreading. A range of variations on this theme are shown in Fig. 7.1. The CD rotaxane precursor necessarily lacks one or both blocking end groups so that the CD may be threaded, and such species are often referred to as pseudorotaxanes shown as **7.1** and **7.2**. The blocking end groups are then attached to form the CD rotaxane **7.3**.

A simple nomenclature system devised by Schill [1,2], has been widely adopted for the rotaxanes, and is followed here. The number of mechanically joined molecular components in the rotaxane is written first and is enclosed in square brackets. This is followed by the names of the threading molecule and the macrocycle, each in square brackets, and ends with rotaxane also in square brackets. Thus, **7.3** is named [2]-[threading molecule]-[CD]-[rotaxane]. Should there be two CDs simultaneously threaded, the name becomes [3]-[threading molecule]-[CD]-[CD]-[rotaxane] and so on. The pseudorotaxanes **7.1** and **7.2** are both named [2]-[threading molecule]-[CD]-[pseudorotaxane] under this system. As the threading molecule becomes longer and more complex, large numbers of CDs may be threaded in different ways as illustrated by the polyrotaxanes **7.4-7.6**.

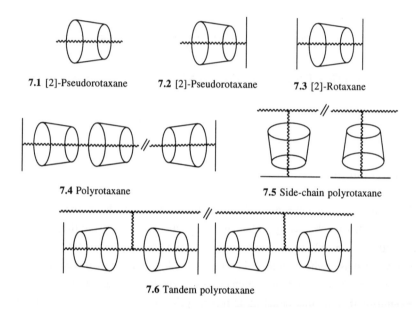

7.1 [2]-Pseudorotaxane   **7.2** [2]-Pseudorotaxane   **7.3** [2]-Rotaxane

7.4 Polyrotaxane   **7.5** Side-chain polyrotaxane

7.6 Tandem polyrotaxane

Fig. 7.1. Representations of CD rotaxanes where the truncated cones represent CDs, the wavy lines represent lengthy molecular segments and the straight lines represent blocking end groups. The orientations of the CDs may be selective or random depending on the interactions occurring with the threading segment, the end groups and neighbouring CDs.

A simple catenane consists of two interpenetrating macrocycles as illustrated by **7.7** in Fig. 7.2. The Schill nomenclature system also applies to catenanes. Thus, **7.7** is named [2]-[macrocycle]-[CD]-[catenane], **7.8** is named [4]-[macrocycle]-[CD]-[CD]-[CD]-[catenane] and so on. As the threading macrocycle increases in size so increasing numbers of CDs may be threaded to give polycatenanes such as **7.9**.

Many rotaxanes and catenanes have been studied and their chemistry has been extensively reviewed [3-9]. A considerable variety of participating macrocycles has been explored as is exemplified by crown ethers, cyclophanes and CDs. A wide range of threading molecules has also been explored. Rotaxane formation sometimes involves the formation of the macrocyclic component as an integral part of the synthetic procedure, a process which is not available in the formation of CD rotaxanes and thereby restricts their formation to the passing of the threading molecule through the CD annulus. Necessarily, in the formation of CD catenanes the ends of the threading molecule must link to form the second macrocycle during the catenane synthesis.

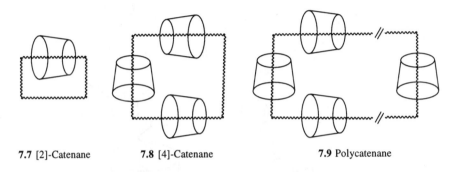

7.7 [2]-Catenane   **7.8** [4]-Catenane   **7.9** Polycatenane

Fig. 7.2. Representations of CD catenanes where the truncated cone represents a CD and the threading macrocycle is represented by the rectangle traced by the wavy line. The orientations of the CDs may be selective or random depending on the interactions occurring with the threading macrocycle and with neighbouring CDs.

7.2. Cyclodextrin Pseudorotaxanes

Pseudorotaxanes of the **7.1** type differ little from other CD complexes except that the threading molecule entering the CD annulus is significantly longer than the CD longitudinal axis so that end groups may be attached to form a rotaxane. In some cases, the threading molecule may possess end groups which fit the CD annulus quite closely so that the pseudorotaxane has quite a high stability but still allows the CD to dethread. Similar considerations apply to the **7.2** type pseudorotaxanes, but now it is only necessary to attach one additional end group to form a rotaxane.

A well characterised pseudorotaxane is represented by [2]-[1,10-bis(1-(4-*tert*-butylpyridinium)-decane)]-[αCD]-[pseudorotaxane], whose self-assembly is shown in Fig. 7.3 [10]. The rapid, but weak, complexation of the *tert*-butylpyridinium end group of 1,10-bis(1-(4-*tert*-butylpyridinium)-decane) (**7.10**) to form the initial complex (**7.11**) is characterised by K_{11} (298.2 K) = 18 dm^3 mol^{-1} in D$_2$O 0.1 mol

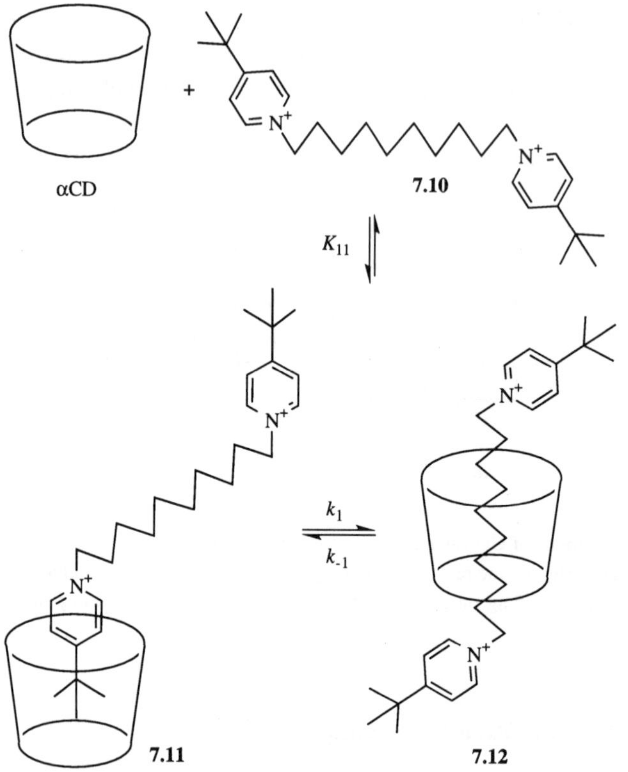

Fig. 7.3. Self-assembly of [2]-[1,10-bis(1-(4-*tert*-butylpyridinium))-decane]-[αCD]-[pseudorotaxane] **7.12**, by a slippage mechanism.

dm^{-3} in NaCl as shown by ^{1}H NMR studies. (Although complexation in **7.11** is shown occurring through the secondary face of αCD, it may occur through the primary face or through both routes.) The slow slippage of αCD over the pyridinium moiety to give the [2]-pseudorotaxane **7.12** is characterised by k_1 (298.2 K) = 4.2 × 10^{-4} s^{-1}, $\Delta H^{\ddagger}$ = 109 kJ mol^{-1} and $\Delta S^{\ddagger}$ = 57 J K^{-1} mol^{-1}, and the dethreading, process by k_{-1} (298.2 K) = 4.2 × 10^{-6} s^{-1}, $\Delta H^{\ddagger}$ = 99 kJ mol^{-1} and $\Delta S^{\ddagger}$ = -17 J K^{-1} mol^{-1}. The similarity of $\Delta H^{\ddagger}$ for both processes is consistent with the formation of **7.12** being largely entropy driven. Thus, $\Delta S^{\ddagger}$ = 57 J K^{-1} mol^{-1} for the threading process may partially result from dehydrating the cationic pyridinium moiety as it enters the αCD annulus in the transition state, while $\Delta S^{\ddagger}$ = -17 J K^{-1} mol^{-1} for dethreading may reflect its partial rehydration as it leaves the annulus. When the 1,10-bis(1-(4-*tert*-butylpyridinium))-decane end groups are replaced by bulkier moieties, such as 3,5-dimethylpyridinium or quinuclidinium, pseudorotaxane analogues of **7.12** are not formed.

A series of α,ω-alkanedicarboxylates **7.13** form pseudorotaxanes **7.14** with αCD in D_2O at pD 8 [11]. Both one- and two-dimensional ^1H NMR studies are consistent with αCD being located close to the centre of the alkyl chain in the general [2]-pseudorotaxane structures. No analogous βCD and γCD pseudorotaxanes are detected. This suggests that the closeness of fit of the threading α,ω-alkanedicarboxylate to the interior of the CD annulus is an important stabilising factor in these [2]-pseudorotaxanes. In D_2O at pD 13, where an αCD secondary hydroxy group is deprotonated to produce a negative charge, the formation constant, K_{11} (303.2 K) increases in the sequence: 3.1×10^2, 6.3×10^2, 1.4×10^3, 1.5×10^3 and 5.4×10^3 dm^3 mol^{-1} as n increases from 8 to 12. It appears from this that electrostatic repulsion between the threaded and negatively charged αCD and the negatively charged carboxylate end groups may repel the threaded αCD towards the centre of the threading alkane chain of the [2]-pseudorotaxanes and thereby stabilise them. Similar [2]-pseudorotaxanes form with α,ω-alkanedipyridinium threading dications **7.15** [12]. When n = 8, 9, 10 and 12, K_{11} (278.2 K) = 1.1×10^2, 6.4×10^2, 2.2×10^3 and 4.8×10^2 dm^3 mol^{-1}, respectively, for the formation of **7.16** in D_2O.

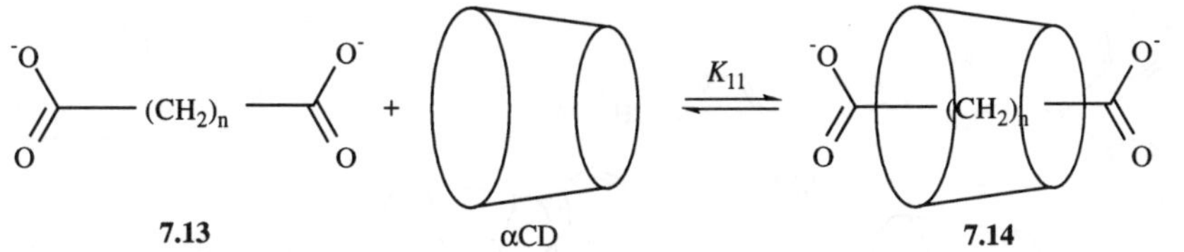

| **7.13** | αCD | **7.14** |

| **7.15** | αCD | **7.16** |

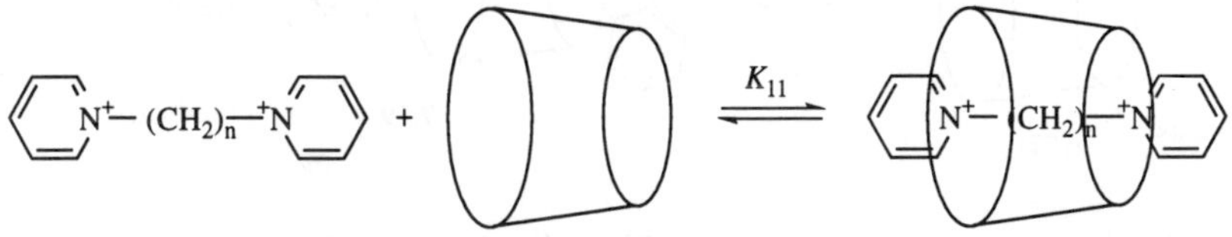

The threading of αCD by carbazole-4,4'-bipyridinium (viologen) linked dications
7.17 to form [2]-pseudorotaxanes **7.19** through the transition states **7.18** shown in
the free energy profile in Fig. 7.4, reveals an interesting variation of stability with
polyalkyl chain length [13]. The carbazole moiety acts as a blocking end group so
that formation of the pseudorotaxane only occurs through αCD threading from the
opposite end of the carbazole-4,4'-bipyridinium dication. When n = 4 or 6 no
complexation is detected, but when n = 8, 10 and 12 the [2]-pseudorotaxanes **7.19**

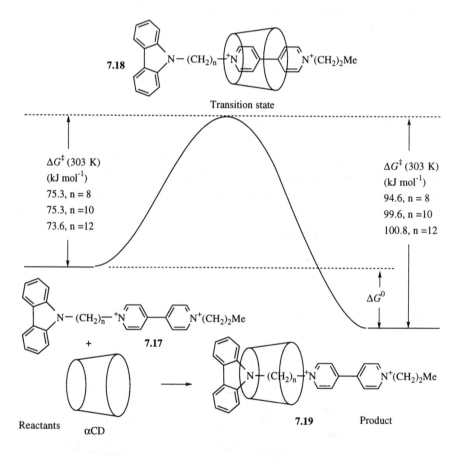

Fig. 7.4. Reaction profile for the threading of αCD by carbazole-viologen linked dications **7.17** to form
the [2]-pseudorotaxanes **7.19**. In **7.17**, the carbazole moiety may fold back over the viologen entity to
allow charge transfer to occur.

are formed as shown by both ^{1}H NMR and uv-visible spectroscopy. The orientation of the αCD annulus shown in Fig. 7.4 is consistent with the structure deduced from NOESY ^{1}H NMR spectra. The stability of **7.19** increases as n increases in the sequence: 8, 10 and 12 as shown by correspondingly increasing $\Delta G^0(303 \text{ K}) = -19.3$, -24.3 and -27.2 kJ mol^{-1}. The formation constants for **7.19**, $K_{11} = 2.1 \times 10^3$, 1.5 $\times 10^4$ and 4.9×10^4 dm^3 mol^{-1}, increase with increasing n = 8, 10 and 12. This is consistent with the dehydrated viologen entity passing through the αCD annulus with difficulty, and with maximum stability being gained when it rehydrates as αCD passes over it to become threaded by the polyalkyl chain. This is achieved to the greatest extent when n = 12, but it is not possible with the shorter polyalkane chains when n = 4 and 6 and the corresponding [2]-pseudorotaxanes are of relatively low stability.

For βCD, the [2]-pseudorotaxane formed when n = 12 for **7.17** is less stable $\{\Delta G^0 (303 \text{ K}) = -23.4 \text{ kJ mol}^{-1}, K_{11} = 1.5 \times 10^4 \text{ dm}^3 \text{ mol}^{-1}\}$ than is its αCD analogue. Greater insight is gained by comparing $\Delta G^{\ddagger}$ (303 K) for threading (n = 12) αCD = 73.6 kJ mol^{-1} with that for βCD = 48.5 kJ mol^{-1}, and the corresponding dethreading values of 100.8 and 72.0 kJ mol^{-1}, respectively. The greater $\Delta G^{\ddagger}$ values for αCD probably reflect a combination of the distortion of the αCD annulus required to permit passage of the viologen moiety together with its dehydration, while the larger βCD annulus more readily permits passage of the dehydrated viologen moiety so that its dehydration represents the major contribution to $\Delta G^{\ddagger}$. The larger γCD is thought to simultaneous complex the carbazole and the viologen moieties in the folded **7.17** dication.

7.20a n = 4
7.20b n = 12

7.21a n = 0
7.21b n = 1

When n is increased to 16 in the **7.17** polyalkyl chain, the increased length allows the threading of two αCDs. Only the isomer where the αCDs are in a head to tail orientation with their primary faces towards the carbazole end group is observed. This indicates the presence of orienting forces which render the opposite head-to-tail, the head-to-head and the tail-to-tail αCD orientations less stable in the [2]-pseudo-rotaxane [14].

The formation of pseudorotaxanes by other threading molecules with single blocking end groups also occur. Thus, while the phenothiazine-viologen linked species **7.20a** appears to be too short to form a stable rotaxane, the longer **7.20b** forms a pseudorotaxane with αCD characterised by $K_{11} = 4 \times 10^4$ dm^3 mol^{-1} at 303.2 K in D$_2$O and less well characterised complexes with βCD and γCD [15]. The polyacetylenic molecule **7.21a** forms pseudorotaxanes with αCD, βCD and heptakis(2,6-di-*O*-methyl)-βCD (DMβCD) characterised by $K_{11} = 1.3 \times 10^3$, 2.3 × 10^3 and 4.3 × 10^3 dm^3 mol^{-1}, respectively, in D$_2$O at 298.2 K [16]. Analogous pseudorotaxanes characterised by $K_{11} = 6.1 \times 10^3$, 6.5 × 10^3 and 2.7 × 10^4 dm^3 mol^{-1}, respectively, are formed by the longer **7.21b**, and the addition of a second DMβCD is characterised by $K_{21} = 5.3 \times 10^3$ dm^3 mol^{-1}.

The effect of charge on the formation of βCD [2]-pseudorotaxanes by **7.22-7.24**, which undergo sequential one electron reductions of their viologen moieties, has been studied voltammetrically [17]. None of **7.22-7.24** form stable [2]-pseudorotaxanes, however, the reduced monocation and zero charged forms of **7.22** form [2]-pseudorotaxanes characterised by $K_{11} = 5.0 \times 10^1$ and 6 × 10^3 dm^3 mol^{-1}, respectively, in aqueous 0.1 mol dm^{-3} phosphate buffer at pH and 298.2 K.

Similar sequential one electron reductions produce anionic and dianionic **7.23** which form βCD [2]-pseudorotaxanes characterised by $K_{11} = 9.0 \times 10^1$ and 2×10^4 dm³ mol⁻¹, respectively, and anionic and dianionic **7.24** whose βCD [2]-pseudorotaxanes are characterised by $K_{11} = 5.0 \times 10^1$ and 7×10^3 dm³ mol⁻¹, respectively. These stability variations are consistent with βCD being threaded by the viologen moiety which becomes increasingly hydrophobic as its local charge decreases from 2+ to 1+ to 0 and its complexing ability correspondingly increases. The analogous DMβCD pseudorotaxanes show the same stability trend where $K_{11} = 5.0 \times 10^1$ and 2×10^4, 1.0×10^1 and 7×10^4, and 7.0×10^1 and 1×10^4 dm³ mol⁻¹, respectively, characterise the monocation and zero charged forms of **7.22**, and the anionic and dianionic forms of **7.23** and **7.24** [17].

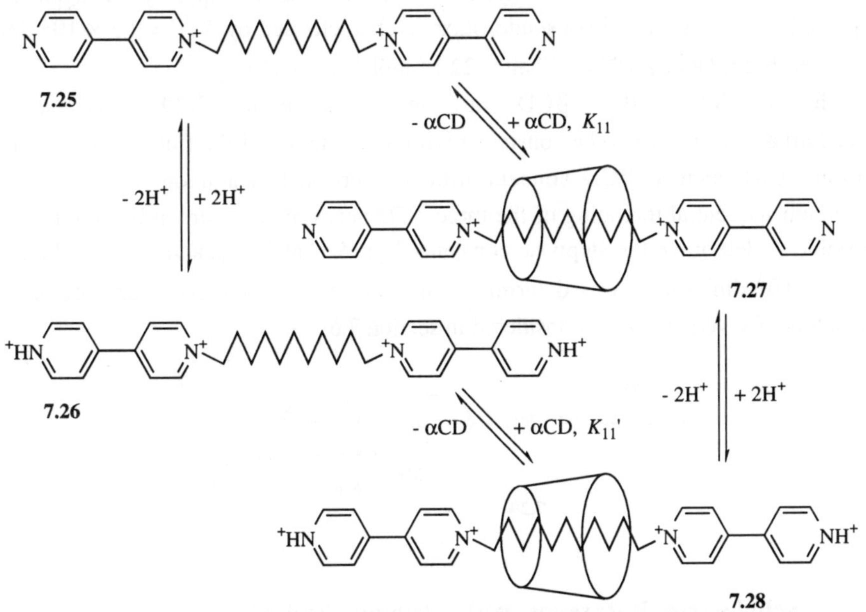

Fig. 7.5. The formation of di- and tetracationic [2]-pseudorotaxanes **7.27** and **7.28**, respectively.

A similar influence of charge is seen in the strong pH dependence of [2]-pseudorotaxane formation from αCD and 1,11-bis(pyridinium)undecane (**7.25**) and its diprotonated form (**7.26**, with pK_as = 2.63 and 3.96) shown in Fig. 7.5 [18]. Thus,

$K_{11} = 4.05 \times 10^3$ dm^3 mol^{-1} with ΔH^0 and $T\Delta S^0$ = -29 and -8.8 kJ mol^{-1}, respectively, and $K_{11}' = 5.8 \times 10^2$ dm^3 mol^{-1} with ΔH^0 and $T\Delta S^0$ = -33 and -17 kJ mol^{-1}, respectively, in aqueous medium at 298.2 K. The greater stability of **7.27**, by comparison with **7.28**, mainly arises from its less negative $T\Delta S^0$ contribution. A decrease in entropy arises from the threading process, but an increase in entropy might be expected from expulsion of water from the αCD annulus. Superimposed on this is the variation in these contributions resulting from the di- and tetracationic natures of **7.25** and **7.26**. The less hydrated viologen moieties in **7.25** probably present a lesser barrier to the threading of αCD than do the more highly hydrated viologen moieties of the tetracationic **7.26** as αCD moves to its ground state position threaded by the polyalkyl chain as shown by ^{1}H NMR spectroscopy. The tetracationic analogue of **7.26**, where the two protons on the bipyridyl nitrogens are replaced by Et$^+$, forms a [2]-pseudorotaxane characterised by $K_{11}' = 2.3 \times 10^2$ dm^3 mol^{-1} with ΔH^0 and $T\Delta S^0$ = -35 and -22 kJ mol^{-1}, respectively.

The threading of three βCD onto the ammonium ion **7.29** to form a [4]-pseudorotaxane appears to be consistent with the variation of the fluorescence of this cationic surfactant as βCD concentration is increased in aqueous solution [19]. Although sequential threading of the three βCD necessarily occurs, it has not proved possible to determine the stepwise constants K_{11}, K_{21} and K_{31}, and their product β_3 = 3.7 × 10^5 dm^9 mol^{-3} was determined instead. More extensive examples of the threading of several CDs are considered in section 7.6.

7.29

7.3. Cyclodextrin Rotaxanes with Organic End Groups

The attachment of organic end groups to the threading moiety of a CD rotaxane is achieved through either covalent bonding or ionic interactions. In Fig. 7.6, an example of the former attachment is shown where the azobenzene diazonium cation **7.30a** adds one end group through reaction with β-naphthol **7.31** [20]. This intermediate product then threads either αCD or βCD to form a [2]-pseudorotaxane which adds a second end group to form [2]-rotaxanes **7.32a** and **7.32b**. Alternatively, **7.30a** may thread αCD or βCD prior to the addition of the first end group. The yield of **7.32a** with αCD and βCD is 12% and 15%, respectively, and the yield of **7.32b** with βCD is 6% when the toluidine diazonium cation **7.30b** is used. As in all CD rotaxane syntheses, the formation of the threading entity complete with two end groups competes with the formation of the [2]-rotaxanes. None of the [2]-rotaxane **7.32b** with αCD is obtained probably because **7.30b** is too large to thread the αCD annulus.

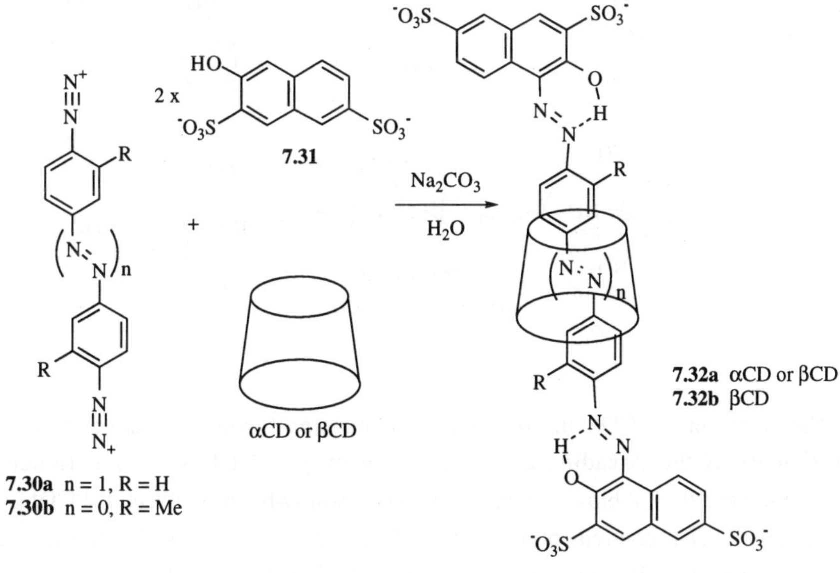

Fig. 7.6. The formation of [2]-rotaxanes **7.32a** and **7.32b** incorporating covalently bound end groups.

Examples of similarly assembled [2]-rotaxanes incorporating covalently attached end groups are represented by **7.33** and **7.34** [21]. The water soluble [2]-rotaxane **7.33** (obtained in 28.1% yield) is the first example where an aromatic chromophore, the stilbene entity, is threaded through a βCD annulus. This facilitates uv-visible absorption and circular dichroic spectroscopic studies from which it is deduced that the stilbene function is within the βCD annulus. The precursor to **7.33** is also a [2]-rotaxane where two chloro groups occupy the place of the two anilino groups. The reaction of 1,12-diaminododecane with two moles of 2,4,6-trinitrobenzenesulfonate and either hexakis(2,6-di-*O*-methyl)-αCD (DMαCD) or hexakis(2,3,6-tri-*O*-methyl)-αCD (TMαCD) gives the corresponding [2]-rotaxane **7.34** in 42% and 48% yield [22]. Both [2]-rotaxanes are insoluble in water, but are soluble in most organic solvents.

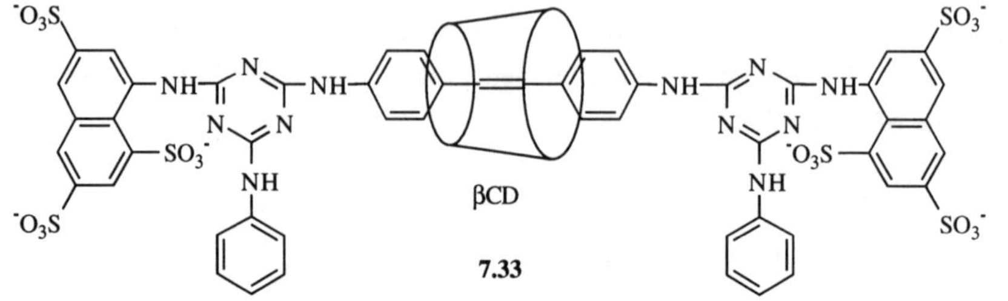

7.33

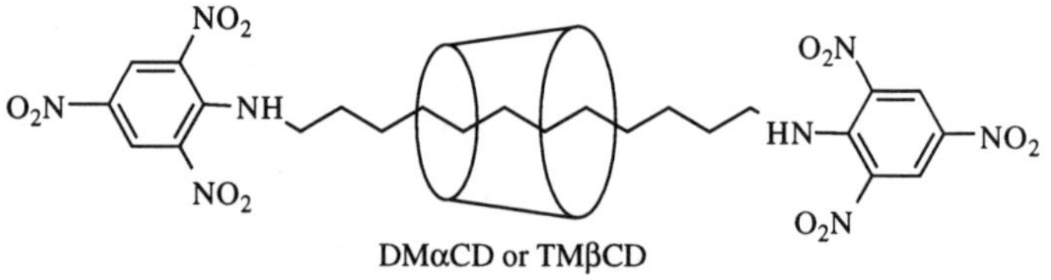

DMαCD or TMβCD

7.34

The position of CDs in rotaxanes can be significantly influenced by the constituents of the threading chain as shown by [2]-[1,1-bis(4-(7,7,7-triphenyl-heptyloxy)-benzyl)-4,4'-bipyridinium]-[heptakis(2,6-di-*O*-butyl-3-*O*-acetyl)-βCD] [23]. Thus, 1-[4-(7,7,7-triphenylheptyloxy)-benzyl]4,4'-bipyridinium ion (**7.35**) forms [2]-pseudorotaxanes **7.36** and **7.37** with heptakis(2,6-di-*O*-butyl-3-*O*-acetyl-βCD), DBMβCD, as shown in Fig. 7.7. Proton NMR spectroscopy shows **7.37** to be the more stable [2]-pseudorotaxane and its subsequent reaction with 4-(7,7,7-triphenyl-heptyloxy)-benzyl bromide **7.38** in 3:1 acetonitrile:chloroform solvent gives the [2]-rotaxane **7.39** in 50% yield. The smaller DBMαCD yields no [2]-rotaxane analogous to **7.39**, probably because its annulus is too small to be threaded. However, the larger DBMγCD gives the analogue of **7.39** in 15% yield in 3:1 acetonitrile/chloroform. Similar studies have been reported for the analogue of **7.39** where both ether oxygens of the threading molecule are replaced by carboxylate groups [24].

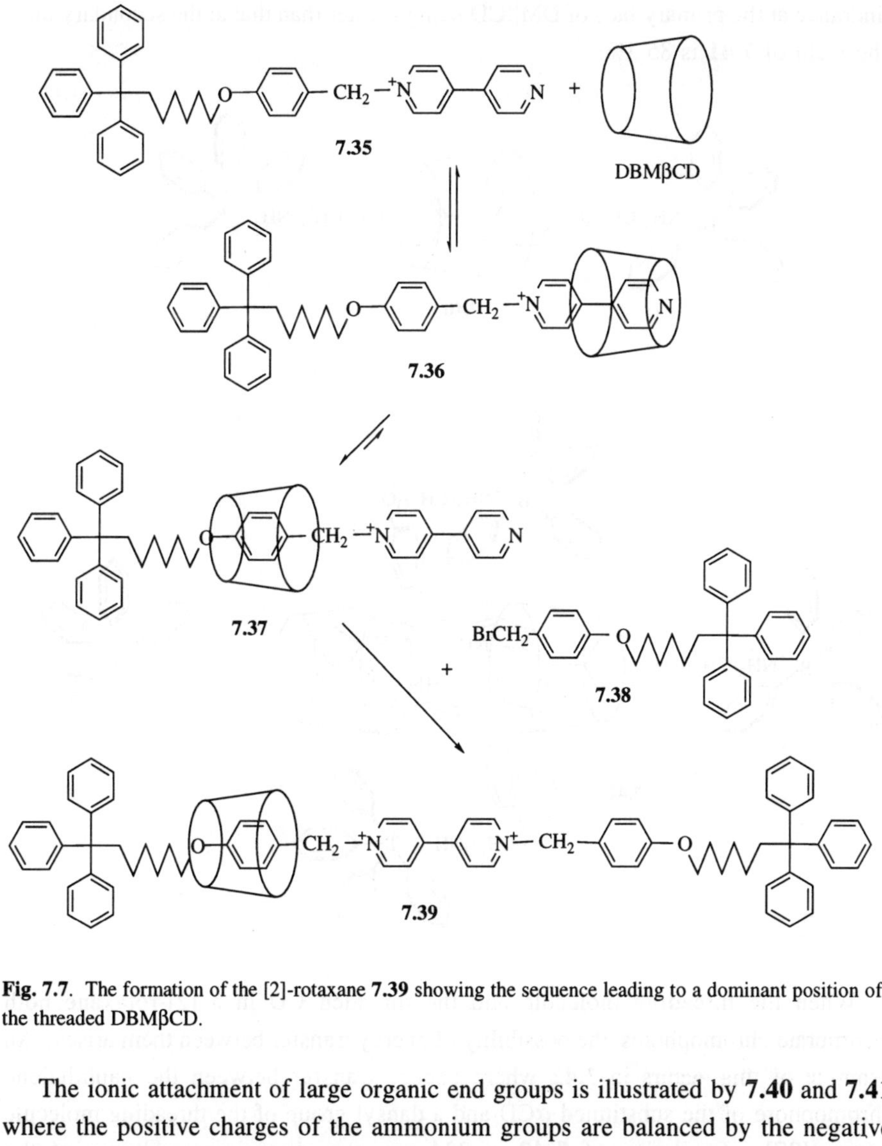

7.35

7.36

7.37

7.38

7.39

Fig. 7.7. The formation of the [2]-rotaxane **7.39** showing the sequence leading to a dominant position of the threaded DBMβCD.

The ionic attachment of large organic end groups is illustrated by **7.40** and **7.41** where the positive charges of the ammonium groups are balanced by the negative charges of the tetraphenylborates [25,26]. The DMβCD [2]-rotaxane **7.40** is obtained in 71.2% yield. The orientation of the two DMβCDs in the [3]-rotaxane **7.41** is shown to be as depicted by ^{1}H NOE NMR spectroscopy, and is attributed to the steric hindrance at the primary face of DMβCD being greater than that at the secondary face. The yield of **7.41** is 85.3%.

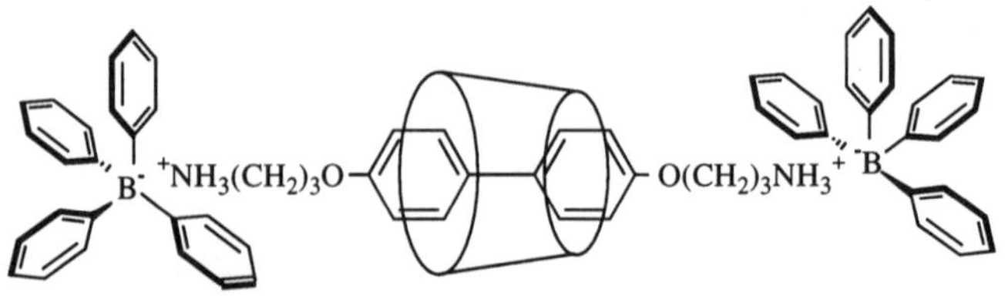

7.40

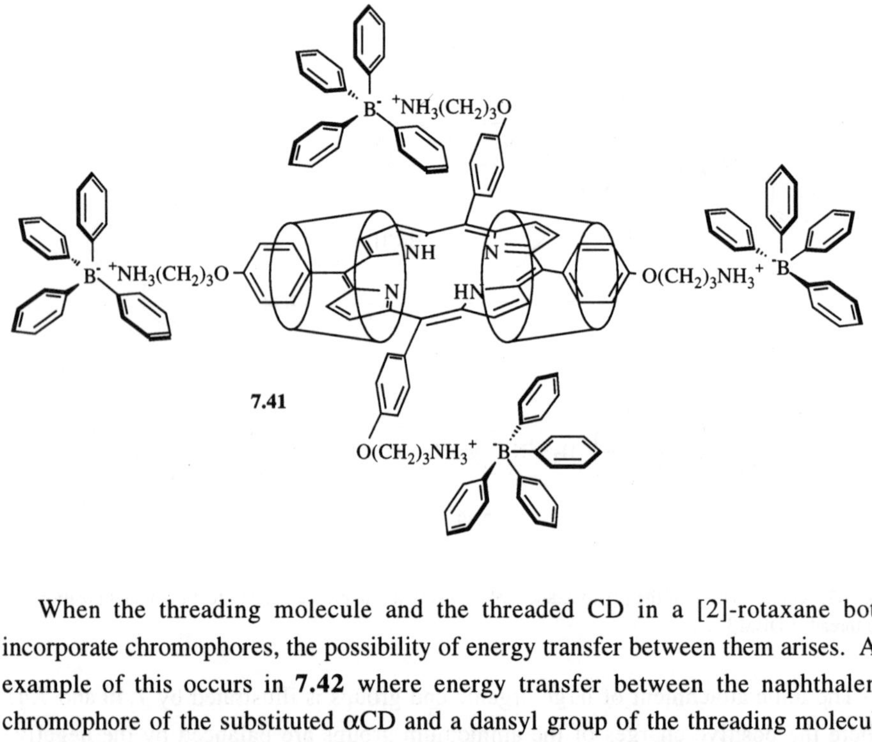

7.41

When the threading molecule and the threaded CD in a [2]-rotaxane both incorporate chromophores, the possibility of energy transfer between them arises. An example of this occurs in **7.42** where energy transfer between the naphthalene chromophore of the substituted αCD and a dansyl group of the threading molecule occurs [27]. Excitation of **7.42** at 234 nm, the absoption maximum of the naphthalene substituent, results in a much reduced fluorescence at 350 nm from the naphthalene group by comparison with that seen for the substituted αCD alone, and a fluorescence at > 450 nm where the dansyl group fluoresces. This is consistent with singlet energy transfer from the naphthalene group to a dansyl group in **7.42**.

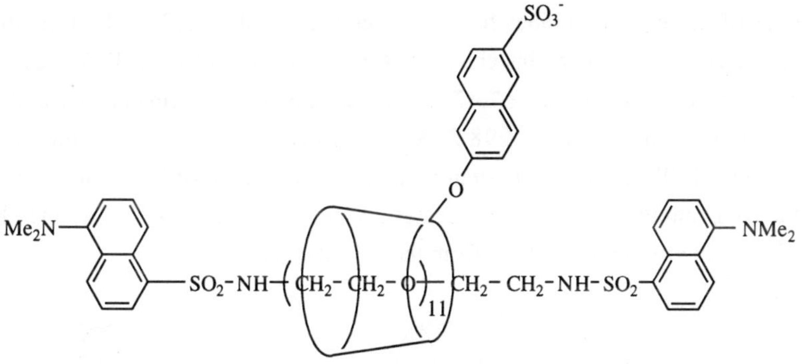

7.42

7.4. Cyclodextrin Rotaxanes with Metal Complex End Groups

The first CD [2]-rotaxanes appeared in an elegant study reported by Ogino in 1981 [28]. These diastereomeric [2]-rotaxanes are composed of βCD threaded onto μ-(1,12-diaminododecane)bis(chlorobis-1,2-diaminoethane)cobalt(III), [(en)$_2$ClCo(NH$_2$-(CH$_2$)$_{12}$)NH$_2$)CoCl(en)$_2$]$^{4+}$. The inert octahedral cobalt(III) moieties are too large to pass through the βCD annulus and act as blocking end groups for the retention of βCD threaded onto the alkyl chain. The starting materials in the self-assembly process shown in Fig. 7.8 are βCD, 1,12-diaminododecane **7.43** and racemic *cis*-dichlorobis-1,2-diaminoethanecobalt(III)chloride, *cis*-[Co(en)$_2$Cl$_2$]Cl, where the reactive constituent is *cis*-[Co(en)$_2$Cl$_2$]$^+$. The reaction may proceed through the [2]-pseudorotaxane **7.44** where βCD is threaded onto **7.43** and a subsequent reaction with *cis*-[Co(en)$_2$Cl$_2$]$^+$ attaches the first end group in **7.45**. Alternatively, the first end group may be attached to **7.43** to form **7.46** which then forms the [2]-pseudorotaxane **7.45**. The attachment of the second end group completes the formation of the [2]-rotaxane **7.47**. Because *cis*-[Co(en)$_2$Cl$_2$]$^+$ may possess either Δ or Λ chirality and the *cis*-[Co(en)$_2$Cl]$^{2+}$ moieties at either end of [(en)$_2$ClCo-(NH$_2$(CH$_2$)$_{12}$)NH$_2$)CoCl(en)$_2$]$^{4+}$ are randomly selected with respect to their chirality, four diastereomeric ΔΔ, ΛΛ, ΔΛ, and ΛΔ [2]-rotaxanes are formed in 7% yield. The preparation of the αCD analogue results in a 19% yield of the combined diastereomers. The analogous 1,10-diaminodecane and 1,14-diaminotetradecane analogues of these [2]-rotaxanes have also been reported [28,29]. The formation of [2]-[μ-(1,12-diaminododecane)bis(chlorobis-1,2-diaminoethane)cobalt(III)]-[αCD]-[rotaxane], the αCD analogue of **7.47**, is accompanied by a volume decrease of about 28 cm^3 mol^{-1} in water at 298.2 K as shown from partial molar volume measurements [30]. This is consistent with the displacement of four water molecules from the αCD annulus as it accommodates six methylene groups as it is threaded by the 1,12-diaminododecane entity to form the [2]-rotaxane.

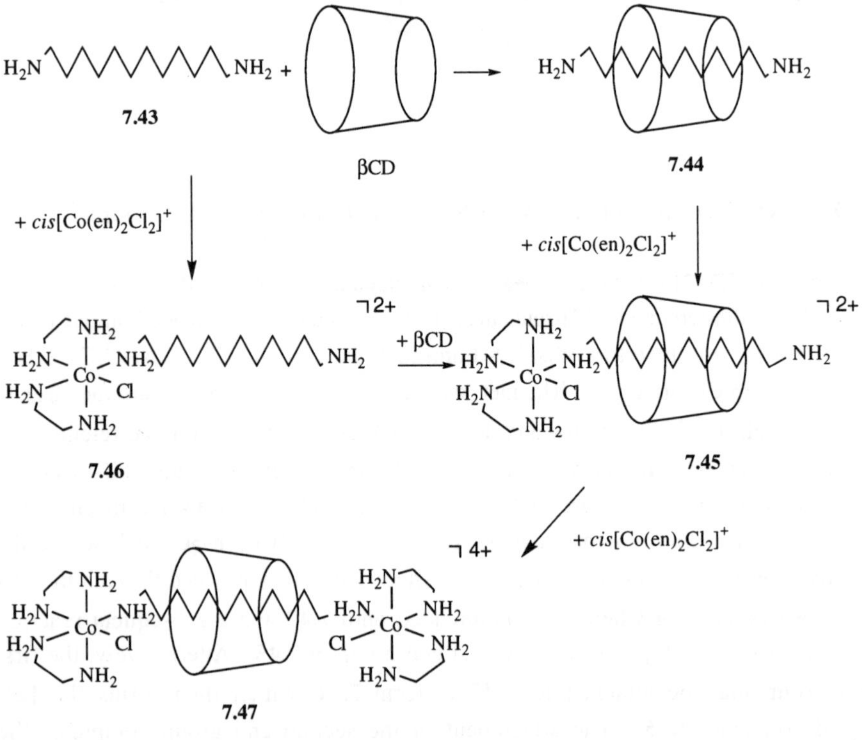

Fig. 7.8. The self-assembly of [2]-[μ-(1,12-diaminododecane)bis(chlorobis-1,2-diaminoethane)cobalt-(III)]-[βCD]-[rotaxane] **7.47**.

7.48 ΔΔ-[2]-rotaxane

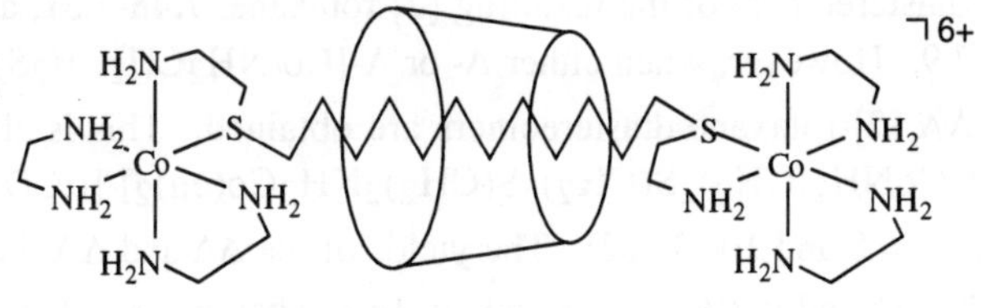

7.49 ΛΛ-[2]-rotaxane

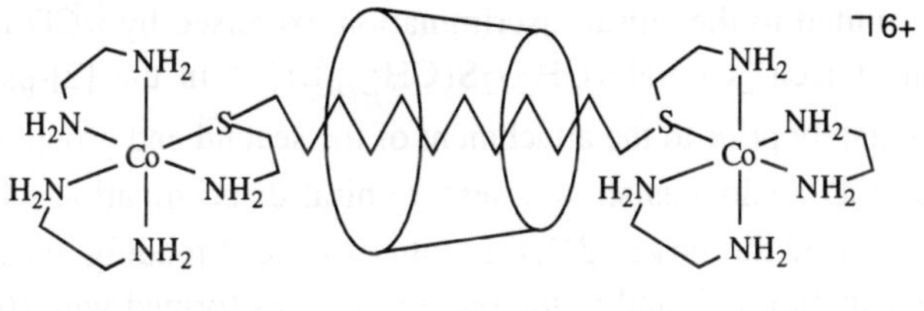

7.50 ΔΛ-[2]-rotaxane

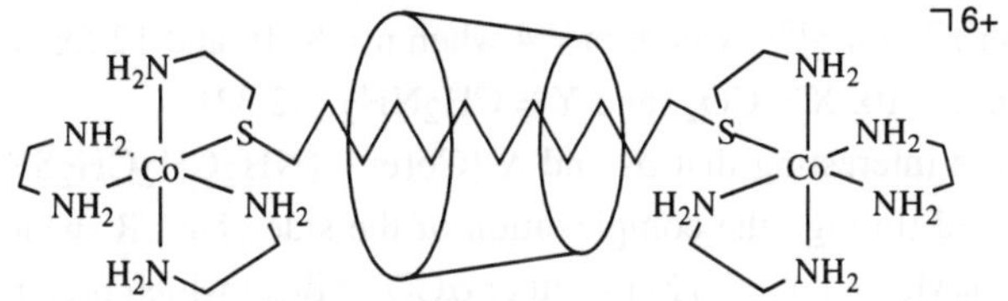

7.51 ΛΔ-[2]-rotaxane

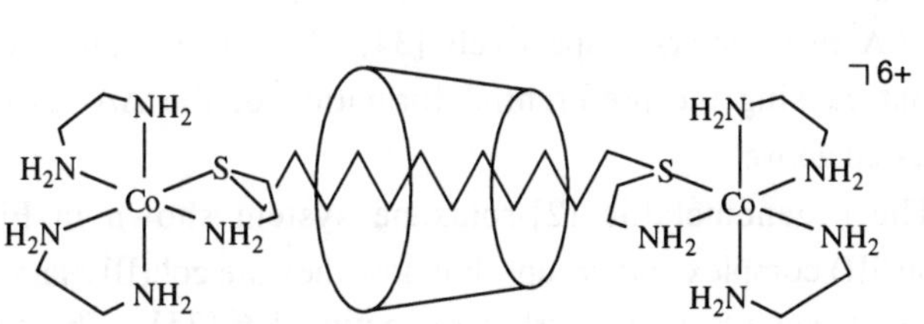

Fig. 7.9. Diastereomers of [2]-[μ-(1,18-diamino-3,16-dithiooctadecane)bis(bis-1,2-diaminoethane)-cobalt(III)]-[αCD]-[rotaxane].

The proportions in which diastereomeric [2]-rotaxanes are formed can be substantially affected by the interaction of the chiral metal complex end groups with the homochiral CD as is exemplified by [2]-[μ-(1,18-diamino-3,16-dithiooctadecane)-bis(bis-1,2-diaminoethane)cobalt(III)]-[αCD]-[rotaxane], [2]-[(en)$_2$CoNH$_2$(CH$_2$)$_2$S-(CH$_2$)$_{12}$S(CH$_2$)$_2$NH$_2$Co(en)$_2$]-[αCD]-[rotaxane] [31]. The preparation of this [2]-rotaxane follows a sequence similar to that shown in Fig. 7.8 through the reaction of [Co(NH$_2$CH$_2$CH$_2$S)(en)]$^{2+}$, which forms the end groups, with 1,12-dibromododecane and αCD. When [Co(NH$_2$CH$_2$CH$_2$S)(en)]$^{2+}$ is a racemate of Δ and Λ enantiomers, four diastereomers of the resulting [2]-rotaxane, **7.48-7.51**, are possible as shown in Fig. 7.9. However, when either Δ- or Λ-$[Co(NH_2CH_2CH_2S)(en)]^{2+}$ is used, the $\Delta\Delta$ and $\Lambda\Lambda$ [2]-rotaxane diastereomers are obtained. This is also the case for the [2]-$[(en)_2CoNH_2(CH_2)_2S(CH_2)_nS(CH_2)_2NH_2Co(en)_2]$-$[\alpha CD]$-[rotaxane] analogues where n = 8 and 10 [31,32]. The yields of the $\Delta\Delta$ and $\Lambda\Lambda$ diastereomers are 3.5 and *ca.* 0%, 21 and < 7%, and 28 and *ca.* 14% when n = 8, 10 and 12, respectively. This is attributed to the chiral discrimination exercised by αCD in complexing the alkyl chain of $[(en)_2CoNH_2(CH_2)_2S(CH_2)_nBr]^{3+}$ in the [2]-pseudorotaxane, the step immediately prior to the attachment of the second end group to form the [2]-rotaxane. (The βCD analogues show a lesser chiral discrimination which is attributed to the looser fit of the larger βCD annulus to the threading species.) A similar chiral discrimination is found in the αCD rotaxanes formed with $[(en)_2CoXCH_2S(CH)_nS-CH_2Y)Co(en)_2]^{m+}$ where m = 4 when n = 8, 10 and 12 for X = Y = CO_2^-; and m = 5 when n = 10, X = CO_2^- and Y = CH_2NH_2 [32,33].

It is interesting that Δ- and Λ-$[Co(en)_2(NH_2CH_2CH_2SR]^{3+}$ have been partially resolved through the complexation of the side chain R (which is either $(CH_2)_nBr$ or $(CH_2)_nMe$ and n = 7-12) by either αCD or βCD which preferentially interact with the Δ and Λ enantiomers, respectively [34]. The origin of this resolution may be related to that causing the preferential formation of the $\Delta\Delta$-[2]-rotaxane diastereomer discussed above.

The organometallic [2]-rotaxane system shown in Fig. 7.10 also involves cobalt(III) complex end groups, but now they are cob(III)alamins derived from vitamin B_{12} that was discussed earlier in section 5.6 [35]. The addition of a solution of aquocob(III)alamin **7.52**, whose detailed structure is shown in Fig. **5.9**, and 1,12-dibromododecane to a solution containing one equivalent of cob(I)alamin **7.53** in the presence of a ten-fold excess of αCD produces 12-bromododecylcob(III)alamin **7.54**, the isomeric [2]-pseudorotaxanes **7.55a** and **7.55b**, and finally the [2]-rotaxane **7.56**, in 50% yield. In **7.55a**, **7.55b** and **7.56** cobalt(III) to carbon bonds are formed and they are accordingly classified as organometallic [2]-pseudorotaxanes and [2]-rotaxanes. The free dimeric μ-(1,12-dodecane)bis(cob(III)alamin threading species **7.57** is obtained as a byproduct.

A labile self-assembly of stable [2]-rotaxanes is represented by the [2]-$[[(NC)_5Fe-\{R(CH_2)_nR'\}Fe(CN)_5]^{4-}]$-$[\alpha CD]$-[rotaxane] systems **7.58-7.60** where R and R' are

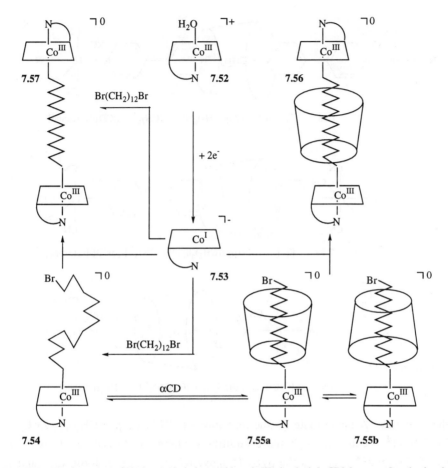

Fig. 7.10. The formation of [2]-[μ-(1,12-dodecane)bis(cob(III)alamin)}-[αCD]-[rotaxane] and related species.

either pyrazinium (pyz) or 4,4'-bipyridinium (bpy), n = 8-12 and αCD is positioned on the alkyl chain section of the threading entity [36,37]. These [2]-rotaxanes self-assemble irrespective of the order of addition of the $[R(CH_2)_nR']^{2+}$ thread, the pentacyanoferrate(II), $[Fe(CN)_5]^{3-}$ end groups and αCD. The [2]-rotaxanes **7.59** and **7.60** have been the subjects of comprehensive kinetic and equilibrium studies. The equilibria represented be Eqs. (7.1-7.7), where R = R' = pyz, are representative of

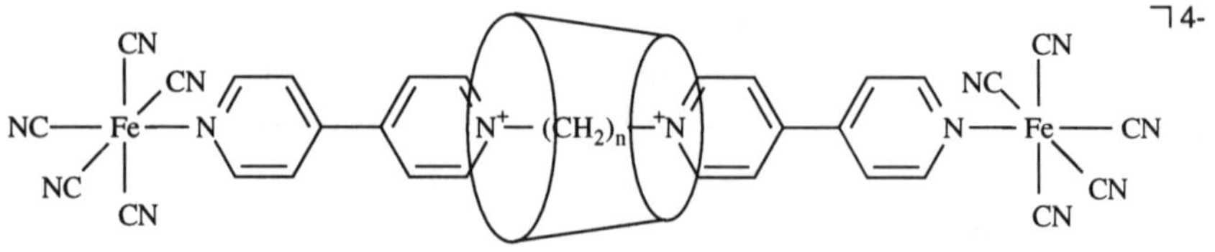

7.58   [2]-[(NC)$_5$Fe(bpy(CH$_2$)$_n$bpy)Fe(CN)$_5$$^{4-}$]-[αCD]-[rotaxane]

7.59   [2]-[(NC)$_5$Fe(pyz(CH$_2$)$_n$bpy)Fe(CN)$_5$$^{4-}$]-[αCD]-[rotaxane]

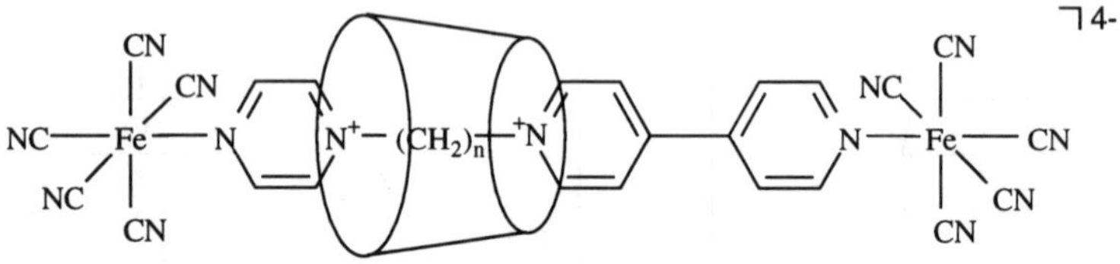

where n = 8-12

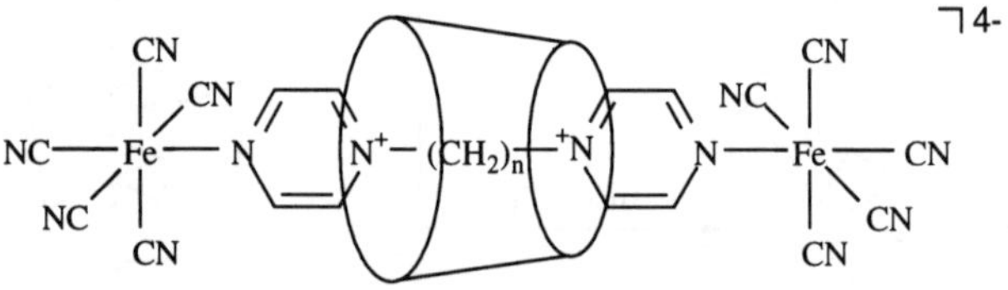

7.60 [2]-[(NC)$_5$Fe(pyz(CH$_2$)$_n$pyz)Fe(CN)$_5$$^{4-}$]-[αCD]-[rotaxane]

these studies. A facile route for the formation of [(NC)$_5$Fe(pyz(CH$_2$)$_n$pyz.αCD)-Fe(CN)$_5$]$^{4-}$ occurs through these equilibria where the threading of αCD by pyz(CH$_2$)$_n$pyz^{2+} produces the pyz(CH$_2$)$_n$pyz.αCD^{2+} [2]-pseudorotaxane and is followed by sequential substitution of the labile water ligands of two [Fe(CN)$_5$OH$_2$]$^{3-}$ by the end nitrogens of pyz(CH$_2$)$_n$pyz^{2+} to give [(NC)$_5$Fe(pyz-(CH$_2$)$_n$pyz.αCD)Fe(CN)$_5$]$^{4-}$ or [2]-[[(NC)$_5$Fe{pyz(CH$_2$)$_n$pyz}Fe(CN)$_5$]$^{4-}$]-[αCD]-[rotaxane]. A slower route to this [2]-rotaxane through equilibria represented by Eqs. (7.5-7.7) involves initial formation of the [(NC)$_5$Fe{pyz(CH$_2$)$_n$pyz}Fe(CN)$_5$]$^{4-}$ dimer which cannot directly thread αCD because of the large [Fe(CN)$_5$]$^{3-}$ end groups. Slow dissociation of a [Fe(CN)$_5$]$^{3-}$ end group produces [(NC)$_5$Fe(pyz(CH$_2$)$_n$pyz)]$^-$ which threads αCD to produce either [(NC)$_5$Fe(pyz(CH$_2$)$_n$pyz.αCD)]$^-$ or [2]-[(NC)$_5$Fe(pyz(CH$_2$)$_n$pyz)$^-$]-[αCD]-[pseudorotaxane]. Subsequent reattachment of a second [Fe(CN)$_5$]$^{3-}$ end group completes the formation of the [2]-rotaxane. Thus, depending on the relative concentrations of the reactants, the equilibria represented by Eqs. (7.1-7.6) feature to a greater or lesser extent in the [2]-rotaxane formation. The self-assembly of **7.58** and **7.59** occur in a similar manner. The $[Fe(CN)_5]^{3-}$ end group has also been used to convert [2]-pseudorotaxanes (where αCD is threaded onto $R(CH_2)_nR^{2+}$ and R is 3- or 4-cyanopyridine and n = 9 or 10) to [2]-rotaxanes where $[Fe(CN)_5]^{3-}$ binds to a cyanopyridine at both ends of the $R(CH_2)_nR^{2+}$ thread in a series of equilbria similar to those shown in Eqs. (7.1-7.6) [38].

$$pyz(CH_2)_npyz^{2+} + \alpha CD \rightleftharpoons pyz(CH_2)_npyz.\alpha CD^{2+} \tag{7.1}$$

$$pyz(CH_2)_npyz.\alpha CD^{2+} + [Fe(CN)_5OH_2]^{3-} \rightleftharpoons$$
$$[(NC)_5Fe(pyz(CH_2)_npyz.\alpha CD)]^- + H_2O \tag{7.2}$$

$$[(NC)_5Fe(pyz(CH_2)_npyz.\alpha CD)]^- + [Fe(CN)_5OH_2]^{3-} \rightleftharpoons$$
$$[(NC)_5Fe(pyz(CH_2)_npyz.\alpha CD)Fe(CN)_5]^{4-} + H_2O \tag{7.3}$$

$$pyz(CH_2)_npyz^{2+} + [Fe(CN)_5OH_2]^{3-} \rightleftharpoons$$
$$[(NC)_5Fe(pyz(CH_2)_npyz)]^- + H_2O \tag{7.4}$$

$$[(NC)_5Fe(pyz(CH_2)_npyz)]^- + [Fe(CN)_5OH_2]^{3-} \rightleftharpoons$$
$$[(NC)_5Fe(pyz(CH_2)_npyz)Fe(CN)_5]^{4-} + H_2O \tag{7.5}$$

$$[(NC)_5Fe(pyz(CH_2)_npyz)]^- + \alpha CD \rightleftharpoons$$
$$[(NC)_5Fe(pyz(CH_2)_npyz \cdot \alpha CD)]^- \tag{7.6}$$

The size selectivity of a metal complex end group is demonstrated by **7.61** (n = 7) which forms the two isomeric [2]-pseudorotaxanes **7.62** and **7.63** with αCD characterised by an overall $K_{11} = 1.2 \times 10^2$ dm^3 mol^{-1} in aqueous 0.05 mol dm^{-3} NaCl at pH 2.6 and 298 K where the carboxylic acid is unionised as shown in Fig. 7.11 [39,40]. The αCD is threaded by the polyalkyl chain (n = 7 or 11) in **7.62** and **7.63** while in their βCD analogues the larger βCD preferentially encapsulates the ferrocene moiety instead of being threaded by the alkyl chain [40,41]. At pH 8.0, where the carboxylic acid is completely ionised, the formation of the carboxylate

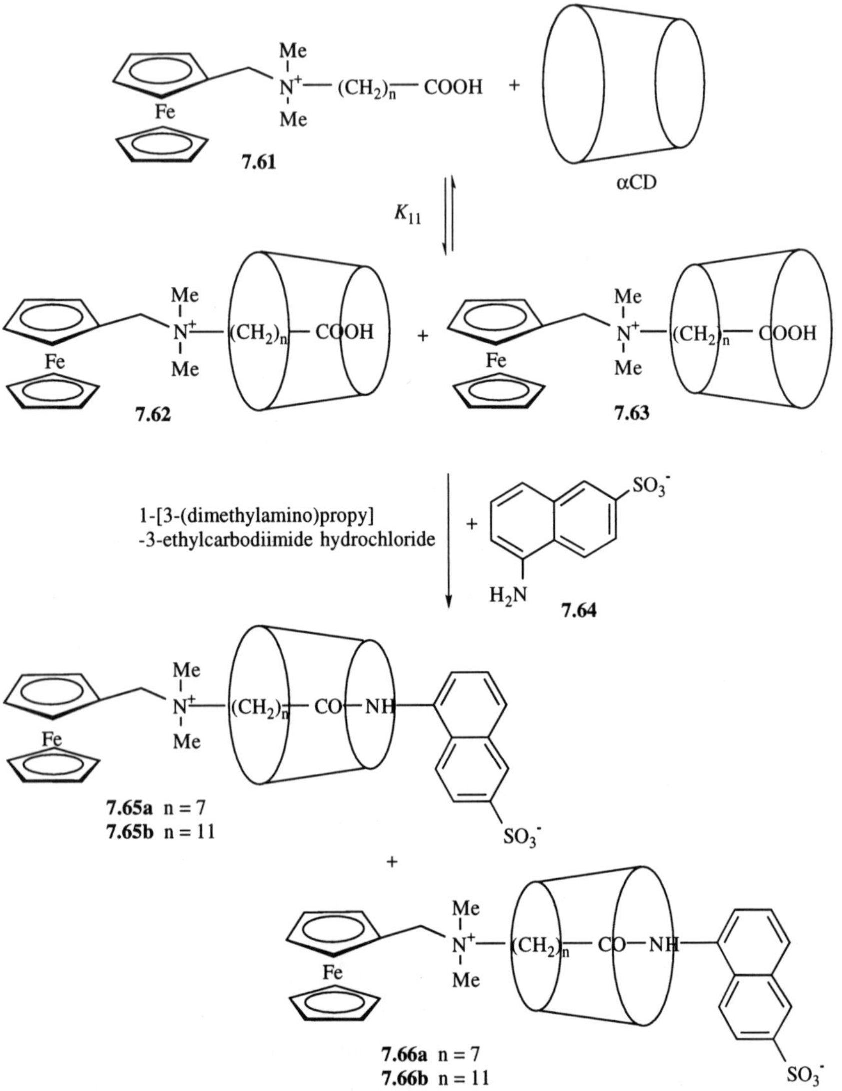

Fig. 7.11. Preparation of the isomeric and zwitterionic [2]-rotaxanes **7.65** and **7.66** from the isomeric [2]-pseudorotaxanes **7.62** and **7.63**.

analogues of **7.62** and **7.63** (n = 7) is weaker (K_{11} = 4.5 × 10^2 dm^3 mol^{-1}) probably because the carboxylate charge decreases the hydrophobic nature of the threading entity and αCD may complex the ferrocenyl moiety instead. Subsequently, reaction of **7.62** and **7.63** with 5-amino-2-naphthalenesulfonate **7.64** yields the isomeric and zwitterionic [2]-rotaxanes **7.65** and **7.66**. The length of the threading alkyl chain has a substantial influence on the ratio of isomers produced which for **7.65a:7.66a** is 2:3 and for **7.65b:7.66b** is 3:2.

The effect of alkyl chain length and CD size in the closely related [2]-pseudorotaxanes **7.67a**, **7.67b** and **7.67c** has been studied by electrochemical and ^{1}H NMR methods [42]. The formation of **7.67a** is characterised by K_{11} = 2.4 × 10^2, 1.9 × 10^3 and 4.0 × 10^2 dm^3 mol^{-1} in aqueous 0.05 mol dm^{-3} NaCl at room temperature for αCD, βCD and γCD, respectively. For **7.67b** the corresponding values are 4.2 × 10^2, 1.4 × 10^3 and 1.7 × 10^2 dm^3 mol^{-1}, consistent with βCD fitting the thread most closely while αCD and γCD are too small and too large for an optimum fit, respectively. However, increasing alkyl chain length increases the stability of the αCD [2]-pseudorotaxane while the opposite trend is established for βCD and γCD. When n is increased to 15 in **7.67c** a [3]-rotaxane is formed where two αCDs are threaded on the polyalkyl chain while βCD and γCD form only [2]-pseudorotaxanes.

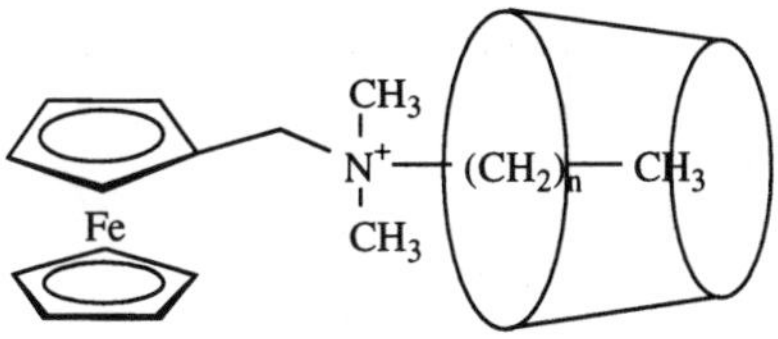

7.67a n = 0
7.67b n = 6
7.67c n = 15

7.5. Cyclodextrin Catenanes

While the first attempt to prepare a CD catenane was reported in 1958 [43], it was not until 1993 that the first successful preparation of CD catenanes was reported

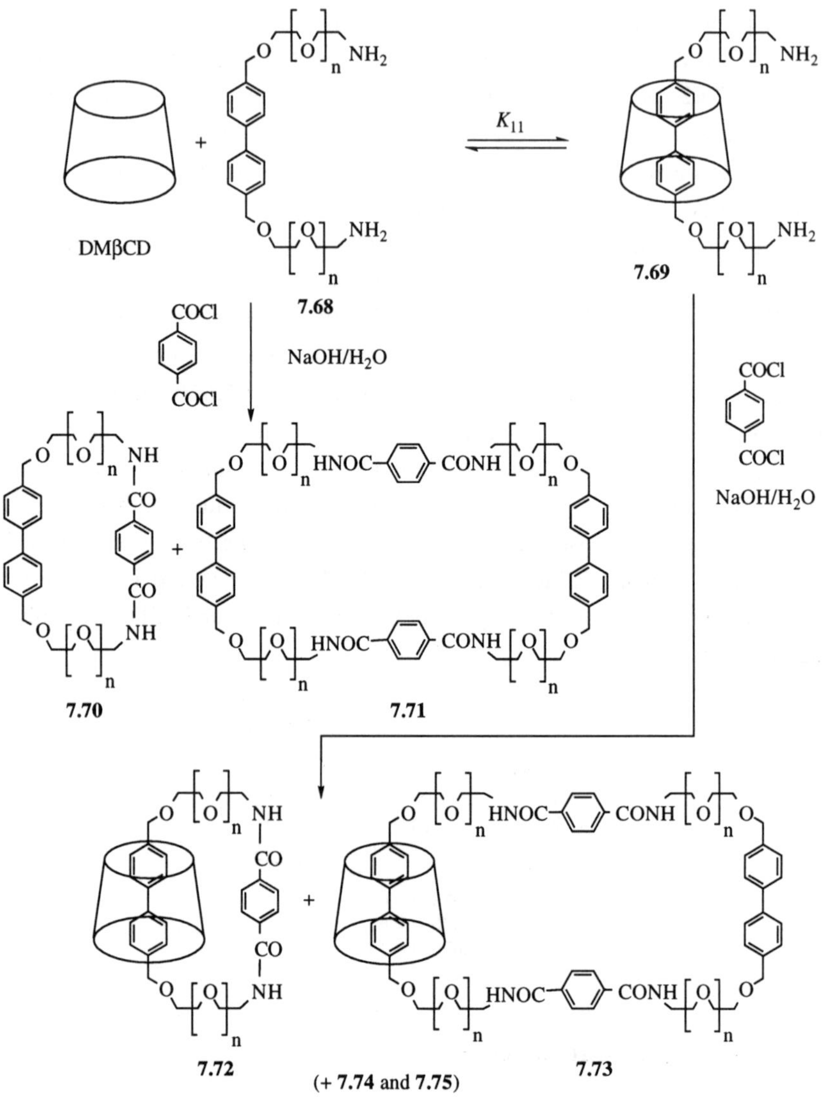

Fig. 7.12. Threading macrocycle formation in the preparation of the DMβCD [2]- and [3]-catenanes **7.72**-**7.75**. (Either n = 3 or n = 4 in each polyether chain.)

7.74

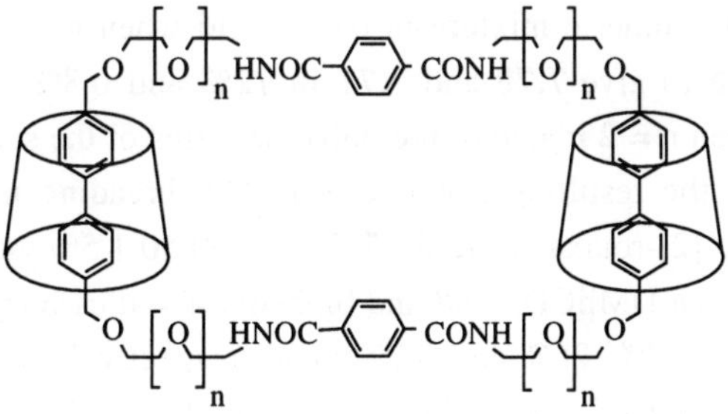

7.75

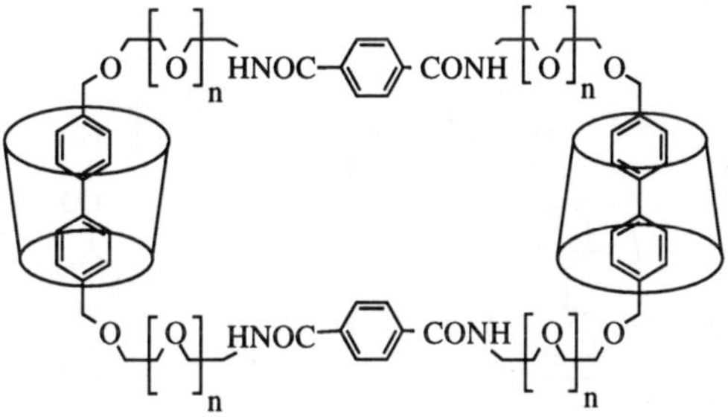

7.76

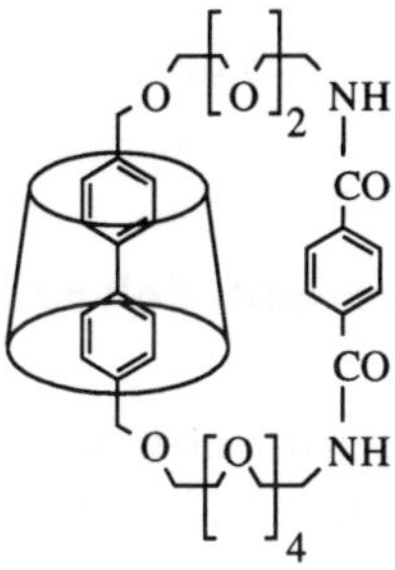

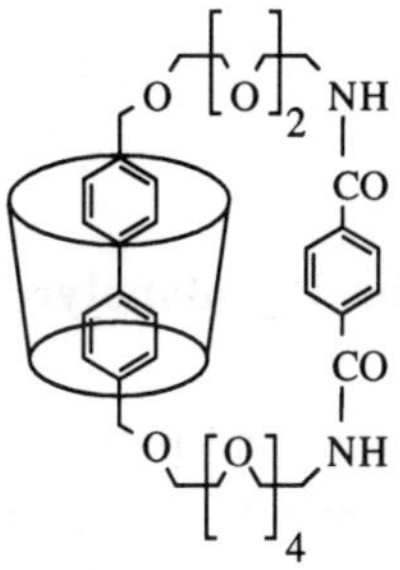

7.77

[9,44,45]. It followed the route shown in Fig. 7.12 where n = 3 in each polyether chain. The reaction is critically dependent on DMβCD and the bitolyl-based diamine **7.68** forming the [2]-pseudorotaxane **7.69** ($K_{11} = 3.28 \times 10^4$ dm^3 mol^{-1} in D$_2$O) with subsequent cyclisation by reaction with terephthaloyl chloride to give the [2]- and [3]-catenanes **7.72-7.75**. The two DMβCD [2]-catenanes **7.72** and **7.73** (n = 3) are obtained in 3.0% and 0.8% yield (2.4% and 0.3% when n = 4), and the two DMβCD [3]-catenanes **7.74** and **7.75** are obtained as a 40:60 isomeric mixture in

1.1% yield (50:50 isomeric mixture in 0.4% yield when n =4). The reaction also cyclises free **7.68** to give **7.70** and **7.71** in 12% and 0.8% yield (8% and 0.6% when n = 4). When n = 2 in one of the polyether arms of the starting diamine, and n = 4 in the other, the resulting asymmetry in the threading macrocycle produces isomeric DMβCD [2]-rotaxanes **7.76 - 7.77** in a 50:50 1.5% yield.

Reaction between DMβCD, **7.68** and biphenyl-4,4'-dicarbonyl chloride produces **7.78-7.81** in 41%, 2.7%, 26% and 0.7% yield, respectively, and also the biphenyl analogue of **7.71** in 2.2% yield. The structures of **7.68-7.81** have been deduced largely from ^{1}H NMR spectroscopy, and the X-ray crystal structure of **7.73** (n = 3) has been determined.

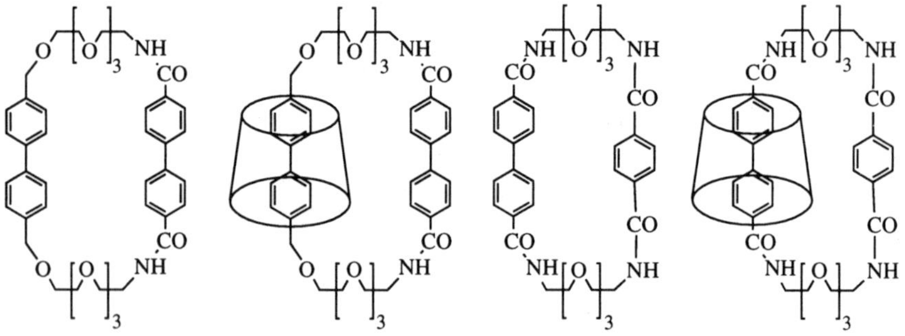

| 7.78 | 7.79 | 7.80 | 7.81 |

7.6.   Cyclodextrin Pseudopolyrotaxanes and Polyrotaxanes

Pseudopolyrotaxanes and polyrotaxanes consist of many CDs threaded onto a polymeric thread and may be prepared by two basic methods in principle. The first method requires the threading of CDs onto a polymer to form a pseudopolyrotaxane which may be converted to a polyrotaxane by the addition of end groups. The second method requires formation of [2]- or [3]-pseudorotaxanes and subsequent polymerisation of the threading molecule. Examples of the application of these and related preparative methods are discussed below.

The stabilities of pseudopolyrotaxanes formed by threading CDs onto polymers exhibit a substantial correlation with the internal diameter of the CD and the cross-sectional area of the polymer chain [46-51]. Thus, for the relatively slender polyethylene glycol polymer, $(-CH_2CH_2O-)_n$, a crystalline pseudopolyrotaxane is obtained in 92% yield with αCD and none is obtained with βCD. With the more bulky polypropylene glycol polymer $(-CH_2CH(Me)O-)_n$ no crystalline pseudopolyrotaxane is obtained with αCD, while with βCD and γCD 96% and 80% yields are obtained, respectively. A further increase in cross-sectional area with the polymethylvinylether polymer $(-CH_2CH(OMe)O-)_n$ produces no crystalline pseudo-polyrotaxane with αCD and βCD but a substantial amount is found with γCD. It appears that the formation of these pseudorotaxanes requires a sufficient closeness of fit to restrict motion of the CDs threaded by the polymer chain if the pseudopoly-rotaxane is to be stable. Such a fit is obtained between polyethylene glycol and αCD, but this polymer is too slender to confer significant stability in the potential pseudopolyrotaxanes of the larger βCD and γCD. While polypropylene glycol is too bulky to enter αCD, it fits βCD and γCD sufficiently closely to form stable polyrotaxanes. The most bulky polymer, polymethylvinylether, only forms a pseudopolyrotaxane with γCD.

In aqueous solution the formation of the αCD polyethylene glycol pseudopolyr-otaxane results in the slow formation of a thick gel preceded by an induction period during which the solution remains clear. This induction period represents the time over which the 3350 average molecular weight polyethylene glycol threads about 20 αCDs to form pseudopolyrotaxanes prior to their aggregation, and is termed the threading time [50]. The threading time increases from about 7 s in an aqueous solution 1.4×10^{-3} mol dm^{-3} in polyethylene glycol and 7.8×10^{-2} mol dm^{-3} αCD to about 3.0×10^3 s in a solution 8×10^{-4} mol dm^{-3} in polyethylene glycol and 4.5×10^{-2} mol dm^{-3} αCD, and generally decreases with increase in temperature. The negative temperature dependence of the threading time is attributed to αCD threading initially being dependent on the attraction between the interior of the αCD annulus and the polyethylene glycol with the subsequent addition of further αCD requiring unfolding of the polyethylene glycol. These effects are thought to be favoured at low temperature where random thermal motion is reduced and hydrogen bonding is more long-lived. In H_2O and D_2O, $\Delta G^{\ddagger} = -50.9$ kJ mol^{-1} and -65.1 kJ mol^{-1}, respectively, which reflects the trend in hydrogen bond strength in these solvents. As $\Delta H^{\ddagger} \sim 0$, $\Delta S^{\ddagger}$ is strongly positive and is attributed to the release of a large number of water molecules on formation of the pseudopolyrotaxane.

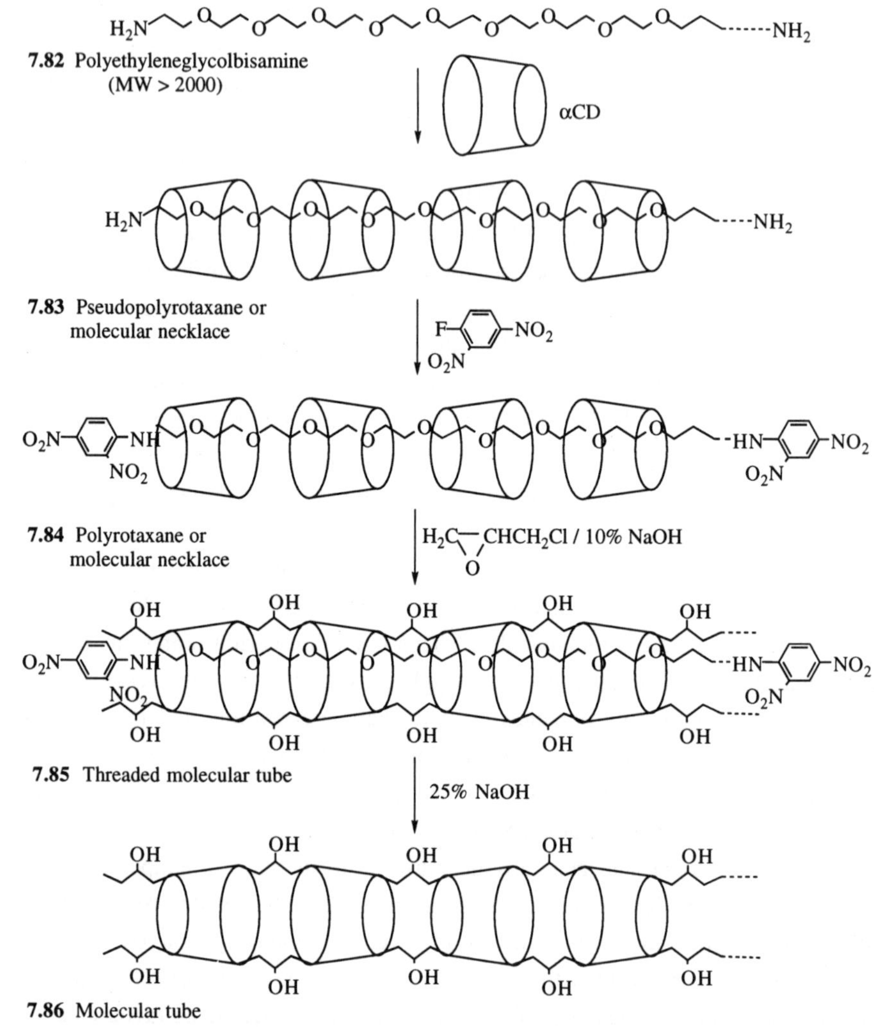

Fig. 7.13. Scheme for the sequential formation of αCD pseudopolyrotaxanes, polyrotaxanes and molecular tubes. For convenience, only two cross-links between each αCD is shown in the molecular tubes.

The composition of the pseudopolyrotaxane formed when αCD is threaded onto polyethylene glycol bisamine polymer (**7.82** in Fig. 7.13) varies significantly with the length of the polymer as is seen from Table 7.1 [49,52,53]. It is evident from these data that the separation between neighbouring αCDs increases as the length of the threading polymer increases. Because the DMαCD and TMαCD, which are completely methylated at (O)2 and (O)6 and at (O)2, (O)3 and (O)6, do not form crystalline pseudopolyrotaxanes with polyethylene glycol it has been concluded that the driving force for the formation of the αCD pseudopolyrotaxane is hydrogen bonding between adjacent αCDs. It is on this basis that the head-to-head arrangement of the αCDs in the pseudopolyrotaxane or molecular necklace **7.83** in Fig. 7.13 is proposed.

Table 7.1. Composition of Pseudopolyrotaxanes Prepared from Polyethylene Gycol Bisamine and αCD.

Polymer MW	Number of ethylene glycol units:		Number of threaded αCDs	Mole ratio of ethylene glycol units to αCD
	inside αCD	outside αCD		
16500	30	5	15	2.3
19000	34	11	17	2.6
20000	36	9	18	2.5
23500	40	37	20	3.9
44000	72	121	36	5.4
89000	140	314	70	6.5

Reaction of **7.83** with 2,4-dinitrofluorobenzene produces the *N,N'*-bis(2,4-dinitrophenyl)polyethylene glycol-threaded αCD polyrotaxane or molecular necklace, **7.84**, where the large 2,4-dinitrophenyl end groups prevent dethreading of αCD [49,52-54]. Reaction of **7.84** with epichlorohydrin produces the cross-linked αCD molecular tube **7.85** and subsequent removal of the 2,4-dinitrophenyl end groups allows dethreading to give the molecular tube **7.86** [49,55,56] This molecular tube acts as a host to I_3^- in solutions of KI/I_2 as evidenced by the development of a deep red colour [56,57]

Different threading polymers **7.87a** and **7.87b** are used in the preparation of the αCD poly(aminoundecamethylene) ($k = l =11$) and poly(aminotrimethylene-aminodecamethylene) ($k = 10$, $l =3$) polyrotaxanes shown in Fig. 7.14 [58]. The pseudopolyrotaxanes **7.88a** and **7.88b** were converted to the polyrotaxanes **7.89a** and **7.89b** by substituting the amino nitrogens with a blocking nicotinyl group at intervals in the polymer, thereby constraining αCD to limited lengths of the polymer thread. Thus, up to 36 αCDs are permanently threaded in the 55000 molecular weight polyrotaxane **7.89b** that is obtained in 43% yield.

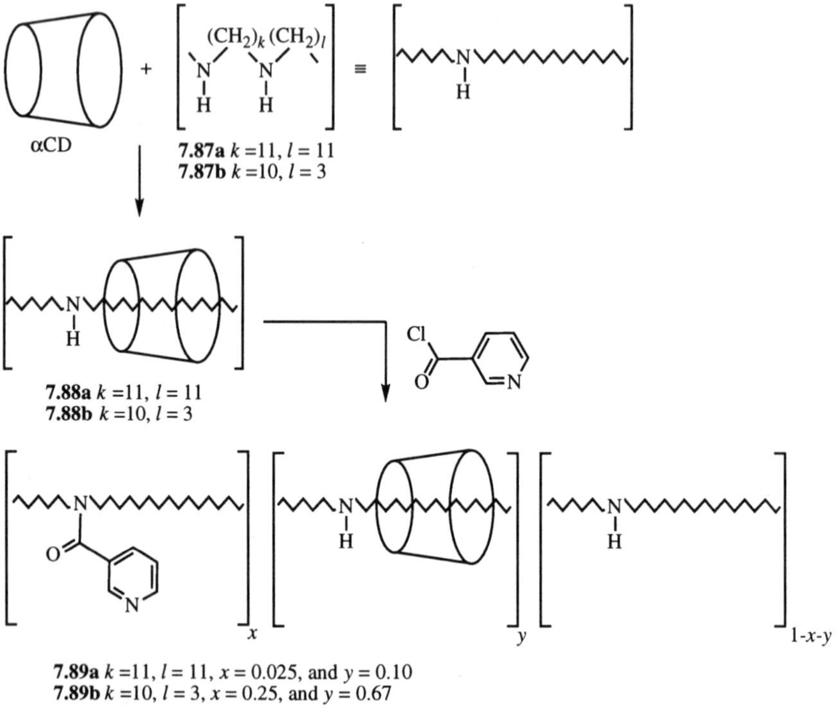
αCD
(CH2)k (CH2)l
N
H
7.87a k =11, l = 11
7.87b k =10, l = 3
7.88a k =11, l = 11
7.88b k =10, l = 3
Cl
O
N
x
y
1-x-y
7.89a k =11, l = 11, x = 0.025, and y = 0.10
7.89b k =10, l = 3, x = 0.25, and y = 0.67

Fig. 7.14. Formation of two αCD poly(aminooligomethylene) based polyrotaxanes.

An interesting study of the generation of blocking groups from the (*E*)- or *trans* stilbene units in the polymer **7.90** through their photochemical conversion to the (*Z*)- or *cis* isomer has been reported [59]. It was anticipated that the formation of a pseudopolyrotaxane by βCD and **7.90** followed by ultraviolet irradiation would produce (*Z*)-stilbene blocking units and a polyrotaxane. While the (*Z*)-stilbene units are produced, they are insufficiently large to prevent dethreading of βCD. However, a pseudorotaxane of composition **7.90.7.91**$_{0.13}$.βCD$_{0.87}$·γCD$_{0.13}$ (**7.92**) undergoes condensation of the stilbene units of **7.90** and **7.91** which are thought to be simultaneously complexed by γCD to produce tetraphenylcyclobutane blocking groups in the polyrotaxane **7.93**.

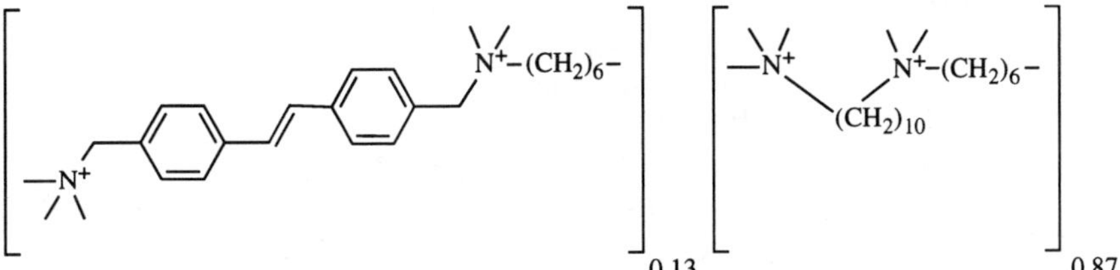

7.90 0.13

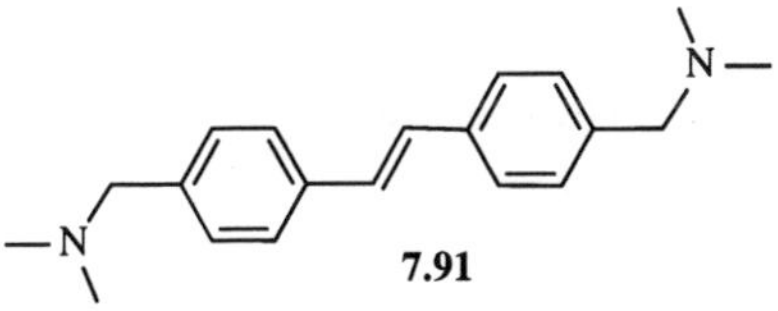

7.91

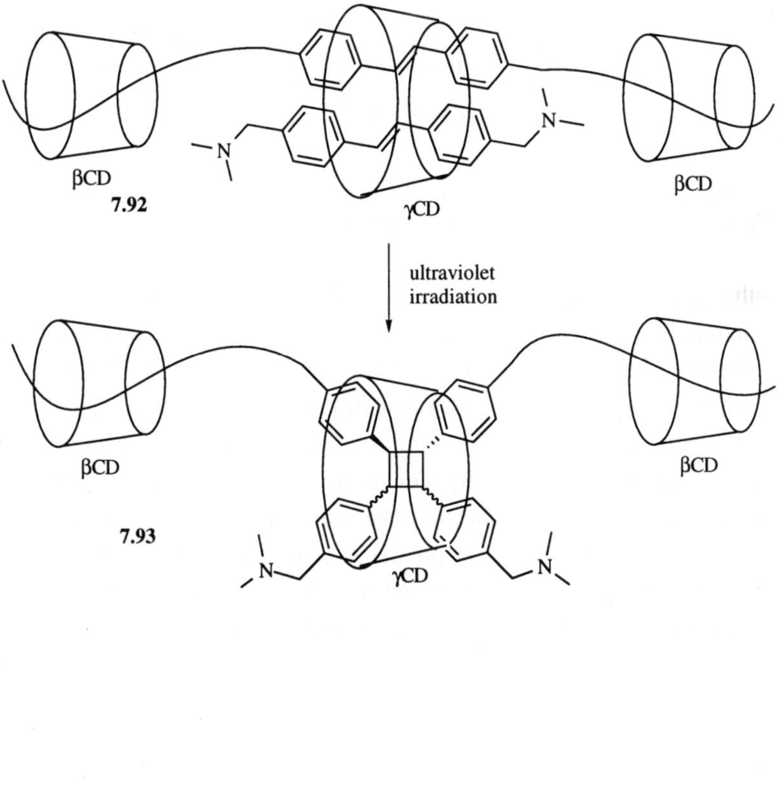

βCD
7.92
γCD

βCD

ultraviolet
irradiation

βCD

7.93
γCD

βCD

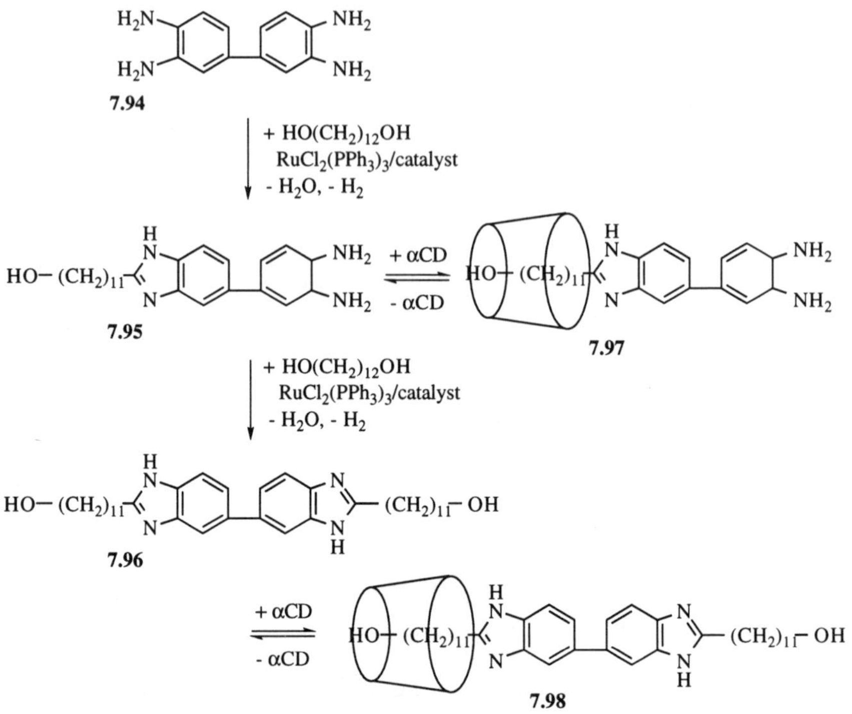

Fig. 7.15. The formation of intermediate polymer units and pseudopolyrotaxanes in the preparation of the αCD poly(alkylbenzimidazole)polyrotaxane **7.99** shown in Fig 7.16.

Another αCD polyrotaxane with blocking groups has been prepared with a poly-(alkylbenzimidazole) thread where the benzimidazole moieties act as blocking groups (X) as shown in Fig. 7.15 [60]. The starting point is the RuCl$_2$(PPh$_3$)$_3$ catalysed reaction of 3,3'-diaminobenzidine (**7.94**) with 1,12-dodecanediol to produce **7.95** or **7.96** which form the αCD [2]-pseudopolyrotaxanes **7.97** and **7.98**. These [2]-pseudopolyrotaxanes subsequently condense with either **7.95** or **7.97** to form rotaxane segments of the polyrotaxane **7.99** shown in Fig. 7.16. This polyrotaxane has the composition ratio of 16:84 for the A and B segments shown in Fig. 7.16 when the RuCl$_2$(PPh$_3$)$_3$ catalysed reaction is carried out in *N*-methylpyrrolidine in a 3,3'-diaminobenzidine (**7.94**):1,12-dodecanediol:αCD mole ratio of 1:3:0.5.

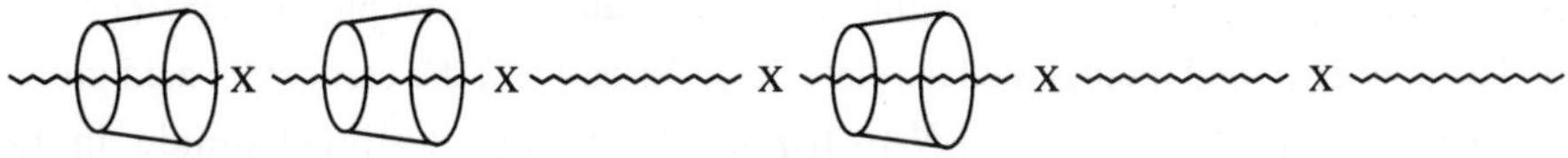
X
X
X
X
X

7.99 Polyrotaxane

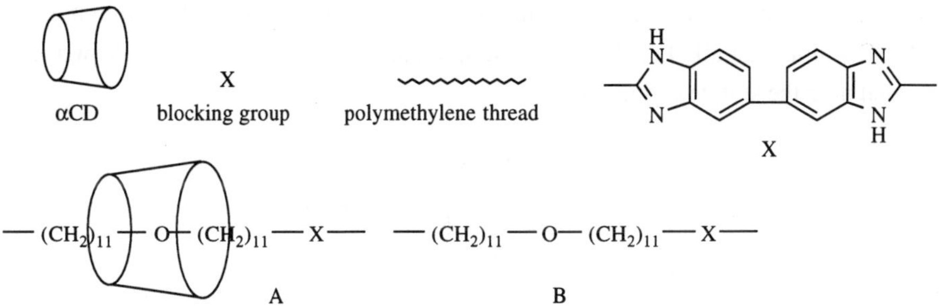
αCD
X
blocking group
polymethylene thread
H
N
N
N
N
H
X
— $(CH_2)_{11}$ — O — $(CH_2)_{11}$ — X —
A
— $(CH_2)_{11}$ — O — $(CH_2)_{11}$ — X —
B

Fig. 7.16. Composition of the αCD poly(alkylbenzimidazole) polyrotaxane **7.99** where the ratio of the A and B segments is 16:84. The orientations shown for the αCDs are illustrative only.

7.100

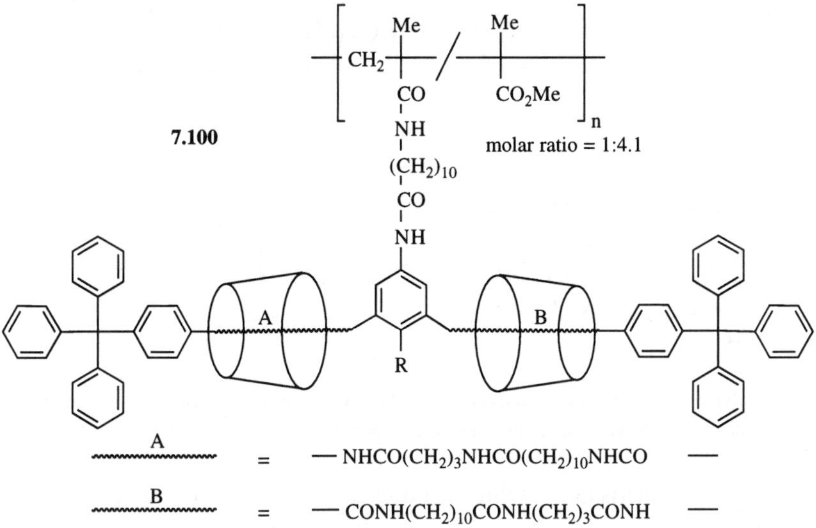
Me
Me
CH_2
CO
CO_2Me
n
NH
molar ratio = 1:4.1
$(CH_2)_{10}$
CO
NH
A
B
R
A
= — $NHCO(CH_2)_3NHCO(CH_2)_{10}NHCO$ —
B
= — $CONH(CH_2)_{10}CONH(CH_2)_3CONH$ —

An example of a CD tandem polyrotaxane is provided by **7.100** [61]. The precursor to **7.100**, where a carboxylic acid group occupies the place of the 4-amino-*N*-[4-(triphenylmethyl)phenyl]butanamide group, is prepared by radical copolymerisation of 5-[(11-methacryloylamino)undecanoylamino]isoisophthaloylbis(12-aza-dodecanoic acid) and methyl methacrylate. Condensation of the carboxylic acid groups of this precursor with 4-amino-*N*-[4-triphenylmethyl)phenyl]butanamide in the presence of DMβCD results in **7.100**. Broadening of the ^{1}H NMR resonances arising from the alkyl threading moieties are consistent with the movement of DMβCD along the alkyl chain at room temperature, while at 263 K the DMβCDs are located near the anilide end groups.

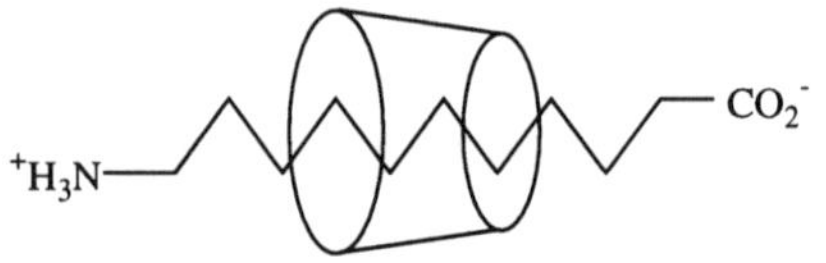

7.101

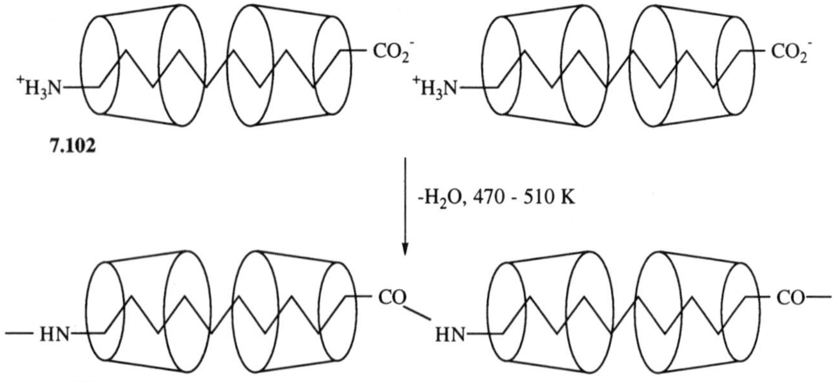

7.102

$-H_2O$, 470 - 510 K

7.103

Fig. 7.17. Polymerisation of 11-aminoundecanoic acid in αCD channels to form a polyamide threaded pseudopolyrotaxane. A head to tail arrangement is also possible in the αCD pairs.

Solid state polymerisation of threaded α,ω-alkylaminocarboxylic acids in [2]- and [3]-pseudorotaxanes proves to be particularly effective [62,63]. This is exemplified by the polymerisation of [2]-[11-aminoundecanoic acid]-[α-cyclodextrin]-[pseudorotaxane] (**7.101**) and [3]-[11-aminoundecanoic acid]-[αCD]-[αCD]-[pseudorotaxane] (**7.102**), for which $K_{11} = 9.58 \times 10^3$ dm^3 mol^{-1} and $K_{12} = 2.34 \times 10^3$ dm^3 mol^{-1} at 298.2

K and pH 4.7 in aqueous solution, respectively, and which are isolated as crystalline solids. (11-Aminoundecanoic acid and similar species exist as α,ω-alkylammonium carboxylates.) When heated under vacuum at 470-510 K in the solid state, **7.102** polymerises as shown in Fig. 7.17 to give a polyamide threaded rotaxane **7.103**, where 19 **7.102** monomer units are polymerised, in 97.4% yield. A similar reaction of **7.101** results in a polyamide threaded rotaxane similar to **7.103**, where 10 **7.101** monomer units are polymerised, in 97% yield. Unlike the polyamide thread alone, **7.103** is water soluble, but a precipitate slowly forms as it loses αCDs, and a similar precipitation occurs with its analogue formed from **7.101**. The extent of polymerisation diminishes with the length of the α,ω-aminocarboxylic acid, and the [2]-[6-aminocarboxylic acid]-[αCD]-[pseudorotaxane] produces no polymer presumably because the 6-aminocarboxylic acid is too short to allow the required condensation in the solid state.

7.7. References

1. G. Schill, in *Catenanes, Rotaxanes and Knots* (Academic Press, New York, 1971).
2. G. Schill, E. Logemann and W. Littke, *Chemie Unserer Zeit.* **18** (1984) 129.
3. J. F. Stoddart, *Angew. Chem., Int. Ed. Engl.* **31** (1992) 846.
4. H. Ogino, *New. J. Chem.* **17** (1993) 683.
5. G. Wenz, *Angew. Chem., Int. Ed. Engl.* **33** (1994) 803.
6. H. W. Gibson, M. C. Bheda and P. T. Engen, *Prog. Polymer Sci.* **19** (1994) 843.
7. D. Philp and J. F. Stoddart, *Angew. Chem., Int. Ed. Engl.* **35** (1996) 1155.
8. M. Belohradsky, F. M. Raymo and J. F. Stoddart, *Collect. Czech. Chem. Commun.* **61** (1996) 1.
9. S. A. Nepogodiev and J. F. Stoddart, *Chem. Rev.* **98** (1998) 1959.
10. D. H. Macartney, *J. Chem. Soc., Perkin Trans.* 2 (1996) 2775.
11. M. Watanabe, H. Nakamura and T. Matsuo, *Bull. Chem. Soc. Jpn.* **65** (1992) 164.
12. H. Saito, H. Yonemura, H. Nakamura and T. Matsuo, *Chem. Lett.* (1990) 535.

13.   H. Yonemura, M. Kasahara, H. Saito, H. Nakamura and T. Matsuo, *J. Phys. Chem.* **96** (1992) 5765.

14.   H. Yonemura, T. Nojiri and T. Matsuo, *Chem. Lett.* (1994) 2097.

15.   H. Yonemura, H. Saito, S. Matsushima, H. Nakamura and T Matsuo, *Tetrahedron Lett.* **30** (1989) 3143.

16.   S. Anderson, R. T. Aplin, T. D. W. Claridge, T. Goodson, A. C. Maciel, G. Rumbles, J. F. Ryan and H. L. Anderson, *J. Chem. Soc, Perkin Trans. 1* (1998) 2383.

17.   A. Mirozoian and A. E. Kaifer, *Chem. Eur. J.* **3** (1997) 1052.

18.   R. Castro, L. A. Godínez, C. M. Criss and A. E. Kaifer, *J. Org. Chem.* **62** (1997) 4928.

19.   X. Shen, M. Belletête and G. Durocher, *J. Phys. Chem. B* **102** (1998) 1877.

20.   S. Anderson, T. D. Claridge and H. L Anderson, *Angew. Chem., Int. Ed. Engl.* **36** (1997) 1310.

21.   M. Kunitake, K. Kotoo, O. Manabe, T. Muramatsu and N. Nakashima, *Chem. Lett.* (1993) 1033.

22.   A. Harada, J. Li and M. Kamachi, *J. Chem. Soc., Chem. Commun.* (1997) 1413.

23.   G. Wenz, F. Wolf, M. Wagner and S. Kubik, *New. J. Chem.* **17** (1993) 729.

24.   G. Wenz, E. van der Bey and L. Schmidt, *Angew. Chem., Int. Ed. Engl.* **31** (1992) 783.

25.   J. S. Manka and D. S. Lawrence, *J. Am. Chem. Soc.* **112** (1990) 2440.

26.   T. Venkata and D. S. Lawrence, *J. Am. Chem. Soc.* **112** (1990) 3614.

27.   M. Tamura and A. Ueno, *Chem. Lett.* (1998) 369.

28.   H. Ogino, *J. Am. Chem. Soc.* **103** (1981) 1303.

29.   H. Ogino and K. Ohata, *Inorg. Chem.* **23** (1984) 3312.

30.   K. Takaizumi, T. Wakabayashi and H. Ogino, *J. Soln. Chem.* **25** (1996) 947.

31.   K. Yamanari and Y. Shimura, *Chem. Lett.* (1982) 1959.

32.   K. Yamanari and Y. Shimura, *Bull. Chem. Soc. Jpn.* **56** (1983) 2283.

33.   K. Yamanari and Y. Shimura, *Bull. Chem. Soc. Jpn.* **57** (1984) 1596.

34.   K. Yamanari and Y. Shimura, *Chem. Lett.* (1982) 1957.

35.   R. B. Hannak, G. Färber, R. Konrat and B. Kräutler, *J. Am. Chem. Soc.* **119** (1997) 2313.

36.   R. S. Wylie and D. H. Macartney, *J. Am. Chem. Soc.* **114** (1992) 3136.

37. D. H. Macartney and C. A. Waddling, *Inorg. Chem.* **33** (1994) 5912.

38. A. P. Lyon and D. H. Macartney, *Inorg. Chem.* **36** (1997) 729.

39. L. A. Godínez, S. Patel, C. M. Criss and A. E. Kaifer, *J. Phys. Chem.* **99** (1995) 17449.

40. R. Isnin and A. E. Kaifer, *J. Am. Chem. Soc.* **113** (1991) 8188.

41. R. Isnin and A. E. Kaifer, *Pure Appl. Chem.* **65** (1993) 495.

42. R. Isnin, C. Salam and A. E. Kaifer, *J. Org. Chem.* **56** (1991) 35.

43. A. Lütteringhaus, F. Cramer, H. Prinzbach and F. M. Henglein, *Liebigs Ann. Chem.* **613** (1958) 185.

44. D. Armspach, P. R. Ashton, C. P. Moore, N. Spencer, J. F. Stoddart, T. J. Wear and D. J. Williams, *Angew. Chem., Int. Ed. Engl.* **32** (1993) 854.

45. D. Armspach, P. R. Ashton, R. Ballardini, V. Balzani, A. Godi, C. P. Moore, L. Prodi, N. Spencer, J. F. Stoddart, M. S. Tolley, T. J. Wear and D. J. Williams, *Chem. Eur. J.* **1** (1995) 33.

46. A. Harada and M. Kamachi, *J. Chem. Soc., Chem. Commun.* (1990) 1322.

47. A. Harada and M. Kamachi, *Macromol.* **23** (1990) 2821.

48. A. Harada, J. Li and M. Kamachi, *Chem. Lett.* (1993) 237.

49. A. Harada, J. Li and M. Kamachi, *Polymer. Adv. Technol.* **8** (1997) 241.

50. M. Ceccato, P. Lo Nostro and P. Baglioni, *Langmuir* **13** (1997) 2436.

51. A. Harada, J. Li and M. Kamachi, *Macromol.* **27** (1994) 4538.

52. A. Harada, J. Li and M. Kamachi, *Nature* **356** (1992) 325.

53. A. Harada, J. Li, T. Nakamitsu and M. Kamachi, *J. Org. Chem.* **58** (1993) 7524.

54. A. Harada, J. Li and M. Kamachi, *J. Am. Chem. Soc.* **116** (1994) 3192.

55. A. E. Kaifer, *Nature* **364** (1993) 484.

56. A. Harada, J. Li and M. Kamachi, *Nature* **364** (1993) 516.

57. M. Ceccato, P. Lo Nostro, C. Rossi, C. Bonechi, A. Donati and P. Baglioni, *J. Phys. Chem. B* **101** (1997) 5094.

58. G. Wenz and B. Keller, *Angew. Chem., Int. Ed. Engl.* **31** (1992) 197.

59. W. Herrmann, M. Schneider and G. Wenz, *Angew. Chem., Int. Ed. Engl.* **36** (1997) 2511.

60. K. Yamaguchi, K. Osakada and T. Yamamoto, *J. Am. Chem. Soc.* **118** (1996) 1811.

61. M. Born and H. Ritter, *Angew. Chem., Int. Ed. Engl.* **34** (1995) 309.

264 Modified Cyclodextrins

62. M. B. Steinbrunn and G. Wenz, *Angew. Chem., Int. Ed. Engl.* **35** (1996) 2139.

63. G. Wenz, M. B. Steinbrunn and K. Landfester, *Tetrahedron* **53** (1997) 15575.

CHAPTER 8

MODIFIED CYCLODEXTRINS

SURFACE AND INTERFACE BEHAVIOUR

8.1. Introduction

The natural CDs each present two parallel planar faces, which are delineated by the rings of primary and secondary hydroxy groups at the narrow and wide ends of the annuli, respectively (see Chapter 1, Fig. 1.1, p.3). These are hydrophilic but either one or both of the faces may be modified, through alkylation or other substitution of the hydroxy groups (as described in Chapter 2), to produce amphiphilic CDs. Such modified CDs possess parallel hydrophobic and hydrophilic surfaces and therefore tend to form layers, in solution, at solvent interfaces and on solid supports. These layers and their applications are discussed in this chapter.

8.2. Amphiphilic Cyclodextrin Monolayers at the Air-Water Interface

Kawabata et al. [1] have reported the formation of monolayers of the amphiphilic βCDs **8.1-8.4** on the surface of water. Examination of surface pressure-molecular area (π-A) isotherms indicates that in each monolayer the CDs are aligned with their hydrophilic faces parallel to the water surface (Fig. 8.1). The molecular area of the CD **8.1** in the monolayer, extrapolated to zero pressure (A_0), was found to be 2.17 nm^2. This corresponds closely to the value of 1.95-2.16 nm^2, calculated for a close-packed structure based on the dimensions of the wide secondary face of βCD. The monolayers of the CDs **8.1-8.4** are each stable up to 55 mN m^{-1} at 290 K.

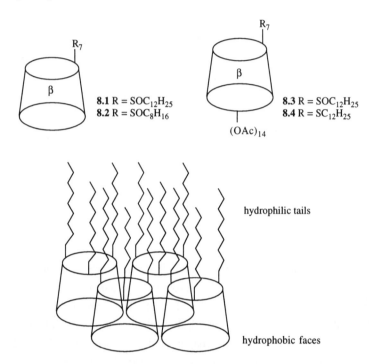

8.1 R = SOC$_{12}$H$_{25}$
8.2 R = SOC$_8$H$_{16}$

8.3 R = SOC$_{12}$H$_{25}$
8.4 R = SC$_{12}$H$_{25}$

hydrophilic tails

hydrophobic faces

Fig. 8.1. Schematic representation of the alignment of the CDs **8.1-8.4** when each forms a monolayer on the surface of water.

A variety of other per-6-substituted CDs also form monolayers at the air-water interface [2-6]. With those formed from the brominated αCD, βCD and γCD **8.5-8.7**, in which each of the primary hydroxy groups is replaced with bromine, the molecular areas (A_0) are 180, 220 and 260 Å^2, respectively, and the corresponding surface collapse pressures are 6, 16 and 19 mN m^{-1} [3]. From this it is clear that long alkyl chains at one face of a CD are not necessarily required to make monolayers, although the monolayers of the bromides **8.5-8.7** are less stable than those of the CDs **8.1-8.4** described above. A second aggregate species is observed for the αCD **8.5** and βCD **8.6**, with areas A_1 of approximately 70 and 100 Å^2, respectively. Since these values are approximately half the A_0 values for the monolayers, it seems likely that these second species are bilayers.

Nucleophilic substitution reactions of the bromides **8.5-8.7** with a range of 4-substituted thiophenols afford the sulfides **8.8-8.10** (where R = H, Br, OC$_4$H$_9$,

C$_5$H$_{11}$ and NO$_2$), and the ability of these compounds to form monolayers has been examined systematically [5]. Hydrophobic substituents on the phenyl ring are found to increase the stability of the monolayers.

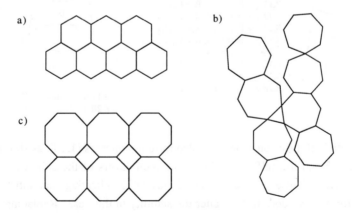

Br$_n$

χ

8.5 χ = α, n = 6
8.6 χ = β, n = 7
8.7 χ = γ, n = 8

R

S   n

χ

8.8 χ = α, n = 6
8.9 χ = β, n = 7
8.10 χ = γ, n = 8

Presumably, CDs form stable monolayers due to their large surface of hydroxy groups available for interaction with the water. On this basis, the stability of the monolayers might be expected to increase with an increase in the number of hydroxy groups, as is observed with the bromides **8.5-8.7**, where the monolayer of the γCD **8.7** is the most stable and that of the αCD **8.5** the least. However, this trend is not seen with the sulfides **8.8-8.10**, and is often not observed, because the

a)

b)

c)

Fig. 8.2. Schematic representation of close-packed structures for the monolayers of a) αCD, b) βCD and c) γCD.

symmetry of the various CDs affects their close-packing arrangements. As discussed in more detail below, αCD and γCD have six- and eight-fold molecular symmetries, respectively, and can therefore form ordered close-packed monolayers, whereas βCD has seven-fold symmetry and can only form disordered dense structures (Fig. 8.2) [7,8], which may be more or less stable.

With the amphiphilic CDs **8.1**, **8.2** and **8.5-8.10** described above, the hydrophobic surface is introduced through substitution of CD primary hydroxy groups. Alternatively, the face of secondary hydroxy groups can be modified. The esters **8.11-8.18** have been prepared, in which all the CD C(2) and C(3) hydroxy groups are acylated [8,9]. Their monolayers at the air-water interface show collapse pressures of approximately 14, 31, 31, 31, 31, 45, 43 and 47 mN m^{-1}, respectively. Obviously the monolayer of the acetylated CD **8.11** is the least stable, but the fact that it forms at all again indicates long alkyl chains are not essential. The CDs **8.11-8.15** each form only one condensed phase, but the molecular area-surface pressure isotherms for the C14 esters **8.16-8.18** are each biphasic, indicating that surface compression results in transitions from a liquid expanded state to a liquid condensed state, and then to a solid state.

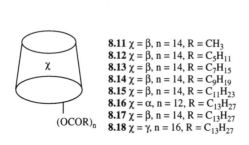

8.11 χ = β, n = 14, R = CH$_3$
8.12 χ = β, n = 14, R = C$_5$H$_{11}$
8.13 χ = β, n = 14, R = C$_7$H$_{15}$
8.14 χ = β, n = 14, R = C$_9$H$_{19}$
8.15 χ = β, n = 14, R = C$_{11}$H$_{23}$
8.16 χ = α, n = 12, R = C$_{13}$H$_{27}$
8.17 χ = β, n = 14, R = C$_{13}$H$_{27}$
8.18 χ = γ, n = 16, R = C$_{13}$H$_{27}$

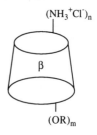

8.19 n = 7, m = 14, R = C$_6$H$_{13}$
8.20 n = 7, m = 14, R = C$_{12}$H$_{25}$

CDs with both faces modified can also form monolayers [1,10], as shown for the sulfoxide **8.3** and the sulfide **8.4** [1]. The monolayers of the sulfoxides **8.1** and **8.3** both have collapse pressures near 55 mN m^{-1}, indicating that substitution of hydroxy for acetoxy does little to alter the stability in this case. Replacing hydroxy substituents with ammonium groups also retains the hydrophilic character of the CD surface, as illustrated by the fact that monolayers of the amine hydrochlorides **8.19** and **8.20** form with collapse pressures of 46 and 50 mN m^{-1}, respectively [11]. The molecular area-surface pressure isotherm is biphasic for the modified CD **8.20**, whereas that of the shorter chain analogue **8.19** shows only a single phase.

As with other types of CDs, much of the interest in the amphiphilic species incorporated in monolayers is related to their ability to form complexes. Heptakis(6-dodecylamino-6-deoxy)-βCD **8.22** has been shown to form a 1:1 complex with *trans*-4-(4-dimethylaminophenyl)azobenzoic acid **8.21** as a monolayer on the surface of water [12]. The surface area occupied per molecule of the CD **8.22** is found to be the same for monolayers formed from the CD **8.22** alone and from a 1:1 mixture of the CD **8.22** with the azobenzoic acid **8.21**. This is convincing evidence that the azobenzoic acid **8.21** is included in the cavity of the CD **8.22**.

Me$_2$N― [benzene ring] ―N=N― [benzene ring] ―CO$_2$H

8.21

(NHC$_{12}$H$_{25}$)$_7$

β

8.22

Similar evidence shows that cholesterol and the γCD **8.25** form a monolayer of the 1:1 inclusion complex [13]. At relatively high surface compression pressures (42 mN m^{-1}), monolayers formed from mixtures of cholesterol with the corresponding αCD **8.23** and βCD **8.24** are not condensed, in that the surface area occupied by a mixture is no less than the sum of that occupied by the individual components at the same pressure. At relatively low surface pressures (10 mN m^{-1}), condensed phases do form, presumably where the cholesterol is incorporated in intermolecular cavities formed by the CD alkyl chains, which are not fully compressed (Fig. 8.3) [13]. An analogous structure has been proposed for monolayers formed from mixtures of dipalmitoylphosphatidylcholine with various amphiphilic CDs [14].

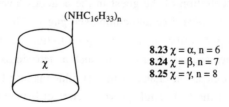

(NHC$_{16}$H$_{33}$)$_n$

χ

8.23 χ = α, n = 6
8.24 χ = β, n = 7
8.25 χ = γ, n = 8

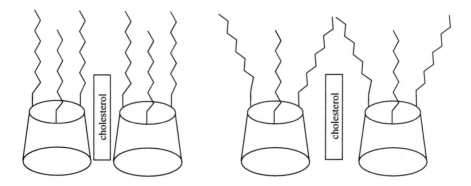

Fig. 8.3. Schematic representation of the monolayers formed from mixtures of cholesterol with the αCD and βCD **8.23** and **8.24** at high (left) and low (right) surface compression pressures.

A potential application of CDs as ion-permeable membranes has been investigated, using the monolayer formed by the amphiphilic CD **8.4** at the air-water interface [15]. With the monolayer maintained at just below the surface collapse pressure, such that the CD molecules are in a close-packed arrangement and the number of intermolecular voids is minimised (Fig. 8.4), and with an electrode in contact with the top surface of the monolayer, the CD annuli serve as well-defined ion-channels between the water subphase and the electrode (Fig. 8.5). Control of ion permeability is then possible by blocking the CD annuli through guest complexation. Thus, when the reduction-oxidation cycle of *p*-quinone (**8.26**) (Scheme 8.1) is examined using this system, cyclohexanol, adenosine, cytidine, benzyl alcohol and adamantan-1-ol are found to complex in the CD annuli and decrease the access of the quinone **8.26** to the electrode. The largest decrease is seen with adamantan-1-ol, which is as expected since the complex of this guest with native βCD is the most thermodynamically favoured. The magnitude of the decrease depends on the concentration of the guest in the aqueous layer, as expected for a dynamic equilibrium between free and complexed guest. The system of control of membrane permeability described here is analogous to ligand control of ion channels that function in biomembranes. Neurotoxins such as tetrodotoxin and saxitoxin act by blocking the biological ion channels in a similar manner to the way in which guest complexation in the CD annuli prevents contact between the water and the electrode.

Fig. 8.4. Illustration of a cross section of a CD monolayer showing that the CD annuli provide membrane channels.

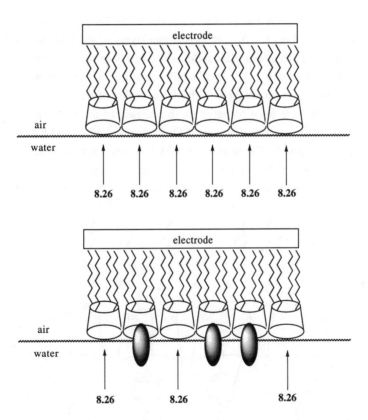

Fig. 8.5. Schematic representation of a CD ion-permeable membrane where the annuli allow passage of the quinone **8.26** from the water to the electrode (above), and where that access is partially restricted by CD guest complexation (below).

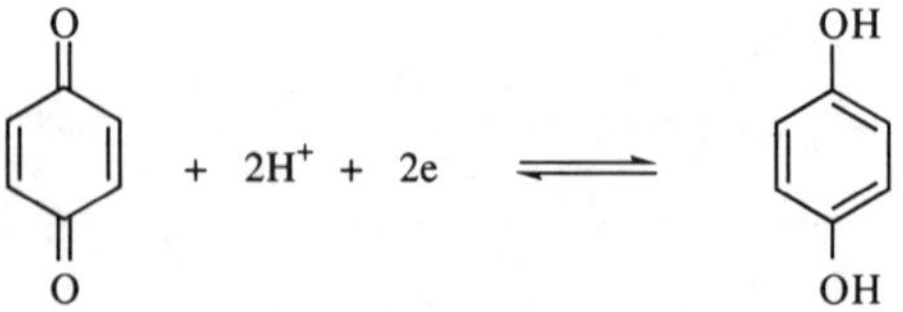

Scheme 8.1. Reversible reduction of *p*-quinone **8.26**.

8.3. Langmuir-Blodgett Films

Monolayers formed at the air-water interface can be deposited as Langmuir-Blodgett films [1,2,4,6,12]. Accordingly, monolayers of the CDs **8.1**, **8.3** and **8.4** have been transferred to quartz glass slides which had been precoated with five layers of cadmium eicosanoate [1]. In a similar fashion, Langmuir-Blodgett films consisting of up to 100 monolayers of the 1:1 complex of the azobenzoic acid **8.21** with the CD **8.22** have been prepared on the surface of quartz [12]. Under these circumstances, the interactions between CDs in the monolayer and between the CD and the surface control the CD orientation, which in turn determines the alignment of the included guest, such that the guest is held in a specific environment and orientation, and isolated from other guest molecules. The orientation of the guest

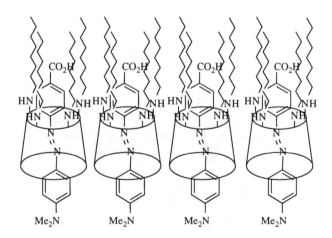

Fig. 8.6. Orientation of the azobenzoic acid **8.21** in monolayers of the CD **8.22**.

8.21 in the CD **8.22** (Fig. 8.6) was elucidated on the basis of guest and host substituent effects, while absorption spectra using polarised light show that the orientation of the guest **8.21** is substantially perpendicular to the film plane.

Irradiation at 360 nm of the Langmuir-Blodgett film of the 1:1 complex of the CD **8.22** with the *trans*-azobenzoic acid **8.21** results in a 60% conversion of the guest to the *cis* isomer (Scheme 8.2) [16,17]. Reversion to the *trans* form occurs spontaneously, and is accelerated at higher temperatures and by irradiation at 254 nm. The azobenzene derivative **8.27**, lacking the electron-donating dimethylamino substituent, shows the same reversible photochemical response to irradiation at 360 and 254 nm, but in this case the *cis* isomer is thermally stable and does not revert spontaneously to the *trans* form. The photochemical cycle may be repeated ten times with very little change to the optical density of the film.

Prior to this work, the photochemical conversion of a *trans*-azobenzene to the *cis* isomer in a Langmuir-Blodgett film had not been observed, although the reverse had been reported. The *cis* isomer is the bulkier form and processes which involve an increase in surface area occupied per molecule are generally difficult to accomplish. In the monolayer of the CD inclusion complex, the space provided by the CD annuli allows the isomerisation to take place because there is no increase in surface area for the complex, despite the increase in surface area of the guest. Reversible photochemical conversion systems of this type may be applicable in the development of optical memory devices and photochemical switches.

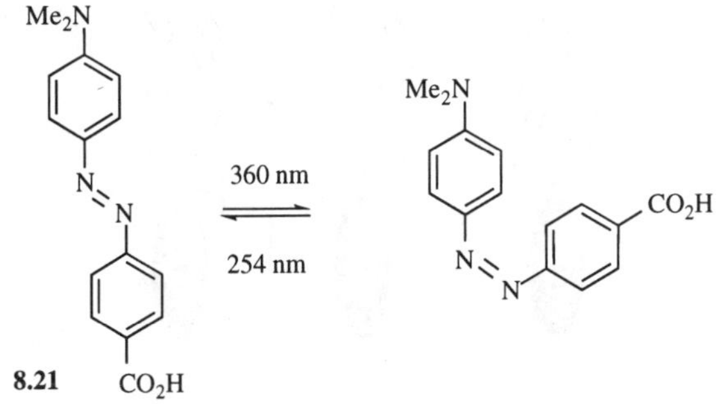

Scheme 8.2. Reversible photochemical conversion of the azobenzoic acid **8.21** included in the CD **8.22** in a Langmuir-Blodgett film.

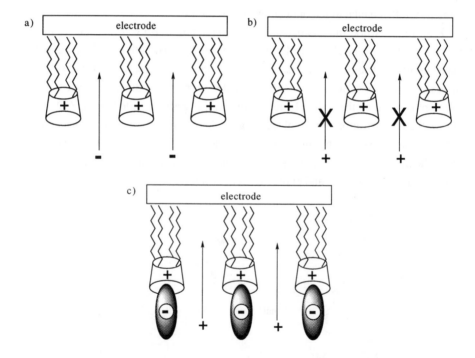

8.27

Another potential application of Langmuir-Blodgett films of modified CDs is in the area of sensor electrodes. The film of the CD **8.25** on the surface of a glassy carbon electrode has already been shown to function as an anionic guest responsive sensor (Fig. 8.7) [18]. When the electrode is immersed in an acidic solution, or the solution is made more acidic, the amino groups of the CD become protonated and

Fig. 8.7. Schematic representation of a Langmuir-Blodgett film of the CD **8.25**, protonated on the surface of an electrode, and the effect of that film on the access to the electrode of a) negative ions, b) positive ions, and c) positive ions when the CD **8.25** complexes anionic guests.

positively charged. This facilitates access of negative ions from the solution to the electrode, though electrostatic attraction, but restricts access of positive ions. The guest anion sensing ability arises from the molecular recognition and complexation of negatively charged guests by the positively charged CD annuli. This reduces the surface charge and therefore allows less access of negative ions and greater access of positive ions to the electrode. In this system, access to the electrode must be *via* intermolecular holes in the monolayer, rather than through the CD annuli, otherwise guest complexation by the CD would limit access of both positive and negative ions.

In general, it seems likely that αCDs and γCDs are likely to be more useful than βCDs in constructing molecular devices based on films and monolayers. As indicated above, monolayers of βCDs tend to show more disorder than those of the corresponding αCDs and γCDs, because the βCD seven-fold symmetry is incompatible with a planar close-packed structure. This is confirmed when monolayers of the CDs **8.16-8.18** which form at the air-water interface are deposited as Langmuir-Blodgett films and examined for roughness, density and thickness using X-ray reflectivity [7]. In particular the roughness of the film-air interface is several Å larger in the case of the βCD **8.17**, indicating a greater variation in the thickness of the monolayer.

8.4. Cyclodextrin Monolayers on Gold and Silver

An alternative approach to the construction of monolayers on electrodes is through chemisorption of thiolated CDs on gold and silver surfaces. The CD **8.28**, which is a mixture of compounds with an average of three 2-mercaptoethylamino substituents in place of primary hydroxy groups on βCD, produces a monolayer on the surface of colloidal silver and a silver electrode, through formation of covalent sulfur-silver bonds. The CD annuli in the monolayers retain their ability to complex the azo-dye Methyl Orange [19]. In a development of this work the corresponding per-6-substituted αCD **8.29** has also been examined as a monolayer on the surface of silver [20] and gold [21]. Complexation of the Methyl Red-phenylethylamine conjugates **8.31-8.34** by the CD annuli in the monolayer on silver is characterised by K_{11} = 2.0 x 10^5, 2.3 x 10^5, 2.0 x 10^4 and 1.3 x 10^4 dm^3 mol^{-1}, respectively, at pH 7.35 and 298 K, while the corresponding values for the monolayer on gold are 7.3

x 10^5, 7.2 x 10^5, 8.7 x 10^6 and 3.9 x 10^6 dm^3 mol^{-1}, respectively. The stereoselectivity displayed in the complexation of the enantiomers **8.33** and **8.34** in the monolayers mirrors that shown by αCD in solution, under the same conditions, although the K_{11} values are much lower for the free CD.

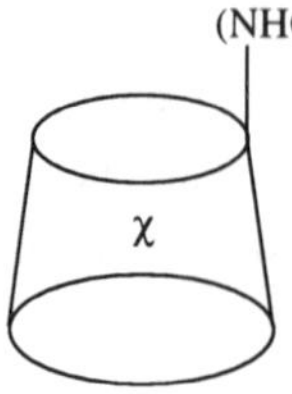

$(NHCH_2CH_2SH)_n$

8.28 χ = β, average degree of substitution, n = 3
8.29 χ = α, n = 6

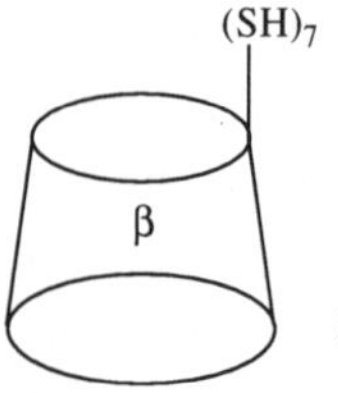

$(SH)_7$

8.30

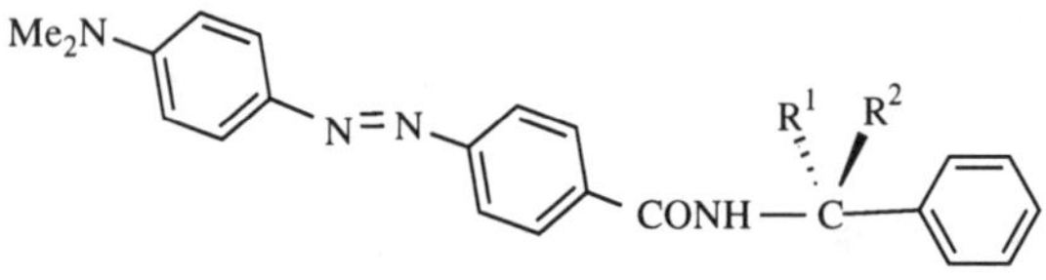

8.31 R^1 = Me, R^2 = H
8.32 R^1 = H, R^2 = Me

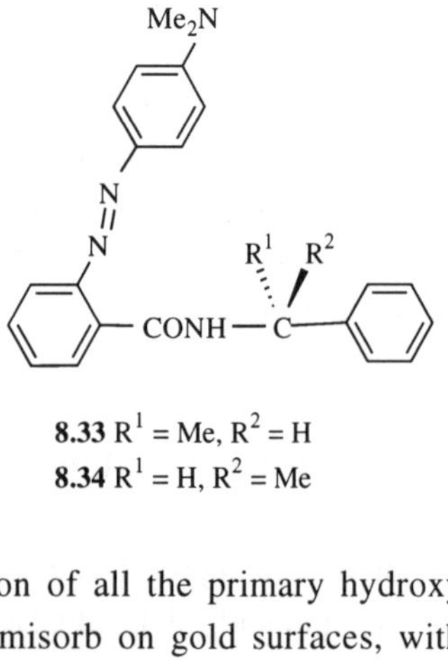

8.33 R^1 = Me, R^2 = H
8.34 R^1 = H, R^2 = Me

The modified CD **8.30**, prepared by substitution of all the primary hydroxy groups of βCD with thiol groups, is found to chemisorb on gold surfaces, with formation of at least six sulfur-gold bonds per CD molecule [22,23]. The geometries of the CD and gold surfaces are incompatible with the formation of sulfur-gold bonds involving all seven CD thiol groups, without distortion of the CD. Even bonding of six requires some deviation from the CD seven-fold symmetry. The CD monolayer is stable, but imperfect, with substantial areas of the gold surface remaining uncovered. To fill the defects, the monolayer has been treated with a mixture of ferrocene and pentanethiol, the latter to chemisorb to the gold surface in the intermolecular voids (Fig. 8.8). Ferrocene is employed to complex in the CD annuli, thereby blocking access of the pentanethiol to the gold surface through the annuli and protecting these monolayer complexation sites. Electrodes formed in this way exhibit the expected voltammetric response for bound ferrocene. The thermodynamic equilibrium of ferrocene complexation in the CD annuli has been confirmed using *m*-toluic acid as a competitive guest, which with increasing concentration displaces greater portions of the ferrocene, with a consequent reduction in the electrode voltammetric response.

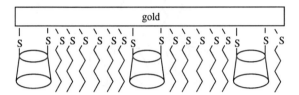

Fig. 8.8. Illustration of the monolayer formed by treating gold with the CD **8.30**, then a mixture of pentanethiol and ferrocene.

The orientation of thiolated CDs on gold surfaces depends on the number of thiol groups, and their mode of attachment to the CD. Molecules of the CD **8.30** are aligned in the monolayers with their faces of secondary hydroxy groups and primary thiol groups parallel with the gold surface. A similar orientation of the mono-thio-substituted βCD **8.35** was observed, whereas the CDs **8.36** and **8.37** having a single thiol group attached to the CD by a long chain, which allows greater flexibility, align with their faces at an angle to the surface (Fig. 8.9) [24,25]. This arrangement is stabilised by intermolecular hydrogen bonding between the CD hydroxy groups and is consistent with a theoretical study [26].

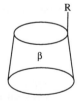

8.35 R = SH
8.36 R = $S(CH_2)_2O(CH_2)_2O(CH_2)_2SH$
8.37 R = $S(CH_2)_{10}SH$

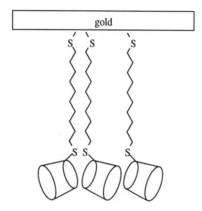

Fig. 8.9. Orientation of the thiolated CD **8.37** chemisorbed on a gold surface.

An alternative way to construct CD layers on gold surfaces is illustrated in Fig. 8.10 [27]. The gold surface is first coated with a mercaptoalkanoic acid such as mercaptopropanoic acid. Self-assembly of CD layers on the modified surface is then induced by electrostatic interactions between the negatively charged carboxylate groups on the surface and the positively charged residues on the CDs **8.38-8.40**. Due to the amphiphilic nature of the CDs **8.38-8.40**, they form multilayers, which are much more organized than those formed by aggregation of the CDs **8.38-8.40** on bare gold electrodes.

As with monolayers of amphiphilic CDs deposited on electrodes, CD monolayers on gold and silver surfaces are suitable for application in the development of optical and electrochemical devices. In this regard, the system illustrated in Fig. 8.11 has been developed, where the CD monolayer on a gold electrode has been used as an

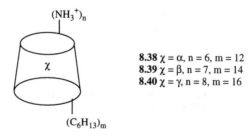

8.38 $\chi = \alpha$, n = 6, m = 12
8.39 $\chi = \beta$, n = 7, m = 14
8.40 $\chi = \gamma$, n = 8, m = 16

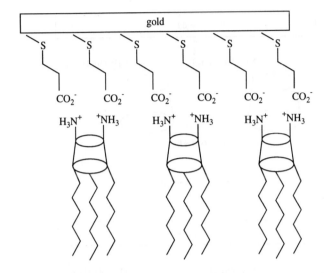

Fig. 8.10. Aggregation of the CDs **8.38-8.40** on a gold surface pretreated with mercaptopropionic acid.

interface for the transduction of optical signals recorded by N-methyl-N'-[1-phenylazobenzyl]-4,4'-bipyridinium ion **8.41** [28]. The azobenzene **8.41** undergoes a one-electron reduction. It also exhibits reversible photochemical *cis-trans* isomerisation (Scheme 8.3), and the *trans*-isomer has a much greater tendency to include in the annuli of a CD. Consequently, irradiation of the *cis*-isomer of the azobenzene **8.41**, with light of wavelength greater than 375 nm, results in conversion to the *trans*-isomer, which complexes in the CD annuli of the monolayer, where it is reduced. Irradiation at 355 nm results in the reversion to the *cis*-isomer, which shows less tendency to complex in the CD annuli and therefore be reduced. In this way the optical input received by the azobenzene **8.41** is converted to an electrochemical output through the CD monolayer on the gold surface.

CD polymers coated as thin films on electrodes function in the same way as amphiphilic CD monolayers deposited on electrodes and thiolated CDs chemisorbed on gold surfaces, in that the CD annuli complex guests in the vicinity of the electrodes, allowing electron transfer processes between the electrodes and the guests to take place. Alternatively, guest complexation in the CD annuli may limit passage

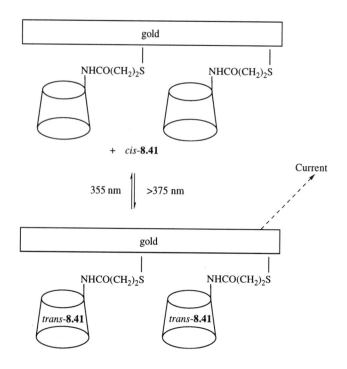

Fig. 8.11. Illustration of the operation of a CD monolayer chemisorbed on the surface of a gold electrode, in the electrochemical transduction of optical signals received by the azobenzene **8.41**.

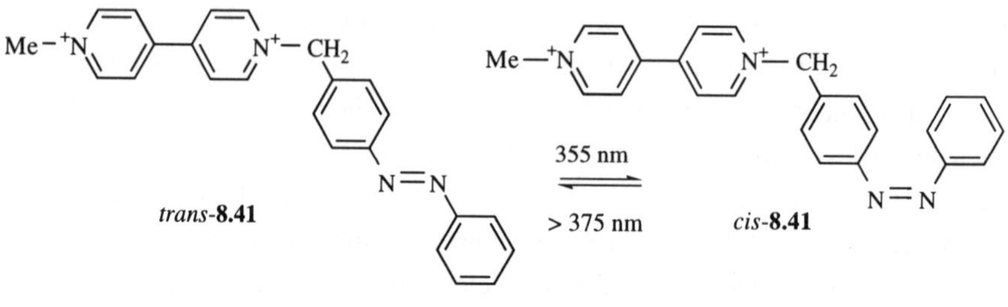

Scheme 8.3. Reversible photochemical isomerisation of the azobenzene **8.41**.

of ions through the annuli and therefore be detectable as an electrode response [29-31]. An example of the use of a CD polymer film on a gold electrode has recently been reported, in the form of a sensor for the determination of the concentration of oxygen in water (Fig. 8.12) [32]. Cobalt tetraphenylporphyrin binds to the CD annuli on the electrode surface, where it electrocatalytically reduces oxygen to hydrogen peroxide. The development of such systems indicates that we may soon see commercial electrochemical devices exploiting CD monolayers and thin films.

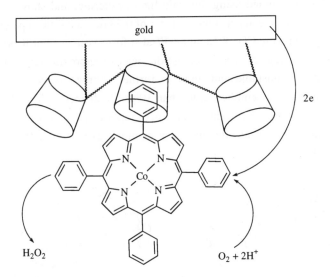

Fig. 8.12. Schematic representation of an oxygen sensor constructed using a gold electrode coated with a film of CD polymer.

8.5. Micelles and Vesicles of Amphiphilic Cyclodextrins

The properties of amphiphilic CDs which lead to their surface behaviour described above also lead to characteristic features in solution. In aqueous solution, amphiphilic CDs form mixed vesicles with phospholipids [33,34] and are of interest in this regard as agents for ion-transport through membranes [35,36]. Alone, amphiphilic CDs adopt lamellar structures in aqueous and non-aqueous solvents, leading to the formation of micelles and vesicles [9]. In water, the critical micelle concentration of the CD **8.11** is 2×10^{-7} mol dm^{-3}. This is orders of magnitude lower than for amphiphilic mono- and di-saccharides, presumably because the planar surface of the CD **8.11** facilitates layer formation.

Micelle and vesicle formation in non-aqueous media is less common but in tetrahydrofuran the CDs **8.12**, **8.15** and **8.17** show behaviour typical of surfactants and consistent with the formation of vesicles, with critical micelle concentrations of 2.8×10^{-2}, 7.1×10^{-3}, and 6.0×10^{-3} mol dm^3, respectively [9]. These micelles of the CD **8.17** were studied using dynamic light scattering, and shown to be stable over 48 hours, with a mean apparent diameter of 0.53 μm and little apparent variation in size. In pyridine, the CDs **8.15** and **8.17** also form vesicles, but these are larger, highly disperse aggregates, with apparent diameter of approximately 4.3 μm. They are also relatively unstable and undergo further aggregation within 2 hours of formation, leading to assemblies larger than 10 μm in diameter. Micellar behaviour of amphiphilic CDs in chloroform has also been reported [37]. The internal structure and organisation of CD micelles has been examine using freeze-fracture electron microscopy [38].

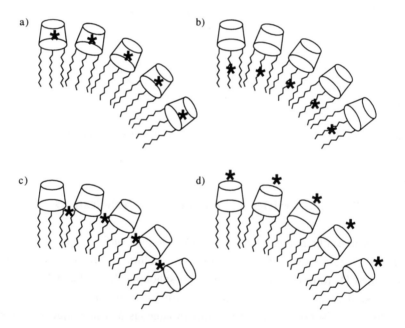

Fig. 8.13. Possible complexation of pharmaceuticals (*) in amphiphilic CD nanospheres; a) in the CD annuli, b) within the alkyl chains, c) between CDs, and d) on the external surface of charged CDs.

CDs incorporated in micelles retain their ability to form complexes. Indeed, aggregated CDs have been reported to show cooperativity in their binding of guests [39,40]. The complexation of pharmaceuticals in CD vesicles has been studied in some detail, with a view to the administration of pharmaceuticals in this manner [41-46]. By dispersing amphiphilic CDs such as **8.12**, **8.15**, **8.17** and **8.42** in aqueous solution, monodisperse nanospheres of 90-300 nm mean diameter have been prepared and characterised, using scanning force microscopy and scanning electron microscopy. The complexation of pharmaceuticals can be accomplished during or after micelle formation. In this manner, progesterone and testosterone are incorporated in vesicles of the CD **8.42**, at 60-80 and 20-30 μg per mg of CD, respectively [45]. In these systems, it is possible that the pharmaceutical complexes either in the CD annuli, within the envelope of the alkyl chains, or between the CD molecules (Fig. 8.13), or through a combination of these processes. Another possible mode of association involves the adsorption of charged molecules on the external surface of charged CDs assembled in nanospheres. Thus, a variety of complexation modes are likely and it should be possible to tailor these to meet specific requirements for the administration of particular pharmaceuticals.

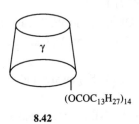

$(OCOC_{13}H_{27})_{14}$

8.42

8.6. References

1. Y. Kawabata, M. Matsumoto, M. Tanaka, H. Takahashi, Y. Irinatsu, S. Tamura, W. Tagaki, H. Nakahara and K. Fukuda, *Chem. Lett.* (1986) 1933.

2. C.-C. Ling, R. Darcy and W. Risse, *J. Chem. Soc., Chem. Commun.* (1993) 438.

3. I. Nicolis, A. W. Coleman, P. Charpin, F. Villain, P. Zhang, C.-C. Ling and C. de Rango, *J. Am. Chem. Soc.* **115** (1993) 11596.

4. D. P. Parazak, A. R. Khan, V. T. D'Souza and K. J. Stine, *Langmuir* **12** (1996) 4046.
5. K. Chmurski, R. Bilewicz and J. Jurczak, *Langmuir* **12** (1996) 6114.
6. Y. Le Bras, M. Sallé, P. Leriche, C. Mingotaud, P. Richomme and J. Møller, *J. Mater. Chem.* **7** (1997) 2393.
7. A. Schalchli, J. J. Benattar, P. Tchoreloff, P. Zhang and A. W. Coleman, *Langmuir* **9** (1993) 1968.
8. P. C. Tchoreloff, M. M. Boissonnade, A. W. Coleman and A. Baszkin, *Langmuir* **11** (1995) 191.
9. P. Zhang, H. Parrot-Lopez, P. Tchoreloff, A. Baszkin, C.-C. Ling, C. de Rango and A. W. Coleman, *J. Phys. Org. Chem.* **5** (1992) 518.
10. M. H. Greenhall, P. Lukes, R. Kataky, N. E. Agbor, J. P. S. Badyal, J. Yarwood, D. Parker and M. C. Petty, *Langmuir* **11** (1995) 3997.
11. H. Parrot-Lopez, C.-C. Ling, P. Zhang, A. Baszkin, G. Albrecht, C. de Rango and A. W. Coleman, *J. Am. Chem. Soc.* **114** (1992) 5479.
12. M. Tanaka, Y. Ishizuka, M. Matsumoto, T. Nakamura, A. Yabe, H. Nakanishi, Y. Kawabata, H. Takahashi, S. Tamura, W. Tagaki, H. Nakahara and K. Fukuda, *Chem. Lett.* (1987) 1307.
13. S. Taneva, K. Ariga, Y. Okahata and W. Tagaki, *Langmuir* **5** (1989) 111.
14. S. Taneva, K. Ariga, W. Tagaki and Y. Okahata, *J. Colloid Interface Sci.* **131** (1989) 561.
15. K. Odashima, M. Kotato, M. Sugawara and Y. Umezawa, *Anal. Chem.* **65** (1993) 927.
16. A. Yabe, Y. Kawabata, H. Niino, M. Tanaka, A. Ouchi, H. Takahashi, S. Tamura, W. Tagaki, H. Nakahara and K. Fukuda, *Chem. Lett.* (1988) 1.
17. H. Tachibana and M. Matsumoto, *Adv. Mater.* **5** (1993) 796.
18. S. Nagase, M. Kataoka, R. Naganawa, R. Komatsu, K. Odashima and Y. Umezawa, *Anal. Chem.* **62** (1990) 1252.
19. Y. Maeda and H. Kitano, *J. Phys. Chem.* **99** (1995) 487.
20. H. Yamamoto, Y. Maeda and H. Kitano, *J. Phys. Chem. B* **101** (1997) 6855.
21. Y. Maeda, T. Fukuda, H. Yamamoto and H. Kitano, *Langmuir* **13** (1997) 4187.
22. M. T. Rojas, R. Königer, J. F. Stoddart and A. E. Kaifer, *J. Am. Chem. Soc.* **117** (1995) 336.

23. A. E. Kaifer, *Isr. J. Chem.* **36** (1996) 389.
24. G. Nelles, M. Weisser, R. Back, P. Wohlfart, G. Wenz and S. Mittler-Neher, *J. Am. Chem. Soc.* **118** (1996) 5039.
25. M. Weisser, G. Nelles, P. Wohlfart, G. Wenz and S. Mittler-Neher, *J. Phys. Chem.* **100** (1996) 17893.
26. J. Qian, R. Hentschke and W. Knoll, *Langmuir* **13** (1997) 7092.
27. L. A. Godínez, J. Lin, M. Muñoz, A. W. Coleman, S. Rubin, A. Parikh, T. A. Zawodzinski, D. Loveday, J. P. Ferraris and A. E. Kaifer, *Langmuir* **14** (1998) 137.
28. M. Lahav, K. T. Ranjit, E. Katz and I. Willner, *J. Chem. Soc., Chem. Commun.* (1997) 259.
29. R. Kataky, P. S. Bates and D. Parker, *Analyst* **117** (1992) 1313.
30. P. S. Bates, R. Kataky and D. Parker, *J. Chem. Soc., Perkin Trans. 2* (1994) 669.
31. A. Gafni, Y. Cohen, R. Kataky, S. Palmer and D. Parker, *J. Chem. Soc., Perkin Trans. 2* (1998) 19.
32. F. D'Souza, Y.-Y. Hsieh, H. Wickman and W. Kutner, *J. Chem. Soc., Chem. Commun.* (1997) 1191.
33. P. Zhang, C.-C. Ling, A. W. Coleman, H. Parrot-Lopez and H. Galons, *Tetrahedron Lett.* **32** (1991) 2769.
34. A. W. Coleman, F. Djedaini and B. Perly, *Proceedings of the 5th International Symposium on Cyclodextrins* (Edition de Santé, Paris, 1990), p. 328.
35. I. Tabushi, Y. Kuroda and K. Yokota, *Tetrahedron Lett.* **23** (1982) 4601.
36. L. Jullien, T. Lazrak, J. Canceill, L. Lacombe and J.-M. Lehn, *J. Chem. Soc., Perkin Trans. 2* (1993) 1011.
37. J. Dey, P. Schwinté, R. Darcy, C.-C. Ling, F. Sicoli and C. Ahern, *J. Chem. Soc., Perkin Trans. 2* (1998) 1513.
38. A. Gulik, H. Delacroix, D. Wouessidjewe and M. Skiba, *Langmuir* **14** (1998) 1050.
39. R. C. Petter and J. S. Salek, *J. Am. Chem. Soc.* **109** (1987) 7897.
40. R. C. Petter, J. S. Salek, C. T. Sikorski, G. Kumaravel and F.-T. Lin, *J. Am. Chem. Soc.* **112** (1990) 3860.
41. M. Skiba, F. Puisieux, D. Dûchene and D. Wouessidjewe, *Int. J. Pharm.* **120** (1995) 1.

42. M. Skiba, D. Dûchene, F. Puisieux and D. Wouessidjewe, *Int. J. Pharm.* **129** (1996) 113.
43. M. Skiba, D. Wouessidjewe, F. Puisieux, D. Dûchene and A. Gulik, *Int. J. Pharm.* **142** (1996) 121.
44. M. Skiba, F. Nemati, F. Puisieux, D. Dûchene and D. Wouessidjewe, *Int. J. Pharm.* **145** (1996) 241.
45. D. Dûchene, *International Pharmaceutical Applications of Cyclodextrins Conference*, Lawrence, Kansas, June 1997.
46. A. M. Dasilveira, G. Ponchel, F. Puisieux and D. Dûchene, *Pharm. Res.* **15** (1998) 1051.

Subject Index